INTRODUCTION TO
MATHEMATICAL MODELING

INTRODUCTION TO
MATHEMATICAL MODELING

MAYER HUMI
WORCESTER POLYTECHNIC INSTITUTE
USA

CRC Press
Taylor & Francis Group
Boca Raton London New York

CRC Press is an imprint of the
Taylor & Francis Group, an **informa** business

A CHAPMAN & HALL BOOK

First published in paperback 2024

First published 2014
by CRC Press
2385 NW Executive Center Drive, Suite 320, Boca Raton FL 33431

and by CRC Press
4 Park Square, Milton Park, Abingdon, Oxon, OX14 4RN

CRC Press is an imprint of Taylor & Francis Group, LLC

ISBN: 978-1-4987-2800-3 (hbk)
ISBN: 978-1-03-247695-7 (pbk)
ISBN: 978-1-315-37030-9 (ebk)

DOI: 10.1201/9781315370309

**Visit the Taylor & Francis Web site at
http://www.taylorandfrancis.com**

**and the CRC Press Web site at
http://www.crcpress.com**

Contents

CHAPTER 1

The Process of Mathematical Modeling

CONTENTS

1.1 WHAT IS MODEL BUILDING?

Definition: modeling is the art of describing in symbolic language a real life system so that approximately correct predictions can be made regarding the behavior or evolution of the system under varied circumstances of interest.

We now elaborate on this definition.

First, note that in this definition "modeling" is referred to as an art. As such one cannot develop rigid preset rules for this task. What can be done, however, is to point out a pattern that is found to be useful in many cases and can help the practitioner to avoid many pitfalls.

Furthermore, a model is described as being able to make "correct predictions" about the system. This usually does not mean 100% accuracy. Predictions of many models have a rather wide error margin. The pertinent question, therefore, is whether these margins are acceptable to the user or not. Moreover, it might turn out that several models are capable of describing the same phenomena with different degrees of accuracy (and complexity).

Another important aspect of the definition is that a model should be "solvable." A sophisticated but "insolvable" model might be less useful from a practical point of view than a simple and straightforward one which is capable of making predictions with acceptable error margins.

One should also bear in mind that every model is constructed with certain limitations on its validity, and these should be borne in mind by the prospective user. Thus, in many practical applications it is not that "the model is incorrect" but it is the application which violates the basic assumptions of the model used.

A classical example to illustrate these points is given by gravitation theory. Here we do have at present two concurrent theories which pertain to modeling the same phenomena viz. Newton's Law of Gravitation and Einstein's theory. Though it is accepted and proven that Einstein's theory is better and more accurate, still it is highly complex and "hard to solve." As a result, in most terrestrial applications we use Newton's Law of Gravitation with "acceptable error margin."

As to the problems which require model construction, their source and scope vary between applied problems in life to attempts to duplicate natural phenomena and (what might seem to be) intellectual curiosity.

Examples and Illustrations:

1. In many cases of daily life, we construct "mini-models" without even paying any attention to these facts; e.g., "How do I get to downtown?" (by car, by bus, by subway, on foot, or otherwise) requires a model which depends on:

   (a) The distance to downtown,

   (b) The time element (How fast do I want to get there?),

   (c) Money considerations,

   (d) Security considerations (Is it safe to ride the subway?),

   (e) Availability of means (How frequently do the buses run?),

   (f) The mood of the person.

2. Consider a truck company operating in the U.S. with "truck depots and service centers" in some major cities.

   A major problem for such a company is how to dispatch trucks to their destinations in the most economical way (saving gas and drivers' time).

3. How can the wheat crop be increased to feed the growing human population?

4. What is the cause of global warming and climate change and how can these effects be mitigated?

5. How can sound and light be recorded in a better way?

6. How can rockets be sent to the Moon or the planets?

7. Why is the sky blue?

1.2   MODELING FRAMEWORK

As we said above, model building is a creative act for which no preset rules apply. We do trace here, however, a series of steps which hopefully will be useful in avoiding costly mistakes and around which one can develop one's skills in this field. We would like to stress, moreover, that model building is a non-sequential process. In some cases, several of these steps will overlap, some might be "missing" (i.e., not needed), and between others "loops" have to be made until one may come up with a reasonable (and acceptable) model for the problem at hand.

We now describe these steps.

STEP 0: Set up as **precisely** as possible the **reasons** for constructing the model and its **objectives**.

We note here that in many projects a clear statement of these reasons and objectives might radically influence the model to be built.

Example: The statement "Build a model to predict the weather" is a rather loose and incomplete statement from a modeling point of view. Thus, if no statement is made for the reasons and the precise objectives of the desired model, the problem should be considered as ill-defined. In fact, as stated above, each of the following might be the actual objective of the model.

1. Predict the weather for the next hour.
   Reason: One wants to go shopping.
   Appropriate model: Just look through the window.

2. Predict the weather tomorrow.
   Reason: Going on a trip one would like to know how to dress.
   Appropriate model: Listen to the weather forecast on the radio or the Internet.

3. Predict the weather next winter.
   Reason: Will it be a good idea to build a ski-motel during the summer?

Discussion of an appropriate model: We note, however, that even after these clarifications the model to be built is not precisely defined since the word "weather" might have different connotations, for example:

1. The model should predict the temperature to within 10° (should I take my sweater with me?).

2. The model should predict the temperature, the condition of the sky, and the possibility of showers (would it be a nice day for a trip? Should I take a raincoat?).

3. The model should predict the possibility of snow (should I take my ski gear with me?).

Another point to remember at this stage is that in many instances a reformulation of the model objectives will be required during the process of model building. This is especially so if the original scope of the model turns out to be too large.

Example: The objective of building a model to "cure cancer" requires many sub-models since there are several types of cancer.

STEP 1: Study the problem as it is in real life.

Example: If one attempts to build a model to improve the performance of a production line, then it is imperative to go to the factory and study *first hand* how this line operates (nothing else will do). In many instances, one might discover that the problem of improved production depends on factors which are independent of the line operation.

STEP 2: Data collection and analysis in real life.

At this stage, one studies the phenomenon and its behavior in real life with the objective of identifying the major factors (i.e., causes) that influence the phenomenon.

Example 1: Analysis of car accidents.

As result of data collection and its statistical analysis, one might conclude that the main factors which have a bearing on the frequency of car accidents are: driver, road, car, and weather. Each of these can be further subdivided into several sub-headings; e.g., to define "driver," we must specify age, sex, height, sight, mental state (e.g., intoxication), etc.

We note that at this step some very crucial decisions have to be made viz. to identify those factors that are most important to the problem at hand. For example, in analyzing car accidents, one might make the (questionable) assumption that the height of the driver is of little importance and hence can be ignored. Once again, it is important to keep in mind the need to strike the balance between model simplicity (i.e., few variables and easy to solve) and effectiveness (i.e., accurate predictions).

Moreover, at this point one must also decide whether to limit the scope of the model to be built or to make its objectives more precise.

Example 1: If we started with the objective of curing cancer, we might decide at this stage to study only the relationship between drug X and lung cancer.

Example 2: If we wanted to study, originally, the performance of a given car, we might decide to limit ourselves to the study of a certain component, e.g., the motor.

STEP 3. Controlled lab studies or simulations.

Studies carried out in the labs enable us to vary the factors that influence the phenomena under study in a controlled manner and thus study the influence of each factor separately.

Example : If the strength of a certain material depends on the temperature and pressure, then lab experiments will enable us to study the strength as a function of one variable only (for a constant value of the other variables), something that is not easy to achieve in real life situations.

STEP 4: Construction of a conceptual qualitative model.

As a result of the studies conducted in the previous steps, one should be able to construct a qualitative model for the phenomena at hand.

Example 1: "The accident rate depends mainly on the driver's age where 25 seems to be the most important in changing driving habits."

Example 2: "Pleasure is the main motivating force in human behavior."

Example 3: "Reliability of a given system decreases with heat and speed but increases with weight."

If such a qualitative model is acceptable to the user, then the author or researcher might feel no further need for a mathematical-quantitative model (or the model might be hard to quantify, e.g., amount of pleasure, motivation, etc.) and therefore may proceed directly to step 8 (bypassing steps 5, 6, and 7, which are needed when a mathematical model is constructed).

In many cases, one is required at this stage to make a creative breakthrough, that is provide a new conceptual framework for the problem under consideration and its solution.

Example 4: "The H_2-molecule can be represented by a rod whose mass is concentrated at the end points." This is Raman Model for the H_2-molecule which earned him the Nobel prize.

STEP 5: Conclusions, predictions, and recommendations that follow for the qualitative model.

Example 1: "Tomorrow will be partly cloudy."
Example 2: "Saccharine is carcinogenic."
Example 3: "Car X is not reliable."
Example 4: "Mr. X has a strong personality."
Example 5: "Travel is a pleasurable experience."

The shortcomings of qualitative models are:

1. Models tend to be limited in scope.

2. Model reliability in making predictions is sometimes questionable.

3. Predictions are qualitative rather than quantitative.

The points of strength of these models, on the other hand, are:

1. Their meaning is clear to almost everybody.

2. They are hard to question and challenge (sometimes these models are accepted solely due to the experience and authority of the person that suggests the model).

3. They are simple in structure.

4. They are based sometimes on long personal experience and intuition.

Example : "Buy stocks in January and sell them in April."

While qualitative models are accepted and used extensively in the social sciences, they are rarely acceptable in the natural sciences or engineering. In these fields, one has to perform the following additional steps which lead to a mathematical-quantitative model and its solutions.

Remark: Not all mathematical models make quantitative predictions, e.g., in the stability theory for solutions (or equilibrium points) of differential equations.

STEP 6: Abstraction and symbolic representation.

This step sometimes requires a lot of insight and creativity as the true variables which control the phenomena might be masked by the data. In the majority of the cases, however, it consists of representing the variables by symbols, naming the functions and identifying the axioms or constraints of the model.

Example 1: "We can treat a car as a point particle whose position is given by the vector x (abstraction and simplification)."

Example 2: "The temperature T is a polynomial function of the speed v and the time t, i.e.,

$$T = p(v, t)$$

where p is a polynomial."

Example 3: "Let the number of trucks at station A be denoted by N_A. Then N_A must satisfy $N_A \leq N$ where N is the total number of trucks operated by the company (symbolic representation and constraint)."

Example 4: "For transportation purposes, it is enough to represent the U.S. map by a set of discrete points whose location coincide with the major cities (abstraction and simplification)."

STEP 7: Derive the equations that govern the phenomena.

Example 1: If $\mathbf{F}$ is acting on point particle of mass m, and the particle acceleration is denoted by $\mathbf{a}$, then

$$\mathbf{F} = m\mathbf{a}.$$

This is Newton's Second Law.

Example 2: Denoting by P, V, and T the pressure, volume, and temperature of a gas, then

$$PV = RT$$

where R is a constant. This is the Ideal Gas Law.

Remarks: Broadly, mathematical models are classified as deterministic versus stochastic (viz. probabilistic). Another possible classification of these models is as continuous (i.e., the variables used are continuous) and discrete. Each of these classifications has its merit within a given context.

STEP 8: Model testing.

To this end, one must solve the model equations and compare the solution with the actual data collected in steps 2 and 3. If there is a bad fit, i.e. non-acceptable deviations, then it will be necessary to redo steps 4, 5, and 6.

In this context, we remark that sometimes new mathematical techniques have been devised to solve a mathematical model. If the model equations remain intractable, then some approximations to the model equations must be made, thereby sacrificing accuracy in favor of easier computability.

Example: If the original model equations are highly nonlinear and hard to solve, then one may find an acceptable linear approximation which might be solved easily.

STEP 9: Model limitations and constraints.

At this point, one must become clearly aware of the limitations that must be imposed on the use of the model and the permissible range of the variables.

Example 1: One cannot use the equations of classical mechanics to predict the motion of a particle whose speed is close to the speed of light.

Example 2: The ideal gas law is a good model for some gases but not for others.

STEP 10: Predictions and sensitivity analysis.

Once a model has been tested and found acceptable, then it can be used to make predictions. Whenever such a prediction is found to be correct, the model is considered to be more reliable (in a way every such prediction is a further test of the model).

One should bear in mind, however, that sensitivity analysis of many models is required before their actual use; i.e., one has to find the extent to which the model predictions are sensitive to small variations of the model parameters. We note that some models are "required" to be highly sensitive while in others insensitivity to such variations is necessary (e.g., if the data contains inherent errors).

Example 1: Ballistic tables.

To construct an "exact" ballistic table, one has to know the exact atmospheric conditions, amount of charge, geographic altitude, and state of the cannon to be fired. In field conditions, however, these variables are known approximately at best. Hence, a good ballistic model must be somewhat insensitive to small variations in these parameters while giving a reasonable prediction about the range of the shot.

Example 2: Models for physical resonances.

Here, sensitivity is highly desirable especially when several such "close by" resonances are involved.

Example 3: Chaotic systems.

When the evolution of a system under consideration displays high sensitivity to the initial conditions, we say that the system is "chaotic." Under these circumstances, it is possible to make only "short time" predictions about the state of the system. This is why weather forecasts are accurate only for a "few days" (at best).

STEP 11: Extensions and refinements.

If a model is found to be correct in some instances but less accurate in others, then a refinement of it is needed to take care of these exceptions.

Example: The ideal gas law needs such a refinement when the gas molecules are "large" (e.g., diatomic gases). The refined model is given by

$$\left(P + \frac{\alpha}{V^2}\right) V = RT$$

where α is a parameter which depends on the gas.

STEP 12: Compounding.

Once a correct and reliable model has been established for some phenomena, then related problems can be modeled by a process of compounding.

Example: Once the equations of motion for the spring-mass system are found, one can compound the model to systems of several masses and springs.

Finally, we present here a schematic overview of the modeling process.

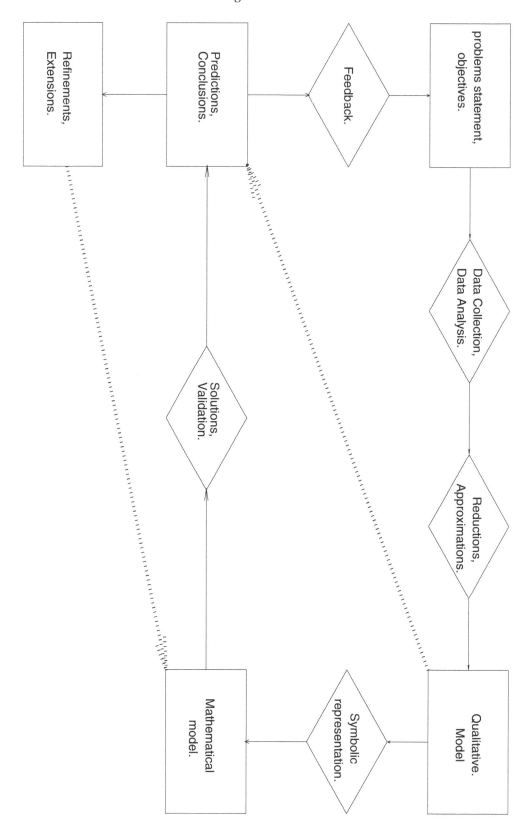

Figure 1.1 A flow chart of the modeling process.

EXERCISES

E. Fermi was one of the great theoretical physicists in the twentieth century. Some of the following "mini-model" questions are attributed to him. We offer these here to sharpen the reader's skill in the modeling of "real world problems."

1. Ignoring oceans and such, how long would it take to walk entirely around the world?

2. How much water per year flows in the Mississippi river?

3. How many dump-truck loads would it take to move Mt. Washington in New Hampshire, USA?

4. Find the dimension of a box that can contain all of the human race (five billion approximately).

5. What is the linear velocity of Earth around the Sun?

6. How many drops of water are in the Pacific ocean?

7. How many books are in a bookstore?

8. How many atoms are in a cell?

9. How many cells are in the human body?

10. How many light-bulbs burn out in one minute throughout the world?

11. What is the actual volume of material in a solid cubic meter of metal (remember atoms are made of nuclei and electrons)?

12. How do you buy the best car for your money?

13. How do you buy the best computer for your money?

1.3 GENES AND BIOLOGICAL REPRODUCTION

Most of the models that we consider in this book will use "continuous variables." Moreover, it will be advantageous in some cases to convert a discrete variable problem into a problem with continuous variables. However, to illustrate the modeling process we consider in this section a model which reduces a "continuous variable problem" into a discrete one.

Motivation and Objective:

A company grows and sells various types of beans (Lima-beans,kidney-beans etc.). It is a known fact that green and smooth texture (Lima-)beans are preferred by the consumers and hence command a higher price. However, on the farms the company grows beans which vary in color (white, red, yellow, green, and anything in between) and texture (from smooth to wrinkled).

It is the objective of this project to understand this phenomenon so that the company will be able to produce larger quantities of the desirable beans and enhance its profits.

Data from the farm:

1. Beans self fertilize.

2. Beans can be divided to a good approximation into the following groups: Green-Smooth (G-S), Green-Wrinkled (G-W), White-Smooth (W-S), White-Wrinkled (W-W), Yellow-Smooth (Y-S), Red-Smooth (R-S), and Red-Wrinkled (R-W).

Remarks: Note that by using the simplification in (2) above, we converted a continuous set of variables for the color and texture into a discrete one. This, sometimes, simplifies the problem considerably. However, sometimes the reverse is true (e.g., in population models).

Experimental Data:

Observations of bean plants grown in seclusion are being made in regard to their crop and their descendants.

Results:

1. It is possible to obtain pure lines of beans, i.e., beans which by self fertilization will always produce descendants of the same type. However, these pure lines are prone to disease and therefore not very desirable from a commercial point of view.

2. Sometimes, a plant from a line producing green smooth beans will produce by self fertilization some descendants which are wrinkled, etc. (so that appearances might be deceptive).

3. If we cross pure lines of G-S beans with G-W, we obtain first generation G-S beans only. However, in the second generation, these G-S beans will give both G-S and G-W beans by approximate ratio of 3:1.

Subproblem:

Build a model to explain texture only; i.e., assume all beans are Green.

Qualitative Model:

1. A bean carries entities which we shall call "genes" which determine whether it is smooth or wrinkled. These will be denoted by S and W.

2. Each bean contains two such genes.

3. A bean is smooth if the combination of genes is SS or SW and wrinkled if WW.

4. In cross fertilization, one gene is accepted (independently) from each parent.

Remark: In such a situation where the combination SW is smooth, we shall say that S is "dominant" with respect to W.

Mathematical Model

Let $R(*, *)$ denote the reproduction function, i.e., the probability distribution of the descendants for a given pair of parents

$$R(p_1, p_2) = (R_1, R_2, R_3)$$

where p_1 and p_2 represent the parents and R_1, R_2, and R_3 are the probabilities of SS, SW, and WW descendants respectively. If $P(*)$ is the probability that a given bean carries a certain gene, we then have

$$R_1(p_1, p_2) = P(S \mid p_1)P(S \mid p_2)$$

$$R_2(p_1, p_2) = P(S \mid p_1)P(W \mid p_2) + P(W \mid p_1)P(S \mid p_2)$$

$$R_3(p_1, p_2) = P(W \mid p_1)P(W \mid p_2).$$

(In these equations $P(S \mid p_1)$ represents the conditional probability that the parent p_1 contributes the S gene to the descendant and so on.)

Model Predictions: In the cross fertilization experiment, we started with two pure lines, i.e. $p_1 = SS$, $p_2 = WW$. As a result, our model predicts for first generation descendants:

$$R_1 = 0 \quad R_2 = 1 \quad R_3 = 0,$$

i.e. all first generation beans are G-S which corresponds to the experimental results. For the second generation, we therefore have

$$p_1 = p_2 = SW$$

and hence,

$$R_1 = \frac{1}{2} \cdot \frac{1}{2} = \frac{1}{4}, \quad R_2 = \frac{1}{2}, \quad R_3 = \frac{1}{4},$$

i.e. $\frac{3}{4}$ of the beans are smooth and $\frac{1}{4}$ are wrinkled, i.e., a ratio of 3:1.

EXERCISES

1. Predict the results of cross fertilization between

   (a) SW and WW beans.

   (b) SW and SS beans.

2. (Compounding) Devise a model for beans which takes color into account.

   Hint: Each bean will now have four genes; S, W for texture and C, c for color.

3. Suppose that beans with (SS, CC) are crossed with (ww, cc) (and C is also dominant with respect to c) and the first generation descendants reproduce by self fertilization. Predict the results for color and texture of the crop.

4. The following are well known facts regarding blood types in humans:

    (a) There are four (major) blood types denoted by A, B, AB and O.

    (b) Each blood cell contains two genes which determine the blood type.

    (c) The O-gene is regressive with respect to the A and B genes, i.e., AO and BO bloods are A and B bloods respectively.

    (d) A and B genes are of "equal strength."

    (e) In the process of reproduction, each parent donates one gene to determine the blood type of the descendant.

Use this data to:

1. Give an explicit representation for the reproduction function of this system.

2. Predict the blood type distribution for the descendants to parents with blood types AO and BO.

Bibliography

[1] R. Aris - Mathematical Modelling Techniques (Ferron-Pitman).

[2] B. Barnes and G.R. Fulford -Mathematical Modeling with Case Studies, 3rd Edition (CRC Press).

[3] Edward A. Bender - An Introduction to Mathematical Modeling (Dover).

[4] C. Dym - Principles of Mathematical Modeling, 2nd Edition (Academic Press).

[5] Lin, C.C, and Segal, L.A.,1974, Mathematics Applied to Deterministic Problems in the Natural Sciences, Macmillan, NY.

[6] Lindsay, R.B., and, Margenau, H.,1955, Foundation of Physics, Dover, NY.

[7] Maki, D.P., and Thompson, M., 2006, Mathematical Modeling and Computer Simulations Brooks/Cole, Belmont, CA, USA

[8] Meerschaert, M. M.,2012 - Mathematical Modeling, 4th Edition, Elsevier, Burlington, MA.

[9] Melnik, R., (Editor), 2015 Mathematical and Computational Modeling: With Applications in Natural and Social Sciences, Engineering, and the Arts Wiley, Hoboken, NJ

[10] Noble, B.,1967, Applications of Undergraduate Mathematics in Engineering, MAA.

[11] Temam, R. and Miranville, R., 2000 Mathematical Modeling in Continuum Mechanics, Cambrige Univ. Press, Cambridge, UK.

CHAPTER 2

Modeling with Ordinary Differential Equations

CONTENTS

The behavior and evolution of many scientific and engineering systems are described by equations which involve unknown functions and their derivatives. These are called differential equations, and methods for their solution play a central role in many disciplines.

Differential equations are classified as ordinary differential equations (ODEs) and partial differential equations (PDEs). ODEs are equations which involve only one independent variable while PDEs involve several independent variables.

To motivate the study of these equations we consider in this chapter problems in various areas which are modeled naturally by ODEs. For some of these models a solution is possible by elementary integration methods. For others more elaborate methods are needed.

For all the models presented in this chapter we illustrate the modeling process by adhering as closely as possible to the modeling framework that was introduced in the previous chapter.

2.1 THE MOTION OF A PROJECTILE

Model Objective and Motivation: Build a prototype model which describes the motion of a small particle in the gravity field of the Earth. Neglect all other forces and the rotation of the Earth.

This study is motivated by the fact that the motion of a projectile in the atmosphere is important in many applications (e.g rockets, cannon shells, etc). As per usual in the modeling process we first consider this problem in its "bare bones" setting and derive a prototype model.

Background: To derive the equations of motion for this problem we need Newton's second law, which states that the external force acting on a point mass is proportional to its acceleration. Thus

$$\mathbf{F} = m\mathbf{a}$$

where $\mathbf{F}, m, \mathbf{a}$ denote respectively the force, mass, and acceleration of the particle.

2.1.1 Approximations and Simplifications

1. We assume that the speed of the particle is small compared to the speed of light. Hence relativistic corrections to Newton's second law can be neglected.

2. The projectile is considered to be a point particle and the motion around its center of gravity is neglected.

3. We assume that the distance covered by the projectile is small compared to the Earth's radius. Consequently the spherical shape of the Earth can be neglected and we consider the flight to be over a flat plane.

4. We neglect the variation of the gravitational force with height and location (which is due to the fact that the Earth is not a perfect sphere). Hence we approximate g - the acceleration due to gravity by a constant.

5. We neglect the influence of the atmosphere on the motion of the projectile. These include air drag and variations in temperature, density, and pressure.

6. We neglect the effect of the Earth's rotation on the projectile motion.

2.1.2 Model:

With the approximations delineated above it follows from Newton's second law that the equation of motion of the projectile is

$$m\frac{d^2\mathbf{x}}{dt^2} = -mg\mathbf{j}. \tag{2.1}$$

where $\mathbf{j}$ is a unit vector in the upward vertical direction. Since the only force acting on the projectile is in the $\mathbf{j}$-direction, we infer also that its motion is constrained to a plane. Without loss of generality we can choose this plane to be the x-y plane with $\mathbf{x} = (x, y)$, (see Fig. 2.1). Equation (2.1) is equivalent then to two scalar equations

$$\ddot{x} = \frac{d^2x}{dt^2} = 0, \quad \ddot{y} = \frac{d^2y}{dt^2} = -g. \tag{2.2}$$

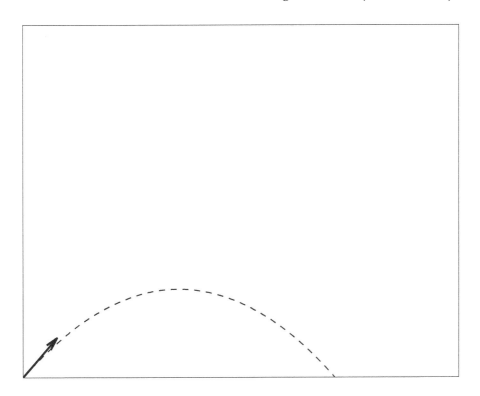

Figure 2.1 Trajectory of a projectile

Since g is constant, we can readily integrate these equations twice to obtain

$$\dot{x} = c_1, \quad \dot{y} = -gt + c_2, \tag{2.3}$$

$$x = c_1 t + c_3, \quad y = -\frac{gt^2}{2} + c_2 t + c_4, \tag{2.4}$$

where c_i, $i = 1, 2, 3, 4$, are constants. To determine these integration constants we need some "initial conditions" which (in this case) must specify the position and velocity of the projectile at some (initial) time. Thus if we assume that at time $t = 0$ the projectile is at the origin and its velocity $\mathbf{v}_0 = (v_0 \cos\theta, v_0 \sin\theta)$, then

$$x(0) = 0, \quad y(0) = 0, \quad \dot{x}(0) = v_0 \cos\theta, \quad \dot{y}(0) = v_0 \sin\theta. \tag{2.5}$$

To use these conditions we substitute $t = 0$ in Equations (2.3), (2.4) to obtain

$$\begin{aligned} c_3 &= c_4 = 0 \\ c_1 &= v_0 \cos\theta, \quad c_2 = v_0 \sin\theta. \end{aligned} \tag{2.6}$$

It follows then that the parametric representation of the trajectory is

$$x = (v_0 \cos \theta)t, \quad y = -\frac{gt^2}{2} + (v_0 \sin \theta)t. \tag{2.7}$$

The nonparametric representation of the trajectory is obtained by eliminating t from Equation (2.7). This leads to

$$y = x \tan \theta - \frac{g}{2} \left(\frac{x}{v_0 \cos \theta} \right)^2. \tag{2.8}$$

Example 2.1.1 *Find the relation between the range of a projectile on Earth and the Moon if they satisfy the same initial conditions.*

Solution 2.1.1 *The range of a projectile is the distance to where it returns to ground zero, i.e., $y = 0$. To find the range R_e on Earth we set $y = 0$ in Equation (2.8) and solve for x.*

We obtain

$$R_e = \frac{v_0^2 \sin 2\theta}{g_e} \tag{2.9}$$

where g_e is the gravitational acceleration on Earth. On the Moon the projectile satisfies the same equation of motion, but the gravitational acceleration is g_m. Hence the range of the projectile on the Moon is

$$R_m = \frac{v_0^2 \sin 2\theta}{g_m}. \tag{2.10}$$

Therefore

$$\frac{R_m}{R_e} = \frac{g_e}{g_m}. \tag{2.11}$$

2.1.3   Model Compounding:

We now compound the prototype model derived above by removing some of its constraints.

Example 2.1.2 *Derive the equation of motion of the projectile when air resistance (drag) has to be taken into consideration.*

Solution 2.1.2 *When the velocity of the projectile is not large, the drag force* $\mathbf{F}_d$ *is (to a good approximation) proportional to the velocity of the projectile*

$$\mathbf{F}_d = -\alpha\mathbf{v}. \tag{2.12}$$

The necessary modifications to Equation (2.1) are given by

$$m\ddot{\mathbf{x}} = -mg\mathbf{j} - \alpha\dot{\mathbf{x}} \tag{2.13}$$

or in scalar form

$$m\ddot{x} = -\alpha\dot{x} \tag{2.14}$$

$$m\ddot{y} = -mg - \alpha\dot{y}. \tag{2.15}$$

Eq. (2.14), (2.15) can be solved by direct integration

$$m\dot{x} = -\alpha x + c_1 \tag{2.16}$$

$$x = \frac{\left[c_2 e^{-bt} + \frac{c_1}{m}\right]}{b}, \quad b = \frac{\alpha}{m}, \quad \alpha \neq 0. \tag{2.17}$$

To solve for y we introduce $\dot{y} = u$. Equation (2.15) becomes

$$m\dot{u} = -mg - \alpha u$$

which then leads to

$$\dot{y} = u = c_3 e^{-bt} - \frac{g}{b} \tag{2.18}$$

$$y = -\frac{1}{b}\left(c_3 e^{-bt} + gt\right) + c_4. \tag{2.19}$$

Once again we need initial conditions in order to solve for the integration constants $c_i, i = 1, \ldots, 4$. We observe that at least formally the solution, Equations (2.16)-(2.19), "looks" totally different from the one obtained when $\alpha = 0$ (however, see ex. 4).

Exercises

1. Derive the equations of motion for a projectile if the variation of the gravitational force with height is to be taken into consideration.

2. Find the maximum height that a projectile will achieve as a function of the initial speed and firing angle. If $v_0 = 1$ km/sec, what will be the maximal change in g along such a trajectory?

3. Solve for the constants $c_i, i = 1, \ldots, 4$, in Equations (2.16)-(2.19) using the initial conditions in Equation (2.5).

4. Use the results of exercise 3 and a first order Taylor expansion for e^{-bt} to show that as $b \to 0$ the solution, Equations (2.16)-(2.19), converges to the one given by Equation (2.7).

5. Write down the equation of motion and initial conditions for the motion of a projectile if there is a wind blowing with velocity $\mathbf{w} = (w_1, w_2)$ where w_1, w_2 are constants (w_1, w_2 are the wind components in the x, y directions). Solve your model.

6. How many firing angles can be used to achieve a given range for a projectile with initial velocity $\mathbf{v}_0$?

7. For a fixed initial speed at what firing angle will a projectile achieve its maximum range?

8. A plane with speed $\mathbf{u}$ is flying from city S to city N, which is at a distance d exactly north of S. A wind of speed $\mathbf{v}$ is blowing in the eastern direction. Find differential equations for the position of the plane if its pilot makes sure that the plane is always aimed towards N. (See Fig. 2.2).

   Hint: Find differential equations for $\dfrac{dx}{dt}, \dfrac{dy}{dt}$ in terms of the position (x, y) and u, v.

2.2 SPRING-MASS SYSTEMS

In this section we model spring-mass systems as well as systems with torsion. Per usual we start with a prototype problem and then compound it to model related systems.

Objective: Build a prototype model which describes the motion (in one-dimension) of a mass attached to a spring whose other end is rigidly fixed (see Fig. 3.3). Neglect gravity and all other external forces.

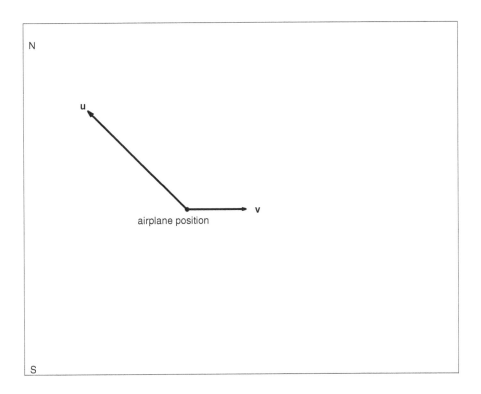

Figure 2.2 A diagram for the plane position and velocity

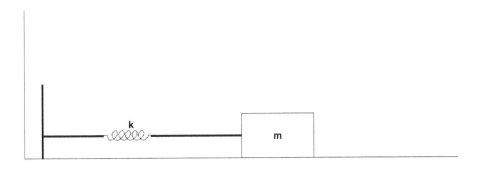

Figure 2.3 Spring Mass system

Background: To model the motion of such a system one usually applies Hooke's law. It states that for small displacements x from the natural length L of the spring the force exerted by it is given by

$$F = -kx, \quad |\frac{x}{L}| \ll 1$$

where $k > 0$ is called the spring-stiffness. However, to understand the limitations and approximations made in its derivation, we state here some of the experiments needed to establish the law.

2.2.1 Data Collection

1. Experiments to measure the force that various springs exert for positive and negative displacements (stretchings and contractions respectively).

2. Experiments to determine to what extent the force exerted by a spring varies with its use.

3. Experiments to find the effects of environmental factors such as temperature, pressure, and location on various springs.

4. Experiments to determine how the force that is exerted by the spring varies as a function of the material, size of the coil, and number and diameter of the loops.

5. Experiments to find how various imperfections in the structure of a spring, e.g. variations in the diameters of the loops and coil or deviations from circular symmetry, affect its performance.

2.2.2 Approximations and Simplifications

As a result of the data collected in the experiments listed above, one is justified in making the following approximations and simplifications for *metal* springs:

1. Small deviations in the structure of a spring minimally affect its performance. Henceforth, we only consider "ideal springs" which are made of homogeneous material, circular coil, and loops whose diameters are constant.

2. The force exerted by a spring depends very weakly on environmental factors and the number of times that the spring is used. Hence we shall neglect the influence of these factors on the performance of the spring.

3. For equal but opposite displacements the magnitude of the force exerted by the spring is equal but in opposite directions.

4. For *small* displacements the force is proportional to the displacement with a negative proportionality constant. The determination of this proportionality constant for a given spring from first principles (i.e. as function of the material, number of loops, etc.) requires a major modeling effort and is in most cases impractical.

2.2.3 Mathematical Model

Let

F	= force exerted by the spring
m	= mass of body attached to the spring
x	= displacement of the mass (or the center of mass) from equilibrium
a	= acceleration of the mass.

Using the approximations to the data introduced above we can now write for $|x| << 1$ that

$$F(x) = -kx \text{ (Hooke's Law)} \qquad (2.20)$$

where $k > 0$ is called the stiffness of the spring.

Using Newton's second law we find that the equation of motion for the mass attached to the spring is given by

$$m\ddot{x} = -kx \qquad (2.21)$$

or

$$m\frac{d^2x}{dt^2} = -kx. \qquad (2.22)$$

2.2.4 Remarks and Refinements

1. If we displace a spring from x to $x + \Delta x$, we infer from Equation (2.20) that

$$\Delta F = F(x + \Delta x) - F(x) = -k\Delta x. \qquad (2.23)$$

This observation, that the additional force exerted by the spring due to a displacement Δx from x is independent of x, is important in many applications.

2. (In preparation for 3) Let there be given an infinite series

$$p(x) = \sum_{n=0}^{\infty} a_n x^n \qquad (2.24)$$

which converges for $\mid x \mid < R, \ \ R > 0$. If $p(x)$ is an odd function, i.e.

$$p(x) = -p(-x), \ \ \mid x \mid < R, \qquad (2.25)$$

then $a_{2m} = 0, \ \ m = 0, 1, \ldots$. In fact we infer from Equations (2.24), (2.25) that

$$2 \sum_{m=0}^{\infty} a_{2m} x^{2m} = 0$$

and since this must be true for all $\mid x \mid < R$, it follows that $a_{2m} = 0$. Similarly if $p(x) = p(-x)$,i.e. $p(x)$ is even, one infers that $a_{2m+1} = 0, \ \ m = 0, 1, \ldots$.

3. In many instances engineers and scientists are called upon to solve or model systems in a short period of time. Under these constraints it is impossible to conduct a thorough set of experiments to establish the laws governing the system's behavior. Instead "mathematical approximations" must be used. We now illustrate this procedure.

   Assume that the only information given about the force exerted by the spring is:

   (a) $F = f(x)$, i.e. the force is a function of the displacement only.

   (b) $F(0) = 0$

(c) $F(x) = -F(-x)$, i.e. F is an odd function of x, where F is some unknown but smooth (i.e. analytic) function.

Since F is analytic, we can expand it in a Taylor expansion around $x = 0$

$$F(x) = F(0) + \frac{F'(0)}{1!}x + \frac{F''(0)}{2!}x^2 + \ldots = \sum_{n=0}^{\infty} \frac{F^{(n)}(0)}{n!}x^n.$$

However, from the fact that F is odd it follows (using the previous observation) that $F^{(2m)}(0) = 0$, $m = 0, 1, 2, \ldots$. Hence

$$F(x) = k_1 x + k_3 x^3 + \ldots \quad .$$

If $| x |$ is small and k_1 is assumed to be nonzero, then we can approximate $F(x)$ by

$$F(x) = -kx, \quad k > 0 \tag{2.26}$$

(the sign can be determined by a simple experiment). Equation (2.26) is called the linear (or first order) approximation to F. It is valid when $F'(0) \neq 0$ and $| x |$ is small. Hooke's law can be interpreted then as representing this approximation. Our analysis, however, goes one step beyond this law. In fact it shows that the next order approximation to the force exerted by the spring (under present assumptions) is not proportional to x^2 but to x^3, i.e.

$$F(x) = -kx \pm k_3 x^3. \tag{2.27}$$

Compounding: Inclusion of external forces.

If an external force besides that of the spring acts on the mass we have from Newton's second law

$$ma = -kx + F_{ext}. \tag{2.28}$$

However if the external force F_{ext} contains frictional forces F_f then it is customary to separate this force from the other external forces so that

$$ma = -kx + F_{ext} + F_f. \tag{2.29}$$

In three dimensions one can obtain the following data about the frictional force: $\mathbf{F}_f$ always acts in the direction opposite of the velocity $\mathbf{v}$ and is a function of $\mathbf{v}$, the material and shape of the body, and the medium in which the body moves.

For a given body moving on a uniform surface, $\mathbf{F}_f = \mathbf{F}_f(v)$, and we infer from the "data" above that

$$\mathbf{F}_f(v) = -\mathbf{F}_f(-v), \quad v = |\mathbf{v}|. \tag{2.30}$$

Hence in one dimension the "first two term approximation" for F_f is given by

$$\mathbf{F}_f(v) = -bv - rv^3 \tag{2.31}$$

where b, r are positive constants.

In three dimensions the equivalent approximation for $\mathbf{F}_f$ is

$$\mathbf{F}_f(\mathbf{v}) = -b\mathbf{v} - r(\mathbf{v} \cdot \mathbf{v})\mathbf{v}. \tag{2.32}$$

For small $|\mathbf{v}|$ we therefore obtain the following equation of motion for the spring mass system in one-dimension.

$$ma + bv + kx = F_{ext} \tag{2.33}$$

or

$$m\ddot{x} + b\dot{x} + kx = F_{ext} \tag{2.34}$$

where dots denote differentiation with respect to time ("standard" notation). The equivalent equation of motion of this system in three dimensions is given by

$$m\ddot{\mathbf{x}} + b\dot{\mathbf{x}} + k\mathbf{x} = \mathbf{F}_{ext}. \tag{2.35}$$

As a particular application of the nonlinear frictional forces given by eq. (2.31) we mention the vibrations in the clarinet tube. Lord Rayleigh, who investigated this problem in the 19th century, modeled these vibrations by the equation

$$m\ddot{x} + kx = b\dot{x} - c(\dot{x})^3, \quad b, c > 0 \tag{2.36}$$

or equivalently

$$m\ddot{x} + [c(\dot{x})^2 - b]\dot{x} + kx = 0 . \tag{2.37}$$

This nonlinear equation is called *Rayleigh equation*.

Solution of Equation (3.84) without friction.

When friction can be neglected and there are no external forces , Equation (3.84) reduces to

$$m\ddot{x} + kx = 0. \tag{2.38}$$

In Equation (2.38) the highest order derivative is $\ddot{x}$, and therefore this is a second order differential equation. This equation is also linear; i.e., if we consider the equation as a polynomial in $x, \dot{x}, \ddot{x}$, etc., then each term is of the first order. Also we observe that the coefficients of the equation are constant.

We now show how Equation (2.38) can be solved by elementary techniques of integration.

Multiplying Equation (2.38) by $\dot{x}$ and observing that $\dot{x}\ddot{x} = \frac{1}{2}\frac{d}{dt}(\dot{x}^2)$. This yields

$$\frac{m}{2}\frac{d}{dt}(\dot{x}^2) + k\dot{x}x = 0. \tag{2.39}$$

Integrating this with respect to t leads to

$$\dot{x}^2 + \omega^2 x^2 = c^2, \quad \omega = \sqrt{k/m} \tag{2.40}$$

where c^2 is a constant of integration (observe that this constant must be non-negative since the left hand side of (2.40) is a sum of squares). Hence

$$\dot{x} = \sqrt{c^2 - \omega^2 x^2}. \tag{2.41}$$

Equation (2.41) can be easily integrated, and we obtain the solution in the form

$$x = A\cos(\omega t + \phi) \tag{2.42}$$

where A, ϕ are constant. Thus the general solution of Equation (2.38) contains two arbitrary constants. These can be determined if the initial conditions $x(0), \dot{x}(0)$ are known.

As expected, the solution, Equation (2.42), represents vibrations with fixed amplitude as there is no friction to damp the motion.

Related Systems:

Example 2.2.1 *Derive the equations of motion for two masses m_1, m_2 which are attached to a spring with stiffness k as in Fig. 2.4.*

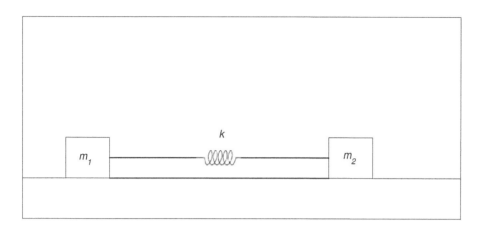

Figure 2.4 Two masses attached to a spring

Solution 2.2.1 *Let the distance between the center of mass of m_1, m_2 at equilibrium be L (if we idealize the system and treat m_1, m_2 as point particles, then L is the natural length of the spring). If these centers of mass at time t are at x_1, x_2 respectively, then either a. $x_2 - x_1 - L > 0$ or b. $x_2 - x_1 - L < 0$.*

In the first case (a) the spring is stretched beyond its natural length, and hence m_1 is pulled to the right and m_2 to the left (by Newton's third law these two forces are equal but in opposite directions.) Hence, using Equation (2.23), we have

$$
\begin{aligned}
m_1 \frac{d^2 x_1}{dt^2} &= k(x_2 - x_1 - L) \\
m_2 \frac{d^2 x_2}{dt^2} &= -k(x_2 - x_1 - L).
\end{aligned}
\tag{2.43}
$$

Similarly in case (b) m_1 is pushed to the left and m_2 to the right. Since $x_2 - x_1 - L < 0$, we infer once again that the equations of motion are given by (2.43). Thus the differential equations which govern the system are the same in both cases.

We observe that this system is modeled by a system of *coupled* ordinary differential equations.

Example 2.2.2 *Derive the equation of motion for a mass in between two springs which are attached to rigid walls whose distance from each other is L, as shown in Fig. 2.5.*

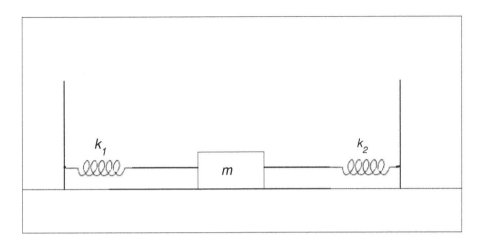

Figure 2.5 A mass and two springs enclosed by rigid walls

Solution 2.2.2 *In problems of this type it is natural to use a coordinate system whose origin coincides with the equilibrium position of the mass (which does not have to be calculated) and obtain a differential equation for the displacement from this position as a function of time. In fact for such a displacement x the* **change** *in the forces acting on m is given by (using Equation (2.23))*

$$F = -k_1 x - k_2 x$$

(regardless of the sign of x). Hence the desired equation of motion is

$$m\ddot{x} = -(k_1 + k_2)x. \tag{2.44}$$

Remark 2.2.1 *To evaluate the position x_{eq} of m at equilibrium we use the fact that in this state $F_{ext} = 0$. Hence if ℓ_1, ℓ_2 are the natural lengths of the springs and m is treated as a point particle we have*

$$k_1(x_{eq} - \ell_1) = k_2(L - x_{eq} - \ell_2)$$

(where we used a coordinate system whose origin is at the left wall of the system). However, note again that Equation x_{eq} is not needed for the derivation of Equation (2.44).

Example 2.2.3 *Derive a model equation for the motion of a mass which is attached to a thin elastic bar and subject to torsional forces ("twists").*

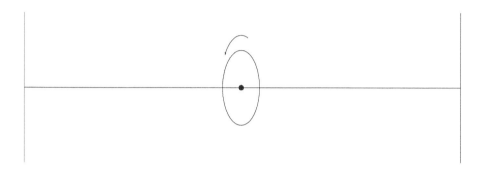

Figure 2.6   A mass attached to a thin elastic bar

Solution 2.2.3 *To model this problem one must conduct the same type of experiments and make the same approximations as in the spring-mass system. For small "twists," i.e., when the twist angle θ is small, the elastic restoring torque due to the bar can be approximated by*

$$T = -k\theta. \tag{2.45}$$

Using Newton's second law for rotating bodies we then have

$$I\ddot{\theta} + k\theta = T_{ext} \tag{2.46}$$

where T_{ext} is the external torque and I is the moment of inertia of m around the axis of rotation which is defined as

$$I = \int_V r^2 \rho(\mathbf{x}) d\mathbf{x}. \tag{2.47}$$

Here r is the distance of $\mathbf{x}$ from the axis of rotation and $\rho(x)$ is the density of the mass attached to the bar.

When frictional forces are also present then for $| \dot{\theta} | \ll 1$, we have

$$F_f = -b\dot{\theta}, \tag{2.48}$$

and the equation of motion for m becomes

$$I\ddot{\theta} + b\dot{\theta} + k\theta = T_{ext}. \tag{2.49}$$

Exercises

1. Find the differential equation which governs the motion of the system shown in Fig. 2.7:

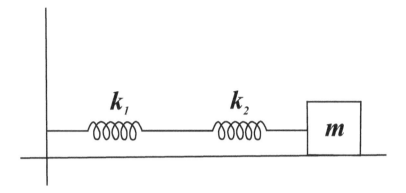

Figure 2.7 A mass attached to two springs "in series"

   Hint: Apply Newton's second law to the massless point P at which the two springs are connected.

2. Repeat Ex. 1 for the system shown in Fig. 2.8.

   Assume that m is always "gliding" on the x-axis.

3. What is the equivalent stiffness for the two springs in the systems of ex. 1,2;i.e., if one wants to replace the two springs by one, what should its stiffness be to yield exactly the same equation of motion for a mass m attached to it? Compare these results to the addition of resistors in series and parallel in an electric circuit.

4. Find the moment of inertia for a thin homogeneous rod of length L and linear density ρ (i.e. mass/unit length) which is rotating around its

   (a) mid-point

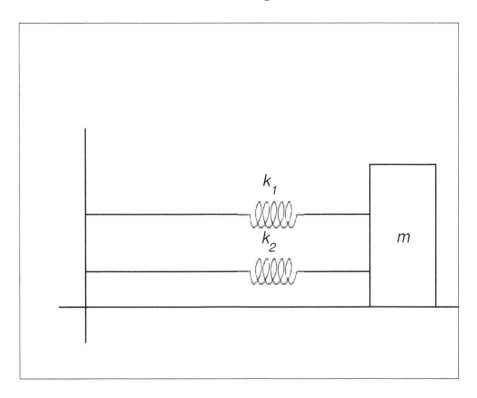

Figure 2.8 A mass connected to two springs in "parallel"

    (b) end point.

5. Model the following systems:

    (a) Two masses attached to three elastic rods (Fig. 2.9).

    (b) Two masses suspended vertically on springs in the gravitational field of the Earth (Fig. 2.10)

Hint: Derive the equation of motion in terms of the displacements x_1, x_2 of the two masses from equilibrium as in example 2.

6. Generalize example 2.2.2 to a system of N masses with $N + 1$ springs as shown in Fig. 2.11.

Hint: Derive equations for the displacements $x_i, i = 1, \ldots, N$ from equilibrium.

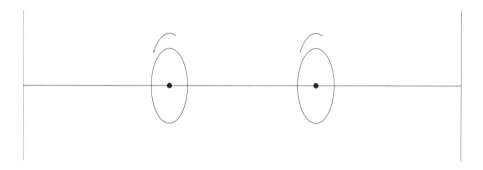

Figure 2.9 Two masses attached to three elastic rods

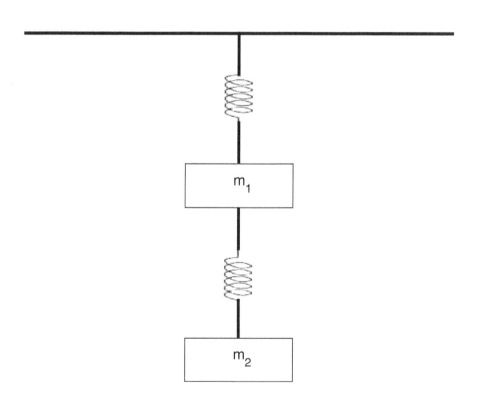

Figure 2.10 Two masses suspended vertically on springs

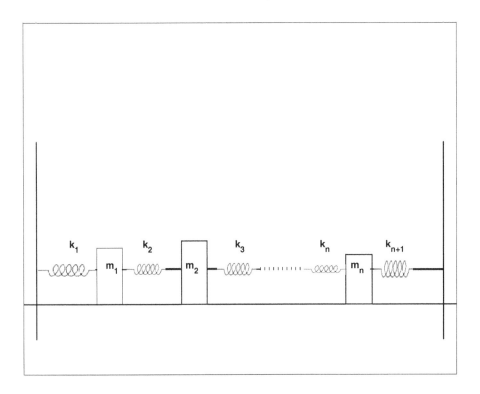

Figure 2.11 N masses attached $N + 1$ springs enclosed by rigid walls

7. Simulate numerically the system in the previous exercise for $n = 2$. Assume $k_i = 1$, $i = 1, 2, 3$, with different mass ratios. For each mass plot x and $\dot{x}$ as a function of time. Plot also $\dot{x}$ versus x versus $\dot{x}$ (phase-diagram). Experiment with different values of k_i to evaluate the impact on the solution.

8. Determine A, ϕ in Equation (2.42) if

   (a) $x(0) = 1$, $\dot{x}(0) = 0$
   (b) $x(0) = 0$, $\dot{x}(0) = 1$.

9. Show that the solution of Equation (2.38) can be written as

$$x = A \cos \omega t + B \sin \omega t, \quad \omega^2 = \frac{k}{m}.$$

10. Explain why Equation (3.84) is linear. What is the order of this equation?

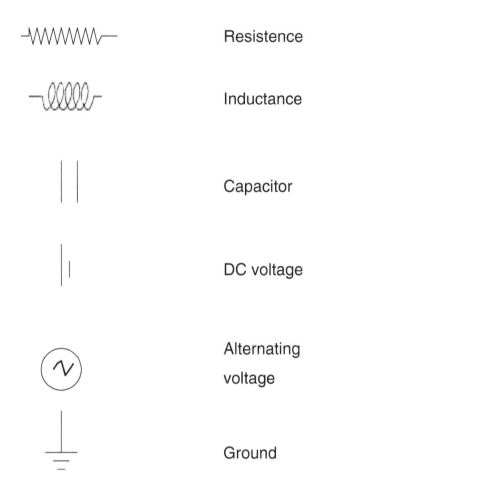

Resistence

Inductance

Capacitor

DC voltage

Alternating
voltage

Ground

Figure 2.12   Components of a RLC circuit

11. Show that when $F_{ext} = 0$ and $b \neq 0$, Equation (3.84) represents a dissipative system. That is, the total energy of the system

$$E(t) = \frac{1}{2}mv^2 + \frac{1}{2}kx^2$$

is a monotonically decreasing function of time.

Solution: Multiply Equation (3.84) by $\dot{x}$ and observe that

$$\ddot{x}\dot{x} = \frac{1}{2}\frac{d}{dt}(\dot{x}^2), \quad x\dot{x} = \frac{1}{2}\frac{d}{dt}(x^2).$$

Hence

$$m\ddot{x}\dot{x} + b(\dot{x})^2 + kx\dot{x} = 0$$

implies

$$\frac{1}{2}m\frac{d}{dt}(\dot{x}^2) + \frac{1}{2}k\frac{d}{dt}(x^2) = -bx^2$$

$$\frac{d}{dt}\left(\frac{1}{2}m\dot{x}^2 + \frac{1}{2}kx^2\right) = -b\dot{x}^2$$

i.e.

$$\frac{dE(t)}{dt} = -b(\dot{x})^2 \leq 0$$

i.e. $E(t)$ is a monotonically decreasing function.

Observe that if $b < 0$ ("negative friction") then $E(t)$ is monotonically increasing.

12. Solve Equation (3.84) with $F_{ext} = 0$ and $b^2 - 4mk > 0$.

    Hint: In some instances one can obtain solutions of differential equations by "inspection" or by making appropriate assumptions on the form of the solution. In this case, assume that $x(t) = Ae^{\alpha t}$ where A, α are constants.

13. Use the results of ex. 12 to show that under present assumptions

    (a) The equation has two independent solutions ϕ_1, ϕ_2

    (b) Show that $\phi = c_1\phi_1 + c_2\phi_2$, where c_1, c_2 are arbitrary constants, is also a solution. This is called the superposition principle. It is true for all linear equations?

    (c) What happens to x as $t \to \infty$?

14. Solve eq. (3.84) with $F_{ext} = 0$ and $b^2 - 4mk < 0$.

15. Consider a spring mass system with a nonlinear spring force

$$F = -kx - \alpha x^3$$

and without friction. Show that for this system

$$E = \frac{1}{2}m\dot{x}^2 + \frac{kx^2}{2} + \frac{\alpha x^4}{4}$$

is constant. Explain why a spring with $\alpha > 0$ is referred to as "hard" while one with $\alpha < 0$ is called "soft."

16. Show that for Rayleigh Equation (2.37) the rate of change in the energy of the system is negative if $\dot{x}^2 \geq b/c$.

    Hint: Multiply this equation by $\dot{x}$.

2.3 ELECTRICAL CIRCUITS

In this section we discuss the modeling of RLC circuits. These circuits contain resistors, capacitors, and inductances.

2.3.1 RLC Circuits

Objective: Build a model which will predict the electric current at any point of an electric circuit.

Background:

The following figure depicts the basic element of an RLC circuit and their universal representation.

Each of these components in a circuit diagram is considered pure; i.e. a resistor has a zero capacity and inductance, etc.

The basic physical quantities of interest in an electric circuit are

1. The electric current $i(x, t)$.

2. The electric potential (or voltage) $V(x, t)$. This potential is measured with respect to a fixed reference point called "ground" – since it is usually the Earth's potential that is used for this purpose.

3. The electric charge Q.

The units (in the MKS system) of the quantities introduced above are as follows:

R - ohms (denoted by Ω)

L - Henries

C - Farads

Q - Coulombs

i - Amperes

V - Volts

Remark: Sometimes one uses a quantity called the conductance G instead of the resistance R where $G = 1/R$. The unit of conductance is called mho (ohm spelled backward).

The relationship between the components and the physical quantities in an electric circuit is as follows:

1. The voltage drop across a resistance is related to the current passing through it by
$$V = Ri. \tag{2.50}$$

2. The potential drop through an inductance L is given by
$$V = L\frac{di}{dt}. \tag{2.51}$$

3. The total electric charge Q on a capacitor is given by
$$Q = CV, \tag{2.52}$$

hence the "virtual current" i in the capacitor is given by
$$i = \frac{dQ}{dt} = C\frac{dV}{dt}. \tag{2.53}$$

Kirchhoff's Laws

Kirchhoff's laws form the basis for the analysis of all electric circuits. To introduce these laws we make the following definitions:

1. A *node* in an electric circuit is a juncture where current can flow along three or more paths (i.e. a point where three or more electric wires are joined together) (see Fig. 2.13).

2. A *loop* in an electric circuit is a sequence of circuit elements which start and end at the same point (i.e. a closed path).

First Kirchhoff law:

The algebraic sum of all the currents at a node is 0. For the node in Fig. 2.13 we have
$$i_1 - i_2 + i_3 = 0.$$

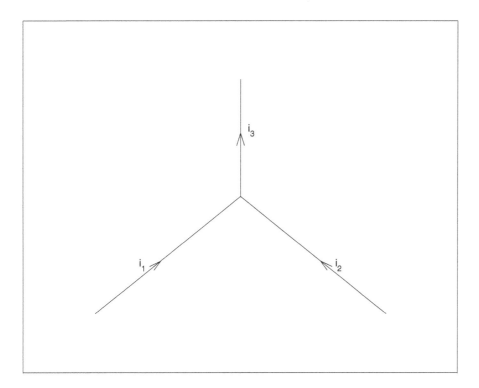

Figure 2.13 Kirchhoff's first law

By convention currents coming to the node are considered positive while those leaving it are negative.

Second Kirchhoff law:

The algebraic sum of the voltage drops around a loop in an electric circuit is equal to the algebraic sum of the external voltage sources in the loop.

2.3.2 Approximations

1. We assume that the resistance, capacitance, or inductance of a given electrical component is independent of the environmental factors (such as temperature, humidity, etc.) and the previous history of the circuit.

2. Cables connecting circuit components have zero resistance, capacity, and inductance.

3. The passage of an electric current through a cable always involves a leakage which leads to a loss of electric energy. For short distances one can

usually ignore this loss. However, over long distances (i.e. transmission lines) one must take these losses into account.

With this data and approximations one can in principle analyze any given circuit. We present a few examples.

Example 2.3.1 *RLC Circuit.*

A simple RLC circuit is illustrated in Fig. 2.14.

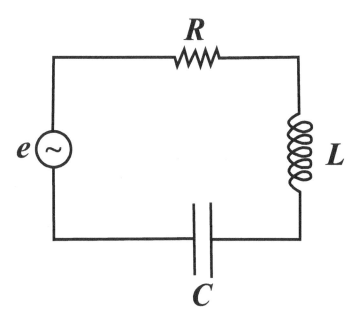

Figure 2.14 RLC circuit

To solve it we first note that there are no nodes in this circuit, and therefore only Kirchhoff's second law applies, hence

$$e(t) = Ri + \frac{Q}{C} + L\frac{di}{dt}. \tag{2.54}$$

Differentiating with respect to t and using (2.53) we obtain

$$\frac{de}{dt} = R\frac{di}{dt} + \frac{1}{C}i + L\frac{d^2i}{dt^2}. \tag{2.55}$$

If $\dfrac{de}{dt}$ is known, then Equation (2.55) constitutes a second order (inhomogeneous) differential equation with constant coefficients for the current i in the circuit.

Example 2.3.2 *Wheatstone Bridge.*

The circuit shown in Fig. 2.14 is used to measure the resistance R_x of a resistor by the use of two fixed resistances R_1, R_2 and a third, variable one, R_3.

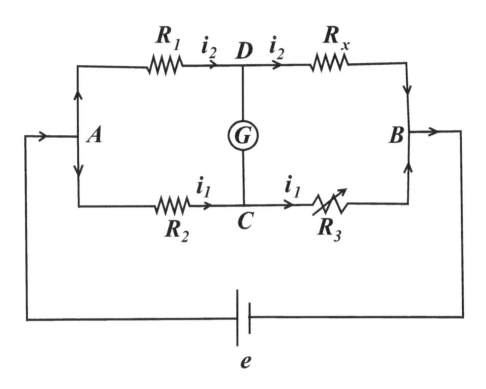

Figure 2.15   Diagram of Wheatstone Bridge

In this circuit e is a low voltage battery and G a galvanometer, i.e. an instrument to measure currents. Once R_x is inserted and the circuit is closed, the operator manipulates R_3 until the current in G is zero (the circuit is then said to be balanced). Apply Kirchhoff's second law to the loops ACD and CBD. In the balanced state we obtain

$$R_1 i_1 - R_2 i_2 = 0, \ \ R_3 i_1 - R_x i_2 = 0, \tag{2.56}$$

hence

$$\frac{R_2}{R_x} = \frac{R_1}{R_3}, \quad \text{i.e. } R_x = \frac{R_2 R_3}{R_1}. \tag{2.57}$$

Example 2.3.3 : *Multicomponent RLC Circuits*

Definition 2.3.1 *We say that a resistance R is equivalent to R_1 and R_2 in a given circuit if the replacement of these resistors by R does not affect the current in the circuit.*

Example 2.3.4 *From Kirchhoff's second law it is easy to see that:*

1. *Two resistors R_1, R_2 in series are equivalent to one resistance $R = R_1 + R_2$*

2. *Two resistors R_1, R_2 in parallel (see Fig. 2.16) are equivalent to one resistor with*

$$\frac{1}{R} = \frac{1}{R_1} + \frac{1}{R_2}.$$

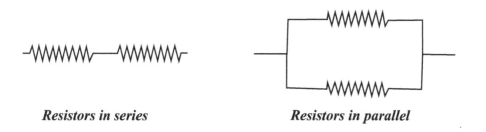

Resistors in series   **Resistors in parallel**

Figure 2.16   Resistors in series and parallel

Example 2.3.5 *Find the equivalent resistance of the "infinite" circuit in Fig. 2.17.*

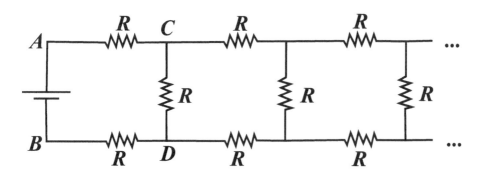

Figure 2.17   Circuit with infinite resistors in parallel

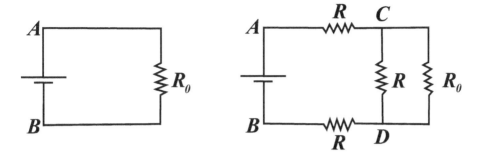

Figure 2.18 Circuits equivalent to the circuit in Fig 2.17

Solution 2.3.1 *Since* "$\infty = \infty - 1$" *the equivalent resistance R_0 of $AB\infty$ is equal to that of $CD\infty$. Hence the circuit $AB\infty$ is equivalent to the one shown in Fig. 2.18 whose total equivalent resistance must also equal R_0.*

Thus we infer

$$R_0 = 2R + \frac{1}{\frac{1}{R} + \frac{1}{R_0}} \text{ i.e. } R_0 = [1 + \sqrt{3}]R.$$

Example 2.3.6 *DC motor.*

A DC motor is an electro-mechanical system consisting of two circuits - the field circuit F, the armament circuit A - and a shaft (Fig. 2.19.).

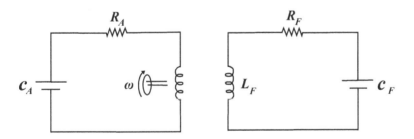

Figure 2.19 A diagram of a DC motor

Background:

The following information about the electro-mechanical coupling in this circuit is needed to model this system.

1. The motion of the shaft causes a potential drop V_s in the armament circuit which is proportional to its angular velocity ω and the field current i_F

$$V_s = c_1 \omega i_F. \tag{2.58}$$

   The proportionality constant c_1 in this equation is called the electromechanical constant of the motor.

2. The torque T exerted on the shaft is proportional to i_A and i_F

$$T = c_2 i_A i_F. \tag{2.59}$$

Model: A mathematical model for this system can be written down now by applying Kirchhoff's second law to the two circuits and Newton's second law to the shaft whose moment of inertia we denote by J.

$$L_F \frac{di_F}{dt} + R_F i_F = e_F \tag{2.60}$$

$$L_A \frac{di_A}{dt} + R_A i_A + c_1 \omega i_F = e_A \tag{2.61}$$

$$J \frac{d\omega}{dt} = c_2 i_A i_F - c_3 \omega \tag{2.62}$$

where c_3 represents frictional damping which is proportional to the angular velocity of the shaft. This is a system of three coupled nonlinear ordinary differential equations of the first order.

Example 2.3.7 *Circuits with nonlinear resistors*

Some electrical circuits contain "vacuum tubes" (or their solid state equivalents). For these elements the resistance is a function of the current, e.g.,

$$R = \mu(i^2 - 1). \tag{2.63}$$

If we substitute this expression for the resistance in Equation. (2.55), we obtain

$$L \frac{d^2 i}{dt^2} + \mu(i^2 - 1) \frac{di}{dt} + \frac{1}{c} i = \frac{de}{dt}. \tag{2.64}$$

Without the forcing term this is equivalent to

$$\ddot{x} + \mu(x^2 - 1)\dot{x} + kx = 0. \tag{2.65}$$

Equation (8.34) is called *Van der Pol equation*. Although this equation was originally derived to model electrical circuits in vacuum tubes, it has been used since then to provide a basic model for the function of nerve cells.

Exercises

1. Prove the statements made in example 2.3.4.

2. Derive model equations for the circuit in Fig. 2.20.

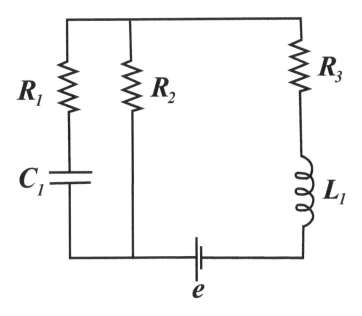

Figure 2.20   Generalized RLC circuit

3. What happens if the direct current source in the previous exercise is replaced by an alternating current source $e(t) = 2\sin(2t)$?

4. Derive model equations for the circuit in the following figure (Fig 2.21).

2.4   POPULATION MODELS

In the first part of this section we introduce the logistic model for the population of one species and the Lotka-Volterra model for two interacting species. Applications to epidemics, chemical reactions, and radioactive decay are discussed in the second part of the section.

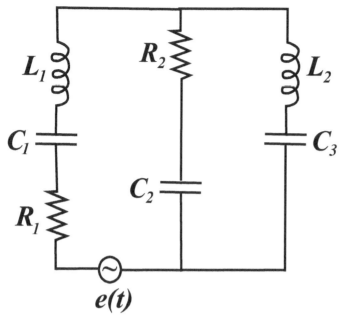

Figure 2.21   Generalized RLC circuit

2.4.1   Logistic Model

Objective: A company which harvests a system of artificial fish pools wants to increase the daily catch but is afraid that this might adversely affect the future population of fish in the pools. As a result a team of biologists and mathematicians is charged with the study and modeling of the fish population.

2.4.2   Prototype Model

Since the pools might contain many species which compete with each other for food and cannibalize each other, it is decided to first study the population of one species of fish in an artificial pool.

2.4.3   Data and Approximations:

1. For the first few generations of fish in the artificial pool the rate of increase of the fish population (i.e. rate of birth - rate of death) was proportional to the population size at that time.

2. As the number of fish in the pool increases, the rate of increase decreases. An investigation of the reasons for this change shows that the

prime reason for this change is the quantity of food available to the fish population.

3. Since the population of the fish, $N(t)$, is large, we can assume that N is a continuous variable rather than discrete.

Model:

To begin with we assume that the rate of birth and death in the fish population is proportional to its size and the time span over which we observe this population

$$N(t + \Delta t) - N(t) = \alpha N(t)\Delta t - \beta N(t)\Delta t = AN(t)\Delta t, \qquad (2.66)$$

where α and β are the birth and death rates respectively. Dividing by Δt and letting $\Delta t \to 0$ we obtain

$$\frac{dN}{dt} = AN(t). \qquad (2.67)$$

Hence

$$N(t) = N(0)e^{At}, \qquad (2.68)$$

where $N(0)$ is the number of fish at time 0.

It follows then that if $A > 0$ (rate of birth is greater than the rate of death), the population will increase exponentially while if $A < 0$ it will decrease exponentially.

To model the competition for food (or resources in general) assume that a "piece of food" is available at some location. The competition for food is represented then by the fact that several fish come together to vie for it. Let us assume, however, that the event where three or more fish coming together at the same time to "grab" the same piece of food is a "rare event" (the population is not "very dense"). The competition will be represented then by a pair of fish coming together to vie for the same piece of food. Hence we can assume that this competition is proportional to the number of pairs in the population which is equal to $\frac{N(N-1)}{2}$. Equation (12.30) will be modified then as

$$N(t + \Delta t) - N(t) = AN(t)\Delta t - \gamma\frac{N(N - 1)}{2}\Delta t, \qquad (2.69)$$

where $\gamma > 0$ is a constant. The minus sign in front of γ represents the adverse effect that food competition has on the population. Combining same terms of N, renaming the constants and letting $\Delta t \to 0$ we finally obtain the *logistic model equation* for the population of one species,

$$\frac{dN}{dt} = aN - bN^2. \tag{2.70}$$

A more formal approach to the derivation of Equation (12.34) is to observe that (12.30) fails to take into account the decrease in $\dot{N}$ as $N(t)$ increases. We can attempt to modify this equation by letting a be a function of N, i.e.,

$$a = a(N).$$

A second order Taylor expansion of $a(N)$ then yields

$$a(N) = a - bN + O(N^2), \; b > 0, \tag{2.71}$$

where the minus sign in front of b in Equation (2.71) is necessary to account for the fact that $\dot{N}$ is decreasing as N increases.

Substituting Equation (2.71) in Equation (12.30) we obtain

$$\frac{dN}{dt} = aN - bN^2 + O(N^3). \tag{2.72}$$

In this equation the term $-bN^2$ ($b > 0$) is interpreted as representing the competition between fish of the same species for food.

2.4.4 Solution of the logistic equation

Although Equation (2.72) is nonlinear, it can be solved analytically using partial fractions as follows:

$$\frac{dN}{N(a - bN)} = dt$$

Hence

$$\left[\frac{1}{N} + \frac{b}{a - bN} \right] dN = a \, dt$$

$$\ln | N | - \ln | a - bN | = at + k$$

which after some algebra yields (note that $N < 0$ is meaningless)

$$N = \frac{aN_0}{bN_0 + (a - bN_0)e^{-at}} \tag{2.73}$$

where N_0 is the initial population of the fish in the pool. We infer from Equation (2.73) that if $a > 0$ and $N_0 \neq 0$ then

$$\lim_{t \to \infty} N(t) = \frac{a}{b}.$$

This ratio is called the saturation level of the pool. It also represents the *equilibrium state* of the fish population since $\dot{N} = 0$ when $N = a/b$.

Fig. 2.22 displays the evolution of the (normalized) fish population with two different initial values of N. The dashed line represents the equilibrium population.

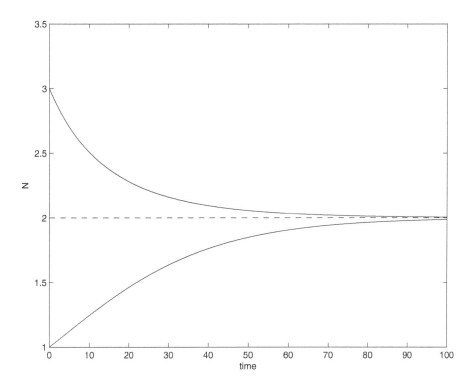

Figure 2.22 Evolution of the normalized fish population in the pool as a function of time for two initial populations

Example 2.4.1 *Predator-prey ecosystem*

Consider a lake in which there are two species of fish. The first of these F feeds on plants while the other P is a predator of F. To write down a mathematical model for this ecosystem we assume that the fish is consumed by

the predator at a rate which is proportional to the population size of the two species (which we denote also by F, P). Hence

$$\frac{dF}{dt} = aF - bF^2 - cFP, \quad b, c > 0. \tag{2.74}$$

As for the predator population we assume that

1. It will become extinct without its prey (rate of death will exceed rate of birth).

2. P increases at a rate which is proportional to F and P (remember as F increases food becomes more abundant for P).

Thus we infer that

$$\frac{dP}{dt} = -kP + eFP. \tag{2.75}$$

The system Equations (2.73) - (2.75) is a special case of Lotka-Volterra equations which model such ecosystems.

Example 2.4.2 *SIR epidemics model*

The SIR model for the spread of disease or epidemics is due to W. O. Kermack and A. G. McKendrick. It assumes that the size of the population remains unchanged (no births and deaths). Since its inception in 1927 the model has been generalized in various ways. Here we consider only the original model.

This model assumes that at time $t = 0$ a part of the population is infected with some infectious disease. We wish to derive equations for the spread of this disease within the population.

To derive these equations the model divides the individuals within the population into three groups:

1. S - individuals susceptible to the disease but not infected as of yet.

2. I - infected individuals who are free to mix in the population at large and transmit the disease.

3. R - individuals who contracted the disease but recovered and are no longer susceptible to the disease.

Subject to the constraint

$$I + S + R = N \tag{2.76}$$

where N is the (fixed) size of the population.

Assuming a free mixing between the S and I groups we infer that:

$$\dot{S} = -aSI \tag{2.77a}$$
$$\dot{I} = -bI + aSI \tag{2.77b}$$
$$\dot{R} = cI \tag{2.77c}$$

where $a, b, c > 0$.

We now use Equations (2.77) to demonstrate how in some instances one can derive inferences about the behavior of a system **even without** solving the differential equations that govern its behavior. In particular we show that if the population size is constant then according to this model:

1. $b = c$.

2. $aS(0) < b$ implies that there will be no epidemics, i.e. $I(t)$ will decrease in time.

3. For all times $S(t) > 0$, thus the population will always contain some healthy individuals.

To prove the first statement we differentiate Equation (2.76) with respect to time

$$\dot{I} + \dot{S} + \dot{R} = 0. \tag{2.78}$$

Substituting Equations (2.77a) - (2.77c) in Equation (2.78) yields $b = c$.

To prove the second statement we first observe that $\dot{S}(t) \leq 0$ since $S, I \geq 0$ in Equation (2.77a). Hence $S(t)$ is a decreasing function of time, i.e. $S(t) \leq S(0)$. Now from Equation (2.77b) we have

$$\dot{I} = (-b + aS)I. \tag{2.79}$$

If $aS(0) < b$ then $aS(t) < b$ for all t. Therefore $\dot{I} < 0$ (i.e. $I(t) \leq I(0)$) and there will be no epidemic.

Finally to prove the third statement we use the chain rule

$$\frac{dS}{dt} = \frac{dS}{dR} \cdot \frac{dR}{dt}. \tag{2.80}$$

Hence (from Equations (2.77a), (2.77b))

$$\frac{dS}{dR} = -\frac{a}{b}S. \tag{2.81}$$

Integrating this equation with respect to R we have

$$S(t) = S(0)e^{-a/bR} > S(0)e^{-a/bN} > 0, \tag{2.82}$$

which proves our statement.

Example 2.4.3 *Chemical reactions*

The basic law which governs the rate of a chemical reaction is the "law of mass action" which states that the rate of a reaction is proportional to the (active) concentration of the reactants. Thus for the reaction

$$\begin{aligned} X \quad + \quad Y &\to Z \\ \frac{d[Z]}{dt} \quad &= \quad k[X][Y] = -\frac{d[X]}{dt} = -\frac{d[Y]}{dt} \end{aligned} \tag{2.83}$$

where $[X], [Y], [Z]$ stand for the (active) concentration of the corresponding chemicals.

If the initial active molar concentration of X, Y is a, b, then at time t

$$\begin{aligned} [X](t) &= a - [Z](t) \\ [Y](t) &= b - [Z](t) \end{aligned} \tag{2.84}$$

since the production of one mole of Z requires one mole of X and Y. Hence

$$\frac{d[Z]}{dt} = k(a - [Z])(b - [Z]). \tag{2.85}$$

This differential equation can be solved by direct integration, and if $a \neq b$ we obtain (using the initial condition $[Z](0) = 0$)

$$k = \frac{1}{t(a - b)} ln \frac{b(a - [Z])}{a(b - [Z])}. \tag{2.86}$$

This relationship is usually used to determine experimentally the rate of the reaction. The solution for concentration of $[Z]$ as a function of time is

$$[Z](t) = \frac{b\exp[k(a-b)(t+c_1)] - a}{\exp[k(a-b)(t+c_1)] - 1}.$$

If $[Z](0) = 0$ then

$$c_1 = \frac{\ln\frac{a}{b}}{k(a-b)}.$$

Example 2.4.4 *Catalytic reactions*

Of particular interest from a chemical point of view are reactions where the addition of some "catalyst" accelerates the rate of a reaction that is "slow going." As an example we consider the oxidation of sulfur dioxide using nitrogen dioxide as a catalyst.

$$NO_2 + SO_2 \xrightarrow{k_1} SO_3 + NO$$
$$NO + \frac{1}{2}O_2 \xrightarrow{k_2} NO_2. \tag{2.87}$$

Observe that the net result of this reaction is

$$SO_2 + \frac{1}{2}O_2 \rightarrow SO_3 \tag{2.88}$$

i.e. the amount of NO_2 in the chemical reactor remains unchanged. Thus a catalyst provides a path for a desired reaction to happen which has a lower activation energy than the uncatalysed reaction.

A model for the reactions in Equation (2.87) consists of five coupled nonlinear differential equations.

$$\frac{d[NO_2]}{dt} = -k_1[NO_2][SO_2] + k_2[NO][O_2]^{1/2} \tag{2.89}$$

$$\frac{d[NO]}{dt} = +k_1[NO_2][SO_2] - k_2[NO][O_2]^{1/2} \tag{2.90}$$

$$\frac{d[SO_3]}{dt} = k_1[NO_2][SO_2] \tag{2.91}$$

$$\frac{d[SO_2]}{dt} = -k_1[NO_2][SO_2] \tag{2.92}$$

$$\frac{d[O_2]}{dt} = -k_2[NO][O_2]^{1/2} \tag{2.93}$$

Example 2.4.5 *Radioactive decay*

The nuclei of many isotopes are not stable and therefore decay over time. In many cases the products of this decay are not stable themselves, and the system then consists of a chain of such reactions. In all these reactions the decay rate is assumed to be proportional to the "population size" viz. to the number of nuclei present.

As a particular example we consider here a chain of such reactions where N_1 decays to N_2, which then decays to a stable nuclei N_3. Here $N_i, i = 1, 2, 3$, represents both the nuclei and their number.

To model these reactions we consider the time interval $[t, t + \Delta t]$. On this interval we have

$$N_1(t + \Delta t) \quad - \quad N_1(t) = -\alpha_1 N_1(t) \Delta t \tag{2.94a}$$

$$N_2(t + \Delta t) \quad - \quad N_2(t) = \alpha_1 N_1(t) \Delta t - \alpha_2 N_2(t) \Delta t \tag{2.94b}$$

$$N_3(t + \Delta t) \quad - \quad N_3(t) = \alpha_2 N_2(t) \Delta t, \alpha_i > 0, i = 1, 2, 3. \tag{2.94c}$$

In Equation (2.94a) the first term on the left hand side represents the number of N_1 nuclei which were converted to N_2, while the second term represents the number of N_2 nuclei which decayed to N_3.

Dividing by Δt and letting $\Delta t \to 0$ we obtain the system

$$\frac{dN_1}{dt} = -\alpha_1 N_1$$

$$\frac{dN_2}{dt} = \alpha_1 N_1 - \alpha_2 N_2$$

$$\frac{dN_3}{dt} = \alpha_2 N_2 \tag{2.95}$$

This is a system of three coupled first order equations. The initial conditions for this system must specify the number of the nuclei N_i, $i = 1, 2, 3$ at some time t_0.

Exercises

1. Derive a model equation for a fish population which consumes plants as well as itself (Hint: Remember that a represents the rate of birth minus the rate of death).

2. In a lake there are three species of fish X, Y, Z. X eats plants that are

highly abundant. Y is a predator of X and Z is a predator of X and Y. Derive a model for this ecosystem.

3. Derive a model for an ecosystem which consists of two species X, Y under the following assumptions:

   a. Both species compete for the same nutrient whose supply is limited.

   b. There is a migration of X (from outside the ecosystem) at a rate r per unit time.

4. Explain why the coefficient of SI in Equations (2.77a), (2.77b) is the same.

5. Develop model equations for the concentrations of X, Y, and A in the reactions

$$A \; + \; X \rightarrow B + 2X \qquad (2.96)$$

$$X \; + \; Y \rightarrow B + 2Y \qquad (2.97)$$

$$A \; + \; Y \rightarrow B. \qquad (2.98)$$

Observe that the net result of these reactions is $2A \rightarrow 3B$.

6. Show that Equation (2.86) is the solution of Equation (2.85).

7. Solve Equation (2.85) when $a = b$.

8. Show (by substitution) that the solution of Equation (2.95) with the initial conditions $N_1(0) = N_0$, $N_2(0) = N_3(0) = 0$ is

$$N_1(t) \;=\; N_0 e^{-\alpha_1 t} \qquad (2.99)$$

$$N_2(t) \;=\; \lambda_2 N_0 \left[e^{-\alpha_1 t} - e^{-\alpha_2 t} \right] \qquad (2.100)$$

$$N_3(t) \;=\; N_0 \left[1 - \lambda_2 e^{-\alpha_1 t} + \lambda_1 e^{-\alpha_2 t} \right] \qquad (2.101)$$

where $\lambda_1 = \dfrac{\alpha_1}{\alpha_2 - \alpha_1}$, $\lambda_2 = \dfrac{\alpha_2}{\alpha_2 - \alpha_1}$.

9. Find the solution of the system, Equation (2.95), when $\alpha_1 \approx \alpha_2$. Assume that $N_1(0) = N_0$ and $N_2(0) = N_3(0) = 0$.

10. Let P, Q be the price and quantity of a certain fuel on the open market. A "population" model for the evolution of these variables was proposed in the form

$$\dot{P} = aP/Q - bP^2$$

$$\dot{Q} = cPQ - dQ^2$$

where $a, b, c, d > 0$. Justify and discuss the meaning of this model.

2.5 MOTION IN A CENTRAL FORCE FIELD

In this section we discuss the equations of motion for a body in a central force field, i.e., when $F = f(r)\mathbf{r}$ where $\mathbf{r}$ is the radius vector from the origin and $r = |\mathbf{r}|$. Then various systems (such as the pendulum) are considered. We begin, however, by introducing the radial coordinate system in R^2.

2.5.1 Radial Coordinate System in R^2

In this coordinate system (which should not be confused with the polar coordinate system) we attach two perpendicular unit vectors (a frame) to any point in the plane. The first of these $\mathbf{e}_r$ is a unit vector along the radius vector connecting the point to the origin of some fixed Cartesian system, and the second $\mathbf{e}_\theta$ is a unit vector orthogonal to $\mathbf{e}_r$ in the counterclockwise direction (see Fig. 2.23).

Using simple trigonometry we infer that the expressions of $\mathbf{e}_r$, $\mathbf{e}_\theta$ in Cartesian coordinates are given by

$$\mathbf{e}_r = \cos\theta\mathbf{i} + \sin\theta\mathbf{j}, \mathbf{e}_\theta = -\sin\theta\mathbf{i} + \cos\theta\mathbf{j} \tag{2.102}$$

where θ is the angle between the radius vector and the positive x axis.

To obtain expressions for the velocity and acceleration in this coordinate system we observe that always

$$\mathbf{x} = r\mathbf{e}_r. \tag{2.103}$$

Hence

$$\dot{\mathbf{x}} = \dot{r}\mathbf{e}_r + r\dot{\mathbf{e}}_r \tag{2.104}$$

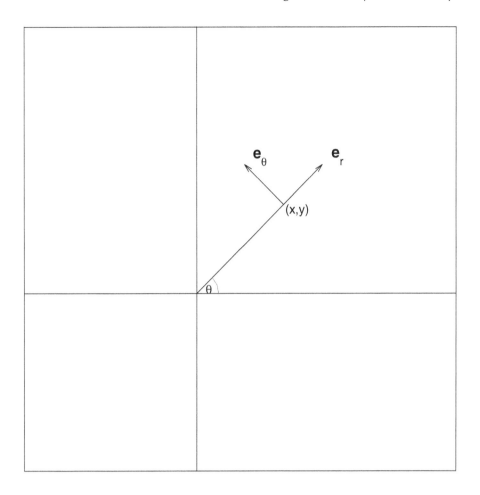

Figure 2.23   Radial coordinate system

and

$$\ddot{\mathbf{x}} = \ddot{r}\mathbf{e}_r + 2\dot{r}\dot{\mathbf{e}}_r + r\ddot{\mathbf{e}}_r \qquad (2.105)$$

but from Equation (2.102) we infer that

$$\dot{\mathbf{e}}_r = \dot{\theta}\mathbf{e}_\theta, \quad \dot{\mathbf{e}}_\theta = -\dot{\theta}\mathbf{e}_r \qquad (2.106)$$

$$\ddot{\mathbf{e}}_r = \ddot{\theta}\mathbf{e}_\theta - \dot{\theta}^2\mathbf{e}_r. \qquad (2.107)$$

Inserting Equation(2.106) - Equation (2.107) in Equation (2.104)-Equation (2.105) we finally obtain

$$\dot{\mathbf{x}} = \dot{r}\mathbf{e}_r + r\dot{\theta}\mathbf{e}_\theta \qquad (2.108)$$

$$\ddot{\mathbf{x}} = (\ddot{r} - r\dot{\theta}^2)\mathbf{e}_r + (r\ddot{\theta} + 2\dot{r}\dot{\theta})\mathbf{e}_\theta. \qquad (2.109)$$

Remark: It is useful in some applications to introduce radial coordinates in

3-dimensions; i.e., attach to each point in space a triad of orthonormal vectors. If (r, θ, ϕ) are the coordinates of a point in spherical coordinates (where ϕ is the angle between the radius vector and z while θ is the azimuthal angle) then these vectors are

$$
\begin{aligned}
\mathbf{e}_r &= (\sin \phi \cos \theta, \sin \phi \sin \theta, \ \cos \phi) \\
\mathbf{e}_\phi &= (\cos \phi \cos \theta, \ \cos \phi \sin \theta, \ -\sin \phi) \\
\mathbf{e}_\theta &= (-\sin \theta, \ \cos \theta, 0).
\end{aligned}
\tag{2.110}
$$

2.5.2 Linear Pendulum

Objective: Derive the equation of motion of a system composed of a rigid (massless) rod and a mass attached at its end in the gravitational field of the Earth.

For this system it is natural to redefine the angle θ as the angle between the radius vector and the negative y axis (see Fig. 2.24). In this case the expressions for $\mathbf{e}_r, \mathbf{e}_\theta$ are given by

$$
\mathbf{e}_r = \sin \theta \mathbf{i} - \cos \theta \mathbf{j}, \quad \mathbf{e}_\theta = \cos \theta \mathbf{i} + \sin \theta \mathbf{j}.
\tag{2.111}
$$

However the expressions for the velocity and acceleration, Equations (2.108) - (2.109) remain unchanged.

Model: Let $\mathbf{x}(t)$ denote the position of the center of mass of m at time t; then in radial coordinates

$$
\mathbf{x} = L\mathbf{e}_r.
\tag{2.112}
$$

Since L is constant, we obtain from Equation (2.109)

$$
\begin{aligned}
\ddot{\mathbf{x}} &= L\ddot{\mathbf{e}}_r \\
\ddot{\mathbf{x}} &= L(\ddot{\theta}\mathbf{e}_\theta - \dot{\theta}^2\mathbf{e}_r).
\end{aligned}
\tag{2.113}
$$

Expressing the gravitational force in radial coordinates

$$
\mathbf{F}_g = -mg\mathbf{j} = -mg \sin \theta \mathbf{e}_\theta + mg \cos \theta \mathbf{e}_r
\tag{2.114}
$$

and using Newton's second law yields

$$
m\ddot{\mathbf{x}} = mL(\ddot{\theta}\mathbf{e}_\theta - \dot{\theta}^2\mathbf{e}_r) = -mg \sin \theta \mathbf{e}_\theta + mg \cos \theta \mathbf{e_r} - T\mathbf{e}_r
\tag{2.115}
$$

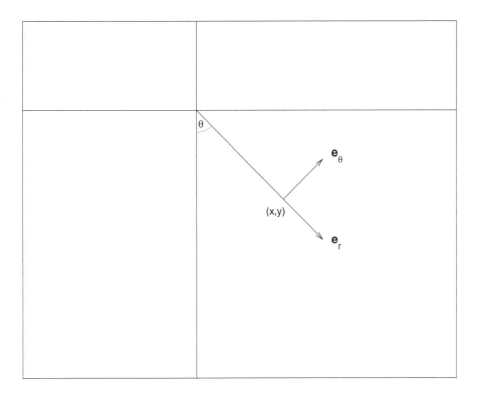

Figure 2.24 Radial coordinate system for the pendulum

where T is the tension in the rod. Rewriting Equation (2.115) in components form we obtain

$$L\ddot{\theta} = -g\sin\theta \tag{2.116}$$

$$mL\dot{\theta}^2 = T - mg\cos\theta. \tag{2.117}$$

The second of these equations can be considered as an equation for T while the first is the desired equation of motion for the pendulum, i.e.

$$\ddot{\theta} = -\frac{g}{L}\sin\theta = -\frac{g}{L}\left(\theta - \frac{\theta^3}{3!} + \dots\right). \tag{2.118}$$

We observe that Equation (2.118) is nonlinear in θ. However for small θ we can approximate Equation (2.118) by

$$\ddot{\theta} + \frac{g}{L}\theta = 0 \tag{2.119}$$

which is formally the same equation as for the spring mass system without friction. The general solution of this equation is $\theta = A\cos(\omega t + \phi)$ where A, ϕ are integration constants and $\omega^2 = \frac{g}{L}$. The period of this pendulum is $P = \frac{2\pi}{\omega}$.

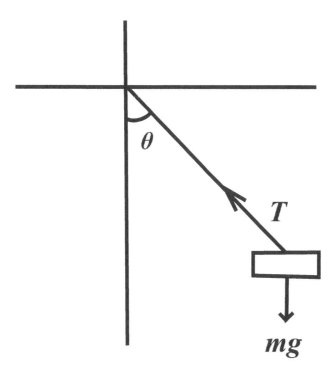

Figure 2.25 Force diagram for the linear pendulum

2.5.3 Nonlinear Pendulum

We now want to treat the nonlinear Equation (2.118) subject to the initial values $\theta(0) = \alpha$ and $\dot{\theta} = 0$. To find the solution of this problem we first multiply eq. (2.118) by $d\theta/dt$. This yields

$$\frac{1}{2}\frac{d}{dt}\left(\frac{d\theta}{dt}\right)^2 = -\omega^2 \sin(\theta)\frac{d\theta}{dt} \tag{2.120}$$

where $\omega^2 = \frac{g}{L}$. We can integrate this equation once on the interval $[0, t]$ to obtain (using the initial conditions):

$$\left(\frac{d\theta}{dt}\right)^2 = 2\omega^2(\cos\theta - \cos\alpha). \tag{2.121}$$

This can be rewritten also as

$$\left(\frac{d\theta}{dt}\right)^2 = 2\omega^2(-1 + \cos\theta + 1 - \cos\alpha)$$

using the identity $1 - \cos(\phi) = 2\sin^2(\phi/2)$ we then obtain

$$\left(\frac{d\theta}{dt}\right)^2 = 4\omega^2(\sin^2\frac{\alpha}{2} - \sin^2\frac{\theta}{2}).$$

We now introduce a new variable ψ which is defined by $\sin(\psi) = \frac{\sin(\theta/2)}{\sin(\alpha/2)}$. The expression above becomes

$$\left(\frac{d\theta}{dt}\right)^2 = 4\omega^2 \sin^2 \frac{\alpha}{2} \cos^2 \psi. \tag{2.122}$$

Using the definition of ψ we obtain (using implicit differentiation) for the left hand side of this equation

$$\left(\frac{d\theta}{dt}\right)^2 = 4 \sin^2 \frac{\alpha}{2} \frac{\cos^2 \psi}{\cos^2 \frac{\theta}{2}} \left(\frac{d\psi}{dt}\right)^2.$$

Equation (3.21) transforms therefore into

$$\frac{d\psi}{dt} = \omega \cos \frac{\theta}{2} = \omega \sqrt{1 - k^2 \sin^2 \psi} \tag{2.123}$$

where $k = \sin(\alpha/2)$. This equation can be integrated now, and we obtain the implicit solution to our problem in the form

$$a + \omega t = \int_0^{\psi(t)} \frac{d\psi}{\sqrt{1 - k^2 \sin^2 \psi}} \tag{2.124}$$

where a is an integration constant. The integral on the right hand side of Equation (2.124) is an elliptic integral of the first kind, since at time 0 $\sin(\psi) = 1$ we infer that $\psi(0) = \pi/2$. Hence

$$a = \int_0^{\frac{\pi}{2}} \frac{d\psi}{\sqrt{1 - k^2 \sin^2 \psi}}.$$

We can rewrite Equation(2.124) in the form

$$t = \frac{1}{\omega} \int_{\frac{\pi}{2}}^{\psi(t)} \frac{d\psi}{\sqrt{1 - k^2 \sin^2 \psi}}. \tag{2.125}$$

The pendulum will complete one period when θ first takes the value $\theta = \alpha$, i.e., $\psi = 2\pi + \frac{\pi}{2}$. Using the following identity for elliptic functions $f(k, \phi + n\pi) = nf(k, \pi) + f(k, \phi)$ (see next subsection) we conclude that the period of the nonlinear pendulum is

$$T = \frac{2}{\omega} f(k, \pi)$$

or explicitly

$$T = \frac{2}{\omega} \int_0^{\pi} \frac{d\psi}{\sqrt{1 - k^2 \sin^2 \psi}}.$$

2.5.4   A Short Introduction to Elliptic Functions

Elliptic integrals of the first and second kinds are defined as

$$f(k, \phi) = \int_0^\phi \frac{d\psi}{\sqrt{1 - k^2 \sin^2 \psi}} \tag{2.126}$$

$$e(k, \phi) = \int_0^\phi \sqrt{1 - k^2 \sin^2 \psi}\, d\psi \tag{2.127}$$

respectively. The variable ϕ is called the amplitude of f or e. Related to these integrals are the functions

$$\begin{align}
sn(k, \phi) &= \sin(f(k, \phi)), \tag{2.128}\\
cn(k, \phi) &= \cos(f(k, \phi)), \tag{2.129}\\
dn(k, \phi) &= \sqrt{1 - k^2 sn(k, \phi)}. \tag{2.130}
\end{align}$$

The expressions $f(\pi/2, k)$ and $e(\pi/2, k)$ are called complete elliptic integrals.

The basic properties of these integrals and functions are:

$$\begin{align}
f(k, \pi) &= 2f(k, \pi/2), \quad e(k, \pi) = 2e(k, \pi/2), \tag{2.131}\\
f(k, \phi + n\pi) &= nf(k, \pi) + f(k, \phi). \tag{2.132}\\
e(k, \phi + n\pi) &= ne(k, \pi) + e(k, \phi). \tag{2.133}\\
sn^2(\phi, k) &+ cn^2(\phi, k) = 1 \tag{2.134}\\
sn(-\phi, k) &= -sn(\phi, k), \quad cn(-\phi, k) = cn(\phi, k). \tag{2.135}
\end{align}$$

Remark: A standard text which contains an extensive list of functions that appear in many applications is

M. Abramowitz and A. Stegun - Handbook of Mathematical functions (Dover Publications).

2.5.5   Motion of a Projectile on a Rotating Earth

Example 2.5.1 *Derive the equations of motion of a projectile taking into account the rotation of the Earth.*

Solution 2.5.1 *When the rotation of the Earth is taken into consideration, it is natural to use an inertial coordinate system fixed at the Earth's center to describe the motion. In this coordinate system the Earth's rotation introduces* *an additional term in Newton's second law which is called Coriolis force. The expression for this force is*

$$F_c = -2m\boldsymbol{\omega} \times \dot{\mathbf{x}} \tag{2.136}$$

where $\boldsymbol{\omega} = (0, 0, \omega)$ is the Earth's angular velocity and $\mathbf{x} = (x, y, z)$ is the position of the particle. The equations of motion of the projectile are then

$$m\ddot{\mathbf{x}} = m\mathbf{g} - 2m\boldsymbol{\omega} \times \dot{\mathbf{x}} \tag{2.137}$$

where in spherical coordinates,

$$\mathbf{g} = -g(\sin\phi\cos\theta, \sin\phi\sin\theta, \cos\phi). \tag{2.138}$$

For a projectile with limited range θ, ϕ can be considered to be constants and the equations of motion in scalar form are

$$\ddot{x} = -g\sin\phi\cos\theta - \omega\dot{y} \tag{2.139}$$

$$\ddot{y} = -g\sin\phi\sin\theta + \omega\dot{x} \tag{2.140}$$

$$\ddot{z} = -g\cos\phi. \tag{2.141}$$

This is a system of three coupled second order differential equations which must be solved simultaneously for x, y, z (however, note that the equation for the z coordinate is decoupled from the other two equations).

2.5.6   A Particle in a Central Force Field

Objective: Derive the equation for the motion of a body of mass m in a force field where the force acting on it is $F = mf(r)\mathbf{e}_r$.
Solution:
  From Newton's second law we know that

$$m\ddot{x} = mf(r)\mathbf{e}_r. \tag{2.142}$$

Using Equation (2.109) we infer that

$$\ddot{r} - r\dot{\theta}^2 = f(r) \tag{2.143}$$

$$r\ddot{\theta} + 2\dot{r}\dot{\theta} = 0. \tag{2.144}$$

The second of these equations can be integrated once (after multiplication by r), and it follows that

$$r^2\dot{\theta} = \text{constant} = h. \tag{2.145}$$

(This is the law of conservation of angular momentum.) Hence we can write formally

$$\frac{d}{dt} = \frac{h}{r^2}\frac{d}{d\theta}$$

and

$$\frac{d^2}{dt^2} = \frac{h}{r^2}\frac{d}{d\theta}\left(\frac{h}{r^2}\frac{d}{d\theta}\right). \tag{2.146}$$

Substituting Equation (9.9) and Equation (2.146) in Equation (2.143) we obtain

$$\frac{h}{r^2}\frac{d}{d\theta}\left(\frac{h}{r^2}\frac{dr}{d\theta}\right) - \frac{h^2}{r^3} = f(r). \tag{2.147}$$

To simplify this equation we introduce $u = \dfrac{1}{r}$ and note that $\dfrac{du}{d\theta} = -\dfrac{1}{r^2}\dfrac{dr}{d\theta}$. This leads to

$$h^2 u^2 \left[\frac{d^2 u}{d\theta^2} + u\right] = -f\left(\frac{1}{u}\right) \tag{2.148}$$

For the Newtonian gravitational field we have $f(r) = -\frac{\mu}{r^2}$ where μ is a constant (for the resulting orbits see Ex. 5).

2.5.7 Motion of a Rocket

Objective: Derive a prototype model for the motion of a one stage rocket.

Background: To derive the required equation we have to use the law of momentum conservation (actually this is a restatement of Newton's second law). This law states that the net change in the momentum of a physical system in time Δt is (approximately) equal to $f_{ext}\Delta t$ where f_{ext} is the sum of the external forces acting on the system.

Approximations:

1. We assume that the flight of the rocket is along the radius vector of the Earth, i.e., "straight up."

2. Neglect the existence of the Earth atmosphere

3. Neglect the rotation of the Earth.

4. Neglect the ignition stage and the stability of the rocket in flight (usually to stabilize this motion the rocket rotates around its axis).

5. Neglect the gravitational field of the Sun and Moon.

6. Assume the Earth is a perfect sphere.

Model: The motion of the rocket is due to the fuel that is being burnt. This generates a stream of gases which leave the rocket and thus exert a forward thrust. To a good approximation the velocity u of these gases is constant (except at ignition). Since the only external force acting on the rocket is the gravitational field of the Earth we obtain by applying the law of momentum conservation (see Fig. 2.26).

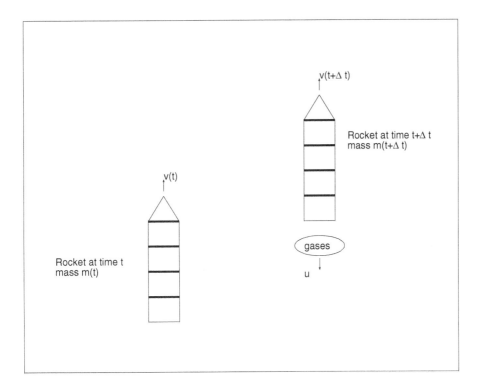

Figure 2.26 Rocket at time t and $t + \Delta t$

Momentum of gases leaving rocket in $[t, t + \Delta t]$	+	Momentum of rocket at $[t + \Delta t]$	-	Momentum of rocket at t	=	Momentum imparted by gravitational field in $[t, t + \Delta T]$

Thus if $v(t)$ is the velocity of the rocket and $m(t)$ is its mass we have

$$\left[-\frac{dm}{dt}\Delta t\right](v(t) - u) + m(t + \Delta t)v(t + \Delta t) - m(t)v(t) = -m(t)g(h(t))\Delta t$$

(2.149)

where $h(t)$ is the altitude of the rocket at t. Dividing by $\Delta t \to 0$ we obtain

$$\frac{d}{dt}(m(t)v(t)) = \frac{dm}{dt}(v(t) - u) - m(t)g(h(t)).$$

(2.150)

If we let $g(h(t))$ be a constant, then the equation of motion reduces to

$$\frac{dv}{dt} = -u\frac{1}{m}\frac{dm}{dt} - g.$$

(2.151)

Integrating with respect to t we obtain

$$v = v_0 + u \ln \frac{m_0}{m(t)} - gt$$

(2.152)

where m_0 is the initial mass of the rocket and v_0 is its initial velocity. We see that if $u \cong 3\text{km/sec}$ and the flight of the rocket is short, the effect of the last term is negligible and therefore

$$v = v_0 + u \ln \frac{m_0}{m(t)}.$$

(2.153)

In a one stage rocket

$$m_0 = m_p + m_f + m_s$$

(2.154)

where m_p is the mass of the payload, m_f the initial mass of the fuel and m_s is the mass of the structures engines and fuel containers. The final velocity of such a rocket (i.e., when the fuel is exhausted) is given then by

$$v_F = v_0 + u \ln \frac{m_0}{m_p + m_s}.$$

(2.155)

Usually $\dfrac{m_s}{m_0} \approx \dfrac{1}{10}$. We see therefore that even if $m_p = 0$ and the effect of gravity is neglected

$$v_F = v_0 + u \ln 10.$$

Letting $v_0 = 0$ and $u = 3$ we obtain $v_F \approx 7.2$ km/sec.

2.5.8   Multistage Rockets

Objective:

A team of engineers, physicists, and mathematicians is asked to examine the desirable gross structure of a rocket capable of putting a satellite in orbit 600 kms above the Earth surface (concentrating on the energy aspects of the problem).

Approximations:

We make here the same approximations as in the previous section but in addition we neglect the final stage of actually putting the satellite in orbit and assume a circular orbit for the satellite.

Reduction of the problem:

I. Determine the speed of the satellite which will keep it in orbit 600 km above ground.

II. Develop and solve a model for the motion of a multistage rocket.

III. Find out the optimal structure of the rocket to be measured by the ratio of payload ($\equiv$ satellite)/total weight.

We discuss each of these points separately.

1. Once the satellite is in orbit, there are two forces which act on it, viz. the gravitational force of the Earth given by $F_G = \frac{\mu m}{r^2}$ and the centrifugal force $F_c = \frac{mv^2}{r}$, where μ is a constant depending on the mass of the Earth and the nature of the gravitation force. We also know that on the surface of the Earth:

$$mg = \mu m / R^2$$

where R is the radius of the Earth (approximately 6400 km). For the satellite to stay in orbit the two forces must cancel each other, i.e.

$$\frac{mv^2}{r} = \mu m / r^2$$

but $\mu = gR^2$ hence $v^2 = gR^2/r$ for $r = 6400 + 600 = 7000$ km. We obtain

$$v \cong 7.6 km/sec.$$

Using the results obtained in the previous section we see that using present technology it is impossible to put a satellite in orbit using a one stage rocket. The reason for the poor performance of this rocket is due to the fact that the engine has to carry not only the payload but also the whole structure up to the final orbit. Ideally, however, one should be able to jettison useless weight as the burning proceeds, i.e. keep the mass ratio between fuel and engines and containers constant.

II. **Motion of a One Stage Ideal Rocket.**

Let us therefore examine the performance of such a rocket. Let $m(t)$ be the rocket mass (without the payload) at time t. Of this the mass of the fuel is $m_f(t) = (1-\lambda)m(t)$ and the mass of the structure is $m_s = \lambda m(t)$. The ratio $\frac{m_f}{m_s} = \frac{1-\lambda}{\lambda}$ will remain constant throughout the flight if λ is independent of time. When λ is constant we have $\lambda m_f = (1-\lambda)m_s$ and hence $m_s = 0$ when the fuel is exhausted (i.e. $m_f = 0$).

We now show that λ will remain constant in time if on the interval $[t, t+\Delta t]$ the mass of the fuel burned equals $\Delta m_f(t) = -(1-\lambda)\frac{dm}{dt}\Delta t$ and the corresponding mass of the structure that is jettisoned away is $\Delta m_s(t) = -\lambda \frac{dm}{dt}\Delta t$ (remember that the total mass change of the rocket on this time interval is $\left(-\frac{dm}{dt}\Delta t\right)$) .

In fact since $\frac{m_f(t)}{m_s(t)} = \frac{1-\lambda}{\lambda}$ and $\frac{\Delta m_f(t)}{\Delta m_s(t)} = \frac{1-\lambda}{\lambda}$ we infer that

$$\frac{m_f(t+\Delta t)}{m_s(t+\Delta t)} = \frac{m_f(t) - \Delta m_f(t)}{m_s(t) - \Delta m_s(t)} = \frac{1-\lambda}{\lambda}.$$

The equation for the conservation of momentum will now read (neglecting gravity)

$$m(t)v(t) = m(t+\Delta t)v(t+\Delta t) - \lambda \frac{dm}{dt}\Delta t v(t) - (1-\lambda)\frac{dm}{dt}\Delta t(v(t)-u).$$

Dividing by Δt and letting $\Delta t \to 0$ we then obtain

$$m\frac{dv}{dt} = -(1-\lambda)u\frac{dm}{dt},$$

therefore

$$v(t) = (1 - \lambda)u \ln \frac{m_0}{m(t)}.$$

At the final state of this rocket we now have $m_f = m_p$ and hence

$$v_F = (1 - \lambda)u \ln \frac{m_0}{m_p}.$$

If we assume that in order to account for air resistance, gravity, etc. we need $v_F = 10.5$ km/sec (rather than 7.6 km/sec) and $\lambda \cong 0.1$ we obtain $m_0/m_p = 50$. Thus the payload in this idealized situation cannot exceed $1/50$ of the initial mass of the whole system.

III. **Optimal Design of a Multistage Rocket.**

Naturally the ideal performance discussed above is impossible to implement in practice. We must therefore approximate it by building a multistage rocket and jettison each stage when its fuel is exhausted.

Let m_i = mass of fuel and structures of the i-th stage. Of this λm_i is the mass of the structure (engines and containers) and $(1 - \lambda)m_i$ is fuel. (We assume same λ for all stages.)

The total mass of a three stage rocket will then be

$$m_t = m_p + m_1 + m_2 + m_3.$$

The equations that govern the motion of this rocket during the "burning" of the first stage are the same as for the one stage rocket. Hence the final speed of the rocket at the end of the first stage is

$$v_1 = u \ln \frac{m_0}{m_p + \lambda m_1 + m_2 + m_3}.$$

At this moment the structural mass λm_1 is dropped and the second stage is ignited. When the fuel in the second stage is exhausted, the final speed of the rocket will be

$$v_2 = v_1 + u \ln \frac{m_p + m_2 + m_3}{m_p + \lambda m_2 + m_3}$$

and similarly at the end of the third stage we have

$$v_F = v_3 = v_2 + u \ln \frac{m_p + m_3}{m_p + \lambda m_3}.$$

Combining all these results we obtain

$$\frac{v_F}{u} = \ln\left(\frac{m_0}{m_p + \lambda m_1 + m_2 + m_3}\right)\left(\frac{m_p + m_2 + m_3}{m_p + \lambda m_2 + m_3}\right)$$
$$\times \left(\frac{m_p + m_3}{m_p + \lambda m_3}\right).$$

We now try to maximize m_p/m_0 for fixed v_F, u, and λ. To this end we introduce the variables

$$x_1 = \frac{m_0}{m_p + m_2 + m_3}, \quad x_2 = \frac{m_p + m_2 + m_3}{m_p + m_3}, \quad x_3 = \frac{m_p + m_3}{m_p}$$

therefore

$$\frac{v_F}{u} = \ln\left\{\left(\frac{x_1}{1 + \lambda(x_1 - 1)}\right)\left(\frac{x_2}{1 + \lambda(x_2 - 1)}\right)\left(\frac{x_3}{1 + \lambda(x_3 - 1)}\right)\right\}.$$

Since m_p appears in the denominators of x_1, x_2, x_3, we would like to minimize the values of these variables. However, by the symmetry of this expression in x_1, x_2, x_3 we must have at minimum $x_1 = x_2 = x_3 = x$. Therefore at minimum

$$\frac{v_F}{u} = \ln\left(\frac{x}{1 + \lambda(x - 1)}\right)^3 \quad \text{and hence} \quad x = \frac{1 - \lambda}{q - \lambda}$$

where $q = \exp(-v_F/3u)$. But

$$m_0/m_p = x_1 x_2 x_3 = \left(\frac{1 - \lambda}{q - \lambda}\right)^3$$

which gives for the values of λ, u, v_F used so far $m_0/m_p = 77$.

2.5.9 Control of a Satellite in Orbit

Consider a satellite in a circular orbit around the Earth which is equipped with thrusters to correct deviations in its orbit. These deviations are in general due to several "secondary effects" such as the gravitation of the Sun and the Moon, friction with the upper atmosphere, etc. We want to derive model equations for the motion of the satellite so that one can find out how to use the thrusters to effect the needed corrections in orbit.

Solution:

a. General Equations of Motion:

If we assume that the satellite always remains in one plane and that the only force acting on it is the Earth's gravitation we deduce from Equation (2.109) that its equations of motion are

$$\ddot{r} = r\dot{\theta}^2 - \frac{GM}{r^2} + u_1 \tag{2.156}$$

$$\ddot{\theta} = -\frac{2\dot{r}\dot{\theta}}{r} + \frac{1}{r}u_2. \tag{2.157}$$

In these equations G is the gravitational constant, M the Earth's mass, and u_1, u_2 are the radial and tangential thrusts.

If initially the satellite is in a circular orbit of radius R and constant angular velocity ω then (since in this orbit the centrifugal and gravitational forces must be equal and opposite)

$$mR\omega^2 = \frac{GMm}{R^2} \quad \text{i.e.} \quad R^3\omega^2 = GM, \tag{2.158}$$

and this is also a solution of Equations (2.156), (2.157) with $u_1 = u_2 = 0$.

b. Equations for Small Deviations in Orbit.

Since we consider only small deviations from the circular orbit described above, it is natural to introduce as variables the deviations of the satellite in position and velocity from the parameters of this orbit:

$$x_1 = \frac{r - R}{R}, \quad x_2 = \frac{\dot{r}}{R} \tag{2.159}$$

$$x_3 = \theta - \omega t, \quad x_4 = \dot{\theta} - \omega$$

Substituting these variables in each of the terms that appear in Equations (2.156) - (2.157) yields

$$R\dot{x}_2 = R(1 + x_1)(\omega + x_4)^2 - \frac{GM}{R^2(1 + x_1)^2} + u_1$$

$$\dot{x}_4 = -\frac{2Rx_2(\omega + x_4)}{R(1 + x_1)} + \frac{u_2}{R(1 + x_1)}.$$

We now use the following expansions

$$\frac{1}{1 + x} = 1 - x + x^2...$$

$$\frac{1}{(1+x)^2} = 1 - 2x - 3x^2...$$

(the second expansion is obtained by differentiating the first) and neglect all the nonlinear terms that appear in these equations (remember: the deviations and the thrusts u_1, u_2 are assumed to be small). Using Equation (2.158) we obtain the following system of differential equations:

$$\frac{d\mathbf{x}}{dt} = A\mathbf{x} + \frac{1}{R}B\mathbf{u} \qquad (2.160)$$

where $x = \begin{bmatrix} x_1 \\ x_2 \\ x_3 \\ x_4 \end{bmatrix}$, $u = \begin{bmatrix} u_1 \\ u_2 \end{bmatrix}$ and

$$A = \begin{bmatrix} 0 & 1 & 0 & 0 \\ 3\omega^2 & 0 & 0 & 2\omega \\ 0 & 0 & 0 & 1 \\ 0 & -2\omega & 0 & 0 \end{bmatrix}, \qquad B = \begin{bmatrix} 0 & 0 \\ 1 & 0 \\ 0 & 0 \\ 0 & 1 \end{bmatrix}. \qquad (2.161)$$

Exercises

1. Enumerate at least five approximations which are made in the derivation of the model equation for the pendulum.

2. Compare the period of the nonlinear pendulum with that of the linear pendulum when $\theta(0) = 5, 10, 20, ..., 85$ degrees and $\dot{\theta}(0) = 0$.

3. Use Equation (2.115) to derive the equation of motion for a pendulum system where the rod has been replaced by a spring (Fig. 2.27).

   Hint: The force exerted by the spring is $-k(L - L_0)\mathbf{e}_r$ where L_0 is the natural length of the spring.

4. Simulate the system of differential equations that was obtained in the previous exercise with proper choices for L and k (using "MATLAB" or

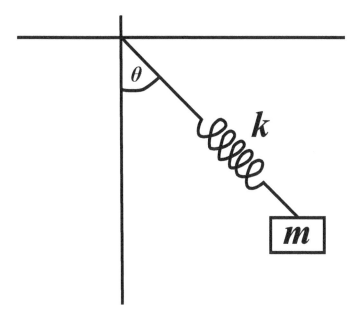

Figure 2.27 A Pendulum with spring attachment

similar). Compare the period and phase diagrams (i.e. plot $\dot{\theta}$ vs. θ) for this pendulum with those for the regular pendulum in Ex 1.

5. Show that when $f(r) = -\dfrac{\mu}{r^2}$ in Equation (2.148) then

$$\frac{1}{r} = \frac{k}{c^2}(1 + e\cos(\theta - \theta_0))$$

where e, θ_0 are constants. Discuss the nature of the orbits described by this equation for different values of e.

Hint: Show that for $e < 1$ the orbit is an ellipse, $e = 1$ a parabola, and for $e > 1$ a hyperbola.

6. Carry out the derivation of Equation (2.160) from Equations (2.156)-(2.157)

7. Verify Equations(2.102)-(2.111).

8. Develop a model for a one stage rocket with air friction. Assume $F_f = \alpha m(t)v(t)$ [try to solve by neglecting gravity]

Hint: $\dfrac{dv}{dt} = \dfrac{dv}{dh}\dfrac{dh}{dt} = v\dfrac{dv}{dh}$

9. Develop model equations for the motion of the rocket if $u = u(t)$ where u is the speed of the gases leaving the rocket.

10. Find the optimal value of m_0/m_p for a two stage and four stage rocket. What happens when the number of stages goes to ∞?

Solution

The ratio $L = \frac{m_0}{m_p}$ for n-stage rocket is given by

$$L = \left[\frac{1 - \lambda}{exp(\frac{v_E}{nu}) - \lambda} \right]^n.$$

To find the limit of this expression as $n \to \infty$ we rewrite it in the form

$$\exp\left\{ \ln \left[\frac{1 - \lambda}{exp(\frac{v_E}{nu}) - \lambda} \right]^n \right\} = \exp\left\{ n \ln \left[\frac{1 - \lambda}{exp(\frac{v_E}{nu}) - \lambda} \right] \right\}.$$

We then have

$$\lim_{n \to \infty} n \left\{ \ln \left[\frac{1 - \lambda}{exp(\frac{v_E}{nu}) - \lambda} \right] \right\} = \lim_{n \to \infty} \frac{\ln \left[\frac{1-\lambda}{exp(\frac{v_E}{nu}) - \lambda} \right]}{1/n}.$$

As $n \to \infty$ both numerator and denominator in the last expression approach zero; therefore, we can compute this limit by replacing n (a discrete variable) by x (a continuous variable) and apply L'Hopital rule:

$$\lim_{x \to \infty} \frac{\ln \left[\frac{1-\lambda}{exp(\frac{v_E}{xu})} - \lambda \right]}{1/x} = \lim_{x \to \infty} \frac{v_F/u}{1 - \lambda exp(\frac{v_E}{xu})} = \frac{v_F/u}{1 - \lambda}.$$

Hence

$$L = \exp\left(\frac{v_F/u}{1 - \lambda} \right).$$

11. A person wants to put a satellite in orbit around the Earth at 64000 km from the Earth's center. A short calculation similar to the one in the text shows that the speed of the satellite in orbit should be (approximately) 2.8 km/sec. Accordingly this person concludes that a one stage rocket is suitable for this purpose. Explain why this conclusion is wrong.

12. Estimate the effect of gravity on the flight of a one stage rocket if its mass is decreasing linearly with time:

$$m = m_0(1 - 5.10^{-4}t)$$

and the flight time is a. 100 sec. b. 190sec. Hint: Use Equation (2.152) with and without the gravity term assuming $g(h)$ to be a constant $0 \leq g \leq 9.8 m/sec^2$.

13. Write down model equations for the motion of a rocket which is launched vertically from under the sea.

Hint: Both below and above sea level the forces acting on the rocket are the thrust $T(h)$, gravity, and the frictional forces which are proportional to the speed (or $(speed)^2$) of the rocket. Since the coefficient of friction is different for $h > 0$ and $h < 0$ we obtain two different equations for these two ranges of h.

14. A person wants to put a satellite in circular orbit around the Earth so that the satellite will always stay over the same point on Earth. Compute the height and velocity of this satellite (these are called "geocentric satellites").

2.6 GREENHOUSE EFFECT

Energy coming from the Sun to Earth is "usually" reflected back to space. Some trace gases such as carbon dioxide (CO_2), methane (CH_4) and others can block part of this energy from going back to space and thereby raise the mean temperature of the Earth. This is called the **Greenhouse Effect**.

There are many models that attempt to gauge the impact of the greenhouse effect on Earth's climate. These models differ in the number of spatial dimensions which are incorporated into the model, their sophistication (in modeling the properties of the Earth surface and atmosphere), and resolution. At the bottom of this "ladder" of models are the 0-dimensional models where all variables depend only on time, i.e., we consider only the mean of these variables as a function of time, e.g., the mean temperature of the Earth.

In this section we consider only 0-dimensional models and start with some qualitative examples which give some insights about the mechanism of this effect.

Example 2.6.1 *Perfect Mirror*

If light is directed at a perfect mirror, then all this light energy will be reflected back and none will be absorbed by the mirror. As a result the mirror temperature remains unchanged.

Example 2.6.2 *Black Body*

By definition a black body absorbs all radiation that is impinging on it. Since radiation is a form of energy, the temperature of this body will rise. However, a black body also emits radiation (per unit area) at a rate which is proportional to the forth power of its temperature. This is the famous Stephan-Boltzmann law. At the equilibrium temperature of such a body the amount of energy absorbed is equal to the amount of the energy emitted and hence

$$P = \sigma T^4,$$

where P is the incident power and T is the equilibrium temperature (usually the energy emitted from such a body is in the infra-red end of the spectrum, i.e. long wave length).

Example 2.6.3 *Grey Body*

A grey body is a body whose properties are in between the two extremes described above. That is, such a body reflects part of the electromagnetic radiation, absorbs part of it, and then emits part of this energy again as thermal energy. The ratio of the reflected radiation to the amount incident upon it is called the "albedo" of the body. Usually the albedo is expressed in percentage form. Current estimates for the albedo of Earth are in the range of $30 - 39$ percent.

For a grey body we therefore have

$$(1 - A)P = \sigma T^4 \tag{2.162}$$

where A is the "albedo" (or grayness) . For a black body $A = 0$.

If Earth had no atmosphere then it would be a good example of such a grey body. It would reflect part of the Sun's energy (this would depend strongly on the ice and snow cover) and would absorb part of it. The part absorbed would lead to a rise in its temperature until the amount of thermal energy radiated into space equals the amount absorbed by it. It is estimated that under these conditions the mean temperature of the Earth would be approximately $-25°C$ (see next section). Thus Earth would be a cold rock covered with ice and snow except (perhaps) for a small belt around the equator.

To understand the effect of Earth's atmosphere on the mean temperature of the Earth we have to explain first how light is absorbed and emitted by atoms (or molecules).

When it comes to the processes of absorption or emission of energy from atoms, light should be considered as made of 'particles' called photons. These photons carry energy of $h\nu$ where h is called the Planck constant and ν is the electromagnetic wave frequency. An atom can absorb such a photon only if it has two energy levels E_1, E_2 so that

$$E_2 - E_1 = h\nu. \tag{2.163}$$

If such an atom is at energy level E_1, then by absorbing a photon with energy $h\nu$ it gets "excited" to an energy level E_2. Such an excited atom will then relax back to its ground energy level E_1 by releasing a photon with energy $h\nu$ (but in a direction which might be different from the one in which the original photon was traveling). Such a photon is then "trapped" in Earth's atmosphere and has a chance of being reabsorbed by Earth and raising its mean temperature.

It turns out that although O_2 and N_2 are the main components of the Earth's atmosphere, they do not have the proper energy levels (or bands) that can absorb the thermal (long wave length) energy that is radiated by the Earth. Therefore they play only a minor role in this process. On the other hand CO_2, methane (CH_4), and other trace gases do have exactly the right energy bands needed for the trapping of this energy, thereby leading to the greenhouse effect.

To develop a 0-dimensional model for the amount of CO_2 (or CH_4) in the atmosphere it is customary to divide the Earth-atmosphere system into several reservoirs, and the number of these reservoirs depends on the fine details and complexity of the model. A popular model which strikes a balance between simplicity and complexity divides the Earth-atmosphere system into seven reservoirs as follows:

1. Troposphere(lower atmosphere up to $\approx 10Km$ above Earth surface)

2. Stratosphere and up (upper atmosphere)

3. Upper ocean layer (up to a depth of 200-300 meters)

4. Deep ocean (below the upper Ocean layer)

5. Short lived vegetation (corn,wheat etc) and animal life

6. Long lived vegetation(trees, forests)

7. Marine life

Furthermore the model contains a forcing term due to human-made effects (burning of fossil fuels, paved tar roads, deforestation, etc.) and natural catastrophes (e.g. volcanic eruptions) (see Fig. 2.28).

These reservoirs interact dynamically (through diffusion and turbulent mixing) in the same way as chemical reactants do in a chemical reaction; i.e., we assume that the rate of CO_2 transfer between the reservoirs is proportional to the concentration of CO_2 in each reservoir. However, not all the reservoirs connect with each other. If we denote the CO_2 concentration in reservoir i by C_i then

$$\frac{dC_i}{dt} = \sum_{i=1}^{7} k_{ij} C_j + F_i, i = 1, \ldots, 7 \qquad (2.164)$$

where k_{ij} are constants and F_i is the forcing (in this model $F_i = 0$ except for the lower atmosphere). Thus the model depends on 49 parameters (some of which are zero). The proper estimation of these coefficients (which remains an outstanding research issue) is one of the important steps that is needed to make accurate predictions about the concentration of CO_2 in the atmosphere.

Once we determine the (mean) concentration of CO_2 in the lower and upper atmosphere, another model is needed to gauge the impact of this concentration on the mean temperature of Earth (see chapter 11).

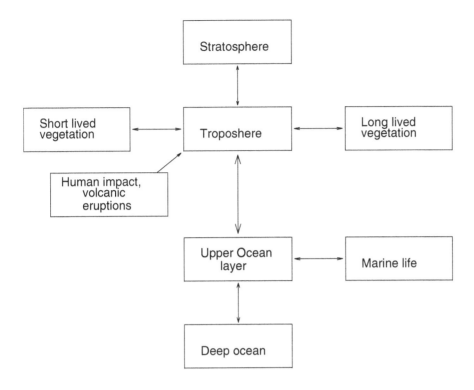

Figure 2.28 Schematic overview of the interaction among the CO_2 reservoirs

2.7 CURRENT ENERGY BALANCE OF THE EARTH.

The mean temperature of Earth has been "stable" for many years (with a small increase in the last decades due to the greenhouse effect). It is reasonable therefore to assume that at the present time the amount of energy incoming to Earth approximately equals the outgoing amount into space. In the following we present the current estimates that lead to this "equilibrium."

Earth receives energy from the Sun in short and visible wavelengths ($\lambda \leq 0.5\mu m$) and emits energy back to space as long wave radiation ($\lambda \sim 10\mu m$). To achieve equilibrium these two processes must balance each other (we neglect energy from geothermal sources as their impact is negligible).

Let the energy flux from the Sun be denoted by $F_s(\approx 1372W/m^2)$. A fraction $A(\approx 30\%)$ of this energy is reflected back to space by the Earth and atmosphere (this fraction is called the albedo of the Earth).

The long wave radiation from Earth can be described by the black-body

Stefan-Boltzmann law

$$P = \sigma T^4, \tag{2.1}$$

where P is the energy flux integrated over all wave lengths emitted by such a body at temperature T and σ the Stefan-Boltzmann constant ($= 5.67 cdot 10^{-8} \frac{Watts}{m^2 K^4}$).

For balance we must have therefore

$$(1 - A)F_s \pi R^2 = 4\pi R^2 \sigma T_e^4, \quad R = \text{ Earth's radius.} \tag{2.2}$$

This leads to $T_e = 255°K$ which is well below the freezing point and $33°$ degrees lower than the observed mean temperature of Earth $T_0 = 288°k$. The difference between T_e and T_0 is due to the greenhouse effect and the structure of the atmosphere and the oceans.

We now examine these issues in greater detail. The average energy flux from the Sun to the Earth is $\bar{F}_s = \frac{F_s}{4} = 343\frac{\text{Watts}}{m^2\text{sec}}$ (the factor "4" represents the ratio of the Earth cross section to its surface area). Of this flux 20 Watts are absorbed by clouds, 48 Watts are absorbed by water, vapor, O_3 and other aerosols, 169 Watts arrive at the surface, and the rest are reflected back to space.

As to long wave radiation flux, the Earth emits $390\frac{\text{Watts}}{m^2\text{sec}}$ in these wavelengths. Of these 22 go to space, 120 are absorbed by clouds, and 248 are absorbed by water vapor, O_3, and other aerosols. However, clouds and aerosols emit $327\frac{\text{Watt}}{m^2\text{sec}}$ back to Earth.

These lead to the following **approximate** budgets:

1. Earth-surface energy flux $\frac{\text{Watt}}{m^2\text{sec}}$

169	327	-390
Short wave radiation	Long wave absorption from atmospheric sources	Long wave emission

This table shows that there is a surplus of $106 \frac{\text{Watt}}{m^2\text{sec}}$ in this flux.

To balance this energy surplus the surface transfers heat to the atmosphere through conduction at a rate of $16 \frac{\text{Watt}}{m^2\text{sec}}$. Furthermore the oceans transfer $90 \frac{\text{Watt}}{m^2\text{sec}}$ through evaporative cooling ("latent heat"). Without these two processes the Earth surface would have had to be $50°K$ warmer to balance this extra energy through black body radiation.

The budget for the atmospheric Energy flux is as follows:

2. Atmospheric flux $\frac{\text{Watt}}{m^2\text{sec}}$

68	+368	-327	-215
Short wave absorption	Long wave absorption (from surface)	Longwave emission to surface	Long wave emission to space

According to this table there is a deficiency in the atmospheric flux which is equal to the surplus of the Earth's energy flux; this keeps the the Earth's total energy in balance. This balance will be disturbed if, due to enhanced greenhouse effect, the long wave emission to space from the atmosphere is diminished.

2.7.1   Critique of the Model

The majority of climate scientists accept the model that man made emissions of CO_2 and other trace gases affect the energy balance of the Earth and raise the mean temperature of the Earth. However there are others who argue that these emissions are small when compared to other natural sources of these gases. There is also some evidence that when these gases diffuse into the upper atmosphere they might have a "cooling effect" by reflecting more of the Sun energy into space. Another major factor that impacts the energy balance of the Earth is the Sun energy output which fluctuates in time. Thus the Earth's climate is influenced of by a large number of variables and the impact of many of these variables is not well understood as of yet. Earth and its climate form a truly complex system due to the interaction of many factors.

2.7.2   Humanity and Energy

Abundant energy at a reasonable cost is a key ingredient for the well being of humanity and its technological progress. Since the advent of the industrial revolution humanity relied on fossil fuels (coal, oil, and gas) to satisfy these needs. Ingenious technologies were developed to tap additional sources of these fuels. For example in the last few decades energy companies were able to access oil and gas reserves buried in the deep ocean. Fracking technology which can unlock gas reserves in certain types of rocks was developed. Extraction of oil from tar sands became feasible.

In spite of all these advances it is obvious that these resources are finite. How long they will last is an open question in view of the growing human population and the energy consumed per capita. Furthermore, fossil fuels contribute to pollution, the rise in the mean temperature of the Earth, and have an impact on the health of humans. (There is some evidence that the rise of respiratory diseases and cancer cases is due to the extensive use of these fuels).

It follows then that in order to ensure sustained energy supply, alternative energy sources have to be used and proper technologies have to be developed to harvest these sources at competitive prices. Foremost among these sources are the wind, using wind turbines, and Sun energy using photovoltaic. However other sources are being considered at the present time. These are biofuels (i.e energy from plants and algae), ocean energy (from the tides and ocean currents), and energy from space which requires the placement of large Sun energy collectors in geocentric orbits around the Earth and beaming this energy to Earth. However the price per energy unit from all of these sources presently exceeds the price of fossil fuels by a wide margin. It should be remembered, however, that fossil fuels have a hidden cost due to their impact on the environment and human health. Furthermore, it can be expected that fossil energy prices will skyrocket as their reserves dwindle.

In the future (and to some extent at the present time) energy supply will come from a basket of sources. Energy policy will be required then to use these different energy sources to optimize the following requirements,

1. Ensure sufficient energy supply to satisfy humanity needs,

2. Minimize the cost per energy unit to the consumer,

3. Minimize the impact on the environment and human health,

4. Ensure the sustainability of the energy supply.

Solutions of Systems of ODEs

CONTENTS

Differential equations, which relate a set of unknown functions with their derivatives, play an important role in many applications of science and engineering. In this chapter, we present some basic analytical and numerical methods for the solution of some classes of systems of differential equations. However, our treatment is not comprehensive. We start with a short review of some basic theory.

3.1 REVIEW

3.1.1 Linear differential equations with constant coefficients

The general form of n-th order linear differential equation with constant coefficients is

$$a_n y^{(n)} + a_{(n-1)} y^{n-1} + \ldots + a_1 y' + a_0 y = f(x) \tag{3.1}$$

where y is a function of x, $\{a_k, \quad k = 0, \ldots n\}$ are constants, and $y^k = \frac{d^k y}{dx^k}$. When $f(x) = 0$, we say that the equation is homogeneous. Otherwise, we refer to it as inhomogeneous (or non-homogeneous).

A unique solution of Equation (10.1) is obtained by specifying n initial conditions (i.e. all the conditions on the desired solution are specified at one value of x) or boundary conditions (where the conditions are specified at different values of x). In the first case, we refer to the problem as an "initial value problem," and in the second case, we refer to it as a boundary value problem."

When $f(x) = 0$, the solutions of Equation (10.1) satisfy the **superposition principle** which states that if y_1 and y_2 are two solutions of Equation(10.1) and c_1, c_2 are constants then

$$y = c_1 y_1 + c_2 y_2$$

is also a solution of Equation (10.1).

Example: The following differential equation

$$ay'' + by' + cy = f(x), \quad y(0) = c_1, \quad y'(0) = c_2 \tag{3.2}$$

where $y = y(x)$ and a, b, c, c_1, c_2 are constants is a second order differential equation with two initial conditions on y at $x = 0$.

When Equation (10.1) is homogeneous, the general solution of this equation y_h is given by

$$y_h = \alpha_1 y_1 + \alpha_2 y_2 + \ldots + \alpha_n y_n \tag{3.3}$$

where $\{y_1, y_2, \ldots, y_n\}$ are n independent solutions of Equation (10.1) and $\alpha_1, \ldots, \alpha_n$ are constants.

When Equation (10.1) is inhomogeneous $(f(x) \neq 0)$, the general solution is obtained if one "particular solution" y_p of the whole equation is known. The general solution of the equation is given by

$$y_G = y_h + y_p.$$

If the general solution of Equation (10.1) is known, then the solution that satisfies a set of initial conditions or boundary conditions can be found by solving a system of algebraic equations for the constants $\alpha_1, \ldots, \alpha_n$.

Example 3.1.1 *Here is an example of a boundary value problem:*

$$y'' + y' - 2y = 0, \ y(0) = 0, \ y(\pi) = 0. \tag{3.4}$$

Example 3.1.2 *Initial value problem:*

For the same differential equation as in the previous example 3.1.1 with $y(0) = 1, \ y'(0) = 2$ *is an initial value problem.*

In the following, we consider only linear second order equations with constant coefficients (as in Equation (3.2)). However, the methods described for these equations hold for higher order equations with constant coefficients.

Equation (3.2) with $f(x) = 0$ is a homogeneous second order equation. It has two independent solutions: $y_1(x)$ and $y_2(x)$. To find these two solutions we use a "trial solution" (or "ansatz") of the form

$$y = e^{kx} \tag{3.5}$$

where k is a constant to be determined. Substituting this expression in (3.2) we obtain

$$(ak^2 + bk + c)e^{kx} = 0 . \tag{3.6}$$

Since $e^{kx} \neq 0$, k must be a root of the "characteristic polynomial"

$$ak^2 + bk + c = 0 . \tag{3.7}$$

This quadratic equation can have

1. Two distinct real roots: k_1, k_2.

2. Double real root: r.

3. Two conjugate complex roots: $k_{1,2} = \alpha \pm i\omega$

We now discuss each of these possibilities separately.

1. Two distinct real roots: $(b^2 - 4ac > 0)$

   In this case, the "trial solution" yielded two independent solutions to the homogeneous equation. The general solution is in the form

   $$y_h = \alpha_1 e^{k_1 x} + \alpha_2 e^{k_2 x}. \tag{3.8}$$

2. Double root: $(b^2 - 4ac = 0)$

   In this case, the characteristic Equation (10.7) yields only one independent solution e^{rx} where $r = -\frac{b}{2a}$. However, a second solution is obtained in the form xe^{rx} and the general solution of Equation. (3.2) is

   $$y_h = e^{rx}(\alpha_1 + \alpha_2 x). \tag{3.9}$$

3. Two complex conjugate roots:

   In this case, we again obtain two independent solutions; the general solution to Equation (3.2) can be written as

   $$y_h = e^{kx}(c_1 e^{i\omega x} + c_2 e^{-i\omega x}) \tag{3.10}$$

   where c_1, c_2 are arbitrary constants. However,

   $$e^{\pm i\omega x} = \cos \omega x \pm i \sin \omega x. \tag{3.11}$$

   Hence,

   $$\begin{aligned} y_h &= e^{kx}[(c_1 + c_2)\cos \omega x + i(c_1 - c_2)\sin \omega x] & (3.12) \\ &= e^{kx}(\alpha_1 \cos \omega x + \alpha_2 \sin \omega x) & (3.13) \end{aligned}$$

   ω is called the natural frequency of the equation. The solution (3.12), can be rewritten in "phase-amplitude" form. To this end we multiply and divide (3.12) by $A = \sqrt{\alpha_1^2 + \alpha_2^2}$. We obtain

$$y_h = Ae^{kx}[\beta_1 \cos \omega x + \beta_2 \sin \omega x] \tag{3.14}$$

where $\beta_1 = \frac{\alpha_1}{A}$ and $\beta_2 = \frac{\alpha_2}{A}$. Since $\beta_1^2 + \beta_2^2 = 1$, we can find an angle ϕ so that

$$\cos \phi = \frac{\alpha_1}{A} \tag{3.15}$$

and rewrite Equation (3.14) as

$$y_h = Ae^{kx} \cos(\omega x - \phi). \tag{3.16}$$

A is the amplitude and ϕ is the phase of the solution.

A Particular Solution

If the general solution of the homogeneous Equation (3.2) (with $f(x) = 0$) is known, then a particular solution to Equation (10.1) (with $f(x)) \neq 0$ can be found by "variation of coefficients." The following (short) table summarizes the trial function one has to use to find y_p for some important cases which appear in applications (this table can be extended). The coefficients A, B, and

Table 3.1 Finding y_p

f(x)	y_p
ae^{kx}	Ae^{kx}
$\sin \beta x, \cos \beta x$	$A \cos \beta x + B \sin \beta x$
$a_n x^n + a_{n-1} x^{n-1} + \ldots + a_0$	$A_n x^n + A_{n-1} x^{n-1} + \ldots + A_0$ '

etc. that appear on the right hand side of the table have to be determined by substituting the trial function in the differential equation. As a result, one obtains a system of algebraic equations which has to be solved for the coefficients.

An exception to this table happens when the forcing function $f(x)$ is one of the solutions of the corresponding homogeneous equation (this is referred to as "resonance").

Remark: Observe that in general y_p will not satisfy the initial conditions. It is only y_G that has to satisfy these conditions.

We illustrate the use of this table by examples.

Example 3.1.3 $f(x) = h = constant$ *in Equation (3.2).*

In this case, y_p is given by

$$y_p = \begin{cases} \frac{h}{c} & c \neq 0 \\ \frac{hx}{b} + \gamma_1 & b \neq 0, c = 0 \\ \frac{hx^2}{2a} + \gamma_1 x + \gamma_2 & b = c = 0 \end{cases} \tag{3.17}$$

where γ_1 and γ_2 are arbitrary constants.

Example 3.1.4 *Consider Equation (3.2) with $b = 0$, $a > 0$, $c > 0$ and*

$$f(x) = b_1 \cos \omega x + b_2 \sin \omega x. \tag{3.18}$$

The general solution of the homogeneous equation is

$$y_h = C_1 \cos \nu x + C_2 \sin \nu x$$

where $\nu = \sqrt{\frac{c}{a}}$ is referred to as the natural frequency of the equation (the frequency of the oscillations when the forcing function $f(x) = 0$). If $\omega \neq \nu$, then to find y_p we try

$$y_p = A \cos \omega x + B \cos \omega x. \tag{3.19}$$

Substituting this in Equation (3.2) we obtain an algebraic equation for A and B

$$\sin \omega x [-aA\omega^2 + cA - b_1] + \cos \omega x [-aB\omega^2 + cB - b_2] = 0. \tag{3.20}$$

However, since $\cos \omega x$ and $\sin \omega x$ are independent functions, we infer that to satisfy this equation each of the brackets in Equation (3.20) must be zero. Hence, we obtain the following system of equations for A and B

$$\begin{pmatrix} -a\omega^2 + c & 0 \\ 0 & -a\omega^2 + c \end{pmatrix} \begin{pmatrix} A \\ B \end{pmatrix} = \begin{pmatrix} b_1 \\ b_2 \end{pmatrix} \tag{3.21}$$

since ω is not the natural frequency of the system $(-a\omega^2 + c) \neq 0$ Equation (3.21) yields nontrivial solutions for A and B.

Example 3.1.5 *Consider (3.2) with $b = 0$, $a > 0$, $c > 0$ and $f(x) = b_1 \cos \omega x + b_2 \sin \omega x$ but with $\omega = \nu$.*

We refer to this situation where the forcing function has frequency equal to the natural frequency of the system as "resonance." (The forcing function $f(x)$ is one of the solutions of the homogeneous equation)

In this case, we use for y_p a trial solution of the form

$$y_p = x(A\cos\omega x + B\sin\omega x). \tag{3.22}$$

Substituting this expression in Equation (3.2) we obtain

$$
\begin{aligned}
\cos\omega x[(2a\omega B - b_1) &\quad + \quad (-a\omega^2 + c)xA] \\
\sin\omega x[(-2a\omega A - b_2) &\quad + \quad (-a\omega^2 + c)xB] = 0.
\end{aligned} \tag{3.23}
$$

Since $\cos\omega x$ and $\sin\omega x$ are independent functions, each bracket in Equation (3.23) must be zero. Moreover, since ω is the natural frequency of the system $-a\omega^2 + c = 0$,

$$2a\omega B - b_1 = 0, \quad 2a\omega A + b_2 = 0. \tag{3.24}$$

We obtain the following solution for A and B.

$$A = -\frac{b_2}{2a\omega}, \quad B = \frac{b_1}{2a\omega}$$

3.2   REVIEW OF LINEAR ALGEBRA

In this section, we review some topics from linear algebra regarding eigenvalues and eigenvectors that are needed fro the solution of systems of differential equations.

3.2.1   Eigenvalues and Eigenvectors

Definition 3.2.1 *Let A be a $n \times n$ matrix with constant entries. The characteristic polynomial of A is*

$$p(\lambda) = \mid A - \lambda I \mid \tag{3.25}$$

where I is the $n \times n$ unit matrix (that is the matrix with 1 along the main diagonal and zero at all other entries).

Definition 3.2.2 *The roots of the characteristic polynomial are called the* **eigenvalues** *of A.*

Thus, when λ is an eigenvalue of A, the matrix $(A - \lambda I)$ is singular. When λ is not an eigenvalue, $(A - \lambda I)$ is not singular and therefore has an inverse.

Now consider the following system of linear equations

$$A\mathbf{x} = \lambda \mathbf{x}. \tag{3.26}$$

This system can be written also as

$$(A - \lambda I)\mathbf{x} = \mathbf{0}. \tag{3.27}$$

It follows then that when λ is not an eigenvalue of A, then $(A - \lambda I)$ is invertible. Therefore, the only solution of this system is the trivial solution $\mathbf{x} = \mathbf{0}$. However, if λ is an eigenvalue of A, the matrix $(A - \lambda I)$ is singular and there is a nontrivial solution of the system, Equation (3.26). That is there exists $\mathbf{v} \neq \mathbf{0}$ that satisfies this equation. We refer to the vector $\mathbf{v}$ as the eigenvector of A that is related to the eigenvalue λ.

Corollary 3.2.1 *If $\mathbf{v}$ is an eigenvector of A, then $c\mathbf{v}$ for all $c \neq 0$ is also an eigenvector of A (with the same eigenvalue λ).*

Therefore, an eigenvector is determined only up to a multiplicative constant.

We note that each eigenvalue has a related eigenvector. However, the number of independent eigenvectors related to a given eigenvalue does not have to equal to its algebraic multiplicity (i.e., the number of times it appears as a root of $p(\lambda)$).

Example 3.2.1 *Find the eigenvalues and eigenvectors of*

$$A = \begin{pmatrix} 2 & 1 & 0 \\ 0 & -2 & 1 \\ 0 & 0 & 2 \end{pmatrix}.$$

Solution 3.2.1 *The characteristic polynomial of A is*

$$p(\lambda) = \begin{pmatrix} 2 - \lambda & 1 & 0 \\ 0 & -2 - \lambda & 1 \\ 0 & 0 & 2 - \lambda \end{pmatrix} = (2 - \lambda)^2(-2 - \lambda). \tag{3.28}$$

Hence, $\lambda = -2$ is a simple root (i.e., multiplicity 1) and $\lambda = 2$ is a double root.

To find the corresponding eigenvectors, we solve Equation (3.26) with each of these eigenvalues.

a. $\lambda = -2$

If $\mathbf{v} = \begin{pmatrix} x_1 \\ x_2 \\ x_3 \end{pmatrix}$, we obtain from the definition the system

$$2x_1 + x_2 = -2x_1$$
$$-2x_2 + x_3 = -2x_2$$
$$2x_3 = -2x_3.$$

Hence, up to a multiplicative constant which can be expected in view of corollary 3.2.1,

$$v = \begin{pmatrix} 1 \\ -4 \\ 0 \end{pmatrix}.$$

b. $\lambda = 2$

Denoting once again the required eigenvector by $v = \begin{pmatrix} x_1 \\ x_2 \\ x_3 \end{pmatrix}$, we obtain the system

$$2x_1 + x_2 = 2x_1$$
$$2x_2 + x_3 = 2x_2$$
$$2x_3 = 2x_3.$$

Hence, $v = \begin{pmatrix} 1 \\ 0 \\ 0 \end{pmatrix}$. Thus, there exists only one independent eigenvector related to $\lambda = 2$ although the multiplicity of this eigenvalue is 2.

Theorem 3.2.1 *If $n \times n$ matrix A has n independent eigenvectors $\mathbf{v}_1, \ldots, \mathbf{v}_n$, then*

$$M^{-1}AM = \begin{pmatrix} \lambda_1 & & \bigcirc \\ & \ddots & \\ \bigcirc & & \lambda_n \end{pmatrix} \qquad (3.29)$$

where $M = (\mathbf{v}_1, \ldots, \mathbf{v}_n)$. Such a matrix A is said to be diagonalizable.

Example 3.2.2 *Verify theorem 3.2.1 for*

$$A = \begin{bmatrix} 1 & 1 & 0 \\ 0 & -2 & 0 \\ 0 & 0 & -2 \end{bmatrix}.$$

Solution 3.2.2 *The eigenvalues of A are $\lambda = 1$ and $\lambda = -2$ (with multiplicity 2). However, contrary to example 3.2.1, A has three independent eigenvectors*

$$v_1 = \begin{pmatrix} 1 \\ 0 \\ 0 \end{pmatrix}, \quad v_2 = \begin{pmatrix} -1 \\ 3 \\ 0 \end{pmatrix} \text{ and } v_3 = \begin{pmatrix} 0 \\ 0 \\ 1 \end{pmatrix}.$$

Hence,

$$M = \begin{pmatrix} 1 & -1 & 0 \\ 0 & 3 & 0 \\ 0 & 0 & 1 \end{pmatrix}, \quad M^{-1} = \begin{pmatrix} 1 & \frac{1}{3} & 0 \\ 0 & \frac{1}{3} & 0 \\ 0 & 0 & 1 \end{pmatrix}$$

and a direct multiplication shows that

$$M^{-1}AM = \begin{pmatrix} 1 & 0 & 0 \\ 0 & -2 & 0 \\ 0 & 0 & -2 \end{pmatrix}.$$

Exercises

1. Find the relationship between the eigenvalues of A and those of A^{-1} and A^T (where A^T is the transpose of A). What about the eigenvectors?

2. Let $p(x)$ be a polynomial. Show that the eigenvalues of $p(A)$ are given by $p(\lambda_i)$, where λ_i are the eigenvalues of A. What can you say about the eigenvectors of $p(A)$?

   Hint: Show that if v is an eigenvector of A, then it is also an eigenvector of A^m.

3. Find the eigenvalues and eigenvectors of the following matrices:

   $$a. \ A = \begin{bmatrix} 1 & -2 \\ -4 & 3 \end{bmatrix} \qquad b. \ A = \begin{bmatrix} 1 & -2 \\ -2 & 4 \end{bmatrix}$$

   $$c. \ A = \begin{bmatrix} 2 & 1 & 0 \\ -1 & -2 & -1 \\ 0 & 1 & 2 \end{bmatrix} \qquad d. \ A = \begin{bmatrix} 1 & 1 & 1 \\ 0 & 0 & 0 \\ -1 & -2 & 3 \end{bmatrix}$$

3.3   REFORMULATION OF SYSTEMS ODES

In the following sections, we consider systems of ordinary differential equations and their solutions. To standardize this treatment, we first show through examples how a coupled system of differential equations of any order can always be rewritten as a system of first order equations.

Example 3.3.1 *Consider the nth order equation*

$$\frac{d^n y}{dx^n} = f\left(x, y, \frac{dy}{dx}, \ldots, \frac{d^{n-1}y}{dx^{n-1}}\right) \tag{3.30}$$

with the initial conditions

$$y(a) = b_0, \quad y'(a) = b_1, \ldots, y^{(n-1)}(a) = b_{n-1}. \tag{3.31}$$

This equation and the initial conditions are obviously equivalent to the system

$$
\begin{aligned}
\frac{dy}{dx} &= y_1 \\
\frac{dy_1}{dx} &= y_2 \left[\equiv \frac{d^2 y}{dx^2}\right] \\
\frac{dy_{n-2}}{dx} &= y_{n-1} \left[\equiv \frac{d^{n-1}y}{dx^{n-1}}\right] \\
\frac{dy_{n-1}}{dx} &= f\left(x, y, y_1, \ldots, y_{n-1}\right)
\end{aligned}
\tag{3.32}
$$

with the initial conditions

$$y(a) = b_0, \quad y_1(a) = b_1, \ldots, y_{n-1}(a) = b_{n-1}. \tag{3.33}$$

Example 3.3.2 *Rewrite the system*

$$
\begin{aligned}
x'' &+ y' + xy = 0 \\
y'' &+ x'y' + 2x^2 y^2 = 0 \\
x(0) &= 0, \quad x'(0) = 1, \quad y(0) = 1, \quad y'(0) = 2
\end{aligned}
$$

(3.34)

(3.35)

where primes denote differentiation with respect to t, as a system of first order equations.

Solution 3.3.1 *The required system is*

$$
\begin{aligned}
x' &= x_1, \quad y' = y_1 \\
x_1' &= -y_1 - xy \\
y_1' &= -x_1 y_1 - 2x^2 y^2 \\
x(0) &= 0, \quad x_1(0) = 1, \quad y(0) = 1, \quad y_1(0) = 2.
\end{aligned}
\tag{3.36}
$$

Example 3.3.3 *Rewrite the system*

$$
x'' - 2y' + x - y = 0 \tag{3.37}
$$

$$
x' - y' + 2x + y = 0 \tag{3.38}
$$

as a system of first order equations.

Solution 3.3.2 *By introducing $x' = z$, the system becomes*

$$
z' - 2y' + x - y = 0 \tag{3.39}
$$

$$
z - y' + 2x + y = 0. \tag{3.40}
$$

Using Equation (3.40) to substitute for y' in Equation (3.39), we finally obtain the system

$$
x' = z \tag{3.41}
$$

$$
z' = 3x + 2z + 3y \tag{3.42}
$$

$$
y' = 2x + z + y. \tag{3.43}
$$

Exercises

1. Rewrite the following as systems of first order ordinary differential equations:

   (a) $x'' + y' - 3xy = t$
   $y'' + x' + 2x - 3y = e^{2t}$
   $x(1) = 1, \quad x'(1) = 0, \quad y(1) = -2, \quad y'(1) = 0$

   (b) $x''' + 5y' + xy^2 = t^3$
   $y'' - 3x''y' - 2y = 0$
   $x(0) = 1, \quad x'(0) = 0, \quad x''(0) = 6, \quad y(0) = 10$
   $y'(0) = 1.$

3.4 LINEAR SYSTEMS WITH CONSTANT COEFFICIENTS

In this section, we describe methods to solve the n-dimensional system

$$\frac{d\mathbf{x}}{dt} = A\mathbf{x} + \mathbf{f}(t), \tag{3.44}$$

with initial conditions

$$\mathbf{x}(a) = \mathbf{C}, \tag{3.45}$$

where A is a matrix with constant entries

$$A = (a_{ij}), \quad \mathbf{x}(t) = \begin{pmatrix} x_1(t) \\ \vdots \\ x_n(t) \end{pmatrix}, \quad \mathbf{f}(t) = \begin{pmatrix} f_1(t) \\ \vdots \\ f_n(t) \end{pmatrix}.$$

If A has n independent eigenvectors, then $\mathbf{x}_G$ can be found by decoupling the system to a set of n independent equations through a change of variables.

Thus, if

$$M = (\mathbf{v}_1, \ldots, \mathbf{v}_n),$$

where $\mathbf{v}_i, i = 1, \ldots n$ are the eigenvectors of A, we define

$$\mathbf{u} = M^{-1}\mathbf{x}. \tag{3.46}$$

Substituting Equation (3.46) in Equation (3.44), we obtain

$$M\frac{d\mathbf{u}}{dt} = AM\mathbf{u} + \mathbf{f}(t).$$

Multiplying both sides of this equation by M^{-1} leads to

$$\frac{d\mathbf{u}}{dt} = M^{-1}AM\mathbf{u} + M^{-1}\mathbf{f}(t), \quad \mathbf{u}(a) = M^{-1}\mathbf{C} \tag{3.47}$$

but

$$M^{-1}AM = D = \begin{pmatrix} \lambda_1 & & \bigcirc \\ & \ddots & \\ \bigcirc & & \lambda_n \end{pmatrix}$$

and, therefore, the system Equation (3.47) is a set of n uncoupled equations. Each equation in this system can be solved independently of the others.

Example 3.4.1 *Solve the system, Equation (3.44) if*

$$A = \begin{pmatrix} 3 & 2 & 4 \\ 2 & 0 & 2 \\ 4 & 2 & 3 \end{pmatrix}, \quad \mathbf{f} = \begin{bmatrix} 1 \\ 0 \\ 1 \end{bmatrix}$$

and $\mathbf{x}(0) = \begin{bmatrix} 1 \\ -1 \\ -1 \end{bmatrix}$.

Solution 3.4.1 *The eigenvalues of A are $\lambda_1 = -1$, $\lambda_2 = -1$, and $\lambda_3 = 8$. The corresponding eigenvectors are*

$$\mathbf{v}_1 = \begin{bmatrix} -1 \\ 2 \\ 0 \end{bmatrix}, \quad \mathbf{v}_2 = \begin{bmatrix} -1 \\ 0 \\ 1 \end{bmatrix}, \quad \mathbf{v}_3 = \begin{bmatrix} 2 \\ 1 \\ 2 \end{bmatrix}.$$

Hence,

$$M = \begin{bmatrix} -1 & -1 & 2 \\ 2 & 0 & 1 \\ 0 & 1 & 2 \end{bmatrix}, \quad M^{-1} = \frac{1}{9} \begin{bmatrix} -1 & 4 & -1 \\ -4 & -2 & 5 \\ 2 & 1 & 2 \end{bmatrix}$$

and from Equation (3.47) we infer that the original system is equivalent to

$$\frac{d}{dt} \begin{pmatrix} u_1 \\ u_2 \\ u_3 \end{pmatrix} = \begin{pmatrix} -1 & 0 & 0 \\ 0 & -1 & 0 \\ 0 & 0 & 8 \end{pmatrix} \begin{pmatrix} u_1 \\ u_2 \\ u_3 \end{pmatrix} + \frac{1}{9} \begin{bmatrix} -2 \\ 1 \\ 4 \end{bmatrix}, \quad \mathbf{u}(0) = \frac{1}{9} \begin{bmatrix} -4 \\ -7 \\ -1 \end{bmatrix}.$$

The solution of this system is

$$u_1(t) = -\frac{2}{9}(1 + e^{-t}), \quad u_2(t) = \frac{1}{9}(-8e^{-t} + 1), \quad u_3 = -\frac{1}{18}(1 + e^{8t}).$$

Finally, we express the solution in terms of the original variables by using equation (3.46) which yields

$$x_1 = -u_1 - u_2 + 2u_3, \quad x_2 = 2u_1 + u_3, \quad x_3 = u_2 + 2u_3.$$

Another way to solve the system, Equation (3.44), is to note that this is a linear system of equations. Therefore, its general solution $\mathbf{x}_G$ is given by

$$\mathbf{x}_G = \mathbf{x}_h + \mathbf{x}_p$$

where $\mathbf{x}_h$ is the general solution of the homogeneous part of Equation (3.44) (that is the system, Equation (3.44), with $\mathbf{f} = 0$) and $\mathbf{x}_p$ is a solution of Equation (3.44).

Theorem 3.4.1 *If λ is an eigenvalue of A, with a corresponding eigenvector $\mathbf{v}$, then $\mathbf{x} = c\mathbf{v}e^{\lambda t}$ (where c is an arbitrary constant) is a solution of the homogeneous system*

$$\dot{\mathbf{x}} = A\mathbf{x}. \tag{3.48}$$

Proof: Substituting $\mathbf{x} = c\mathbf{v}e^{\lambda t}$ on the left-hand side of Equation (3.48), we obtain

$$\dot{\mathbf{x}} = c\lambda\mathbf{v}e^{\lambda t}. \tag{3.49}$$

Substituting $\mathbf{x} = c\mathbf{v}e^{\lambda t}$ on the right-hand side of Equation (3.48) yields

$$A\mathbf{x} = A(c\mathbf{v}e^{\lambda t}) = ce^{\lambda t}(A\mathbf{v}) = ce^{\lambda t}(\lambda\mathbf{v}). \tag{3.50}$$

Hence, $\mathbf{x} = c\mathbf{v}e^{\lambda t}$ is a solution of Equation (3.48).

Since Equation (3.48) is a homogeneous system of equations, the superposition principle holds. Therefore, if all the eigenvalues of A are simple, then the general solution of Equation (3.48) is given by

$$\mathbf{x}_h = \sum_{i=1}^{n} c_i\mathbf{v}_i e^{\lambda_i t}. \tag{3.51}$$

To find a particular solution of the system Equation (3.44) when $\mathbf{f} \neq 0$, we make the ansatz that

$$\mathbf{y}_p = N\mathbf{u}.$$

where N is the matrix

$$N = (\mathbf{x}_1, \ldots, \mathbf{x}_n),$$

with $\mathbf{x}_i = \mathbf{v}_i e^{\lambda_i t}$. This matrix is called the **Fundamental matrix of the system**, Equation (3.44).

Since $\frac{d}{dt}\mathbf{x}_i = A\mathbf{x}_i$ (see Equation (3.48)), it follows that

$$\frac{dN}{dt} = (\frac{d}{dt}\mathbf{x}_1, \ldots, \frac{d}{dt}\mathbf{x}_n) = AN.$$

Therefore, the left hand side of Equation (3.44) becomes

$$\frac{d\mathbf{y}_p}{dt} = N\frac{d\mathbf{u}}{dt} + \frac{dN}{dt}\mathbf{u} = N\frac{d\mathbf{u}}{dt} + AN\mathbf{u}$$

while the right hand side of Equation (3.44) yields

$$A\mathbf{y}_p + \mathbf{f} = AN\mathbf{u} + \mathbf{f}.$$

Hence,

$$N\frac{d\mathbf{u}}{dt} = \mathbf{f},$$

i.e.

$$\frac{d\mathbf{u}}{dt} = N^{-1}\mathbf{f}.$$

Integrating this equation we obtain

$$\mathbf{u}(t) = \int N^{-1}\mathbf{f}(t)\, dt.$$

Thus,

$$\mathbf{y}_p = N \int N^{-1}\mathbf{f}(t)\, dt.$$

Exercises

Solve the following systems of equations:

1. $\dot{x} = x + 2y - 1$
   $\dot{y} = 4x + 3y + (t - 1)$
   $x_1(0) = -1, \quad x_2(0) = 1$

2. $\ddot{x} + 3\dot{x} - 4x = e^{2t}$
   $x(0) = 1, \quad \dot{x}(0) = 0$

3. $\ddot{x} = 2x + 5y - 2t$
   $\dot{y} = \dot{x} - 2y - 1$
   $x(0) = -1, \quad \dot{x}(0) = 0, \quad y(0) = 1$

4. $\ddot{x} + 2\dot{x} + 3x = y$
   $\ddot{y} + 2\dot{y} - 4y = 2x$
   $x(0) = y(0) = 0, \quad \dot{x}(0) = 1, \quad \dot{y}(0) = 2.$

3.5 NUMERICAL SOLUTION OF INITIAL VALUE PROBLEMS

To explore the ideas underlining the numerical solution of initial value problems, we consider the following example.

Example 3.5.1 *Find the value of solution to the following differential equation at $x = 0.2$ and $x = 1$:*

$$y' = x^2 + 2y, \quad y(0) = 0.5. \tag{3.52}$$

Solution 3.5.1 *Assuming $y(x)$ is analytic, it can be expanded in a Taylor series around $x = 0$:*

$$y(x) = y(0) + y'(0)x + \frac{y''(0)}{2!} x^2 + \cdots . \tag{3.53}$$

Thus, inside the radius of convergence of this series, we can evaluate $y(x)$ if we can compute $\{y^{(k)}(0)\}_{k=0}^{\infty}$. To see how this can be carried out, we first note that $y(0) = 0.5$ is given. To compute $y'(0)$, we use the differential equation

$$y'(0) = 0^2 + 2y(0) = 1.$$

To compute $y''(0)$, $y'''(0)$, etc., we differentiate the given differential equation

$$y''(x) = 2x + 2y'(x)$$

$$y'''(x) = 2 + 2y''(x)$$

$$y^{(4)}(x) = 2y'''(x)$$

and in general

$$y^{(n)} = 2y^{(n-1)}, \quad n \geq 3. \tag{3.54}$$

This implies that

$$y''(0) = 2, \quad y'''(0) = 6, \quad y^{(4)} = 12, \quad \text{etc.}$$

Thus,

$$y(0.2) = 0.5 + 0.2 + 0.2^2 + 0.2^3 + \frac{1}{2} 0.2^4 + \ldots$$

and we can obtain a "reasonable" approximation for $y(0.2)$ with a small number of terms in the Taylor expansion. The situation at $x = 1$ is the same *in principle* but a larger number of terms in the Taylor expansion is needed to obtain the same accuracy. For this particular differential equation, this is not a problem since we have a recursive formula for $y^{(n)}(x)$. However, in general, such a recursive formula is not available and this implies that a large number of symbolic computations is necessary to obtain a reasonable approximation for the value of the solution.

3.5.1 Euler Algorithm

From a numerical point of view, we want to utilize only the differential equation and the initial conditions in the computation of the solution. This implies that we must circumvent the need (as described above) for the computation of the higher order derivatives.

To see how this is done, consider the equation

$$y' = f(x, y), \quad y(0) = y_0 \tag{3.55}$$

where the solution is desired on the interval $[a, b]$.

To carry this out, we divide the interval $[a, b]$ into n ("small") sub-intervals of fixed length ($\equiv$ step size) h. We note, however, that under exceptional circumstances (e.g., near a resonance where the solution varies rapidly) a variable step size has to be used. We shall say that the differential equation has been solved numerically if we can compute $y_k = y(x_k)$ at the grid points $x_k = a + kh$, $k = 0, \ldots, n$ (see Fig. 3.1). Obviously, if the value of the solution is needed at any other point $x\epsilon[a, b]$, then interpolation can be used.

To compute $y_1 = y(x_1)$, we now employ a first order Taylor expansion around x_0. This leads to

$$y_1 \cong y(x_0) + hy'(x_0) = y_0 + hf(x_0, y(x_0)) = y_0 + hf(x_0, y_0). \tag{3.56}$$

We now use this (approximate) value of $y(x_1)$ as an initial value for the differential equation on the interval $[x_1, x_2]$. Using the same strategy as before, we obtain

$$y_2 = y(x_1) + hy'(x_1) = y_1 + hf(x_1, y_1) . \tag{3.57}$$

Continuing in this manner, we derive the general formula

$$y_{k+1} = y_k + hf(x_k, y_k), \quad k = 1, \ldots, n - 1. \tag{3.58}$$

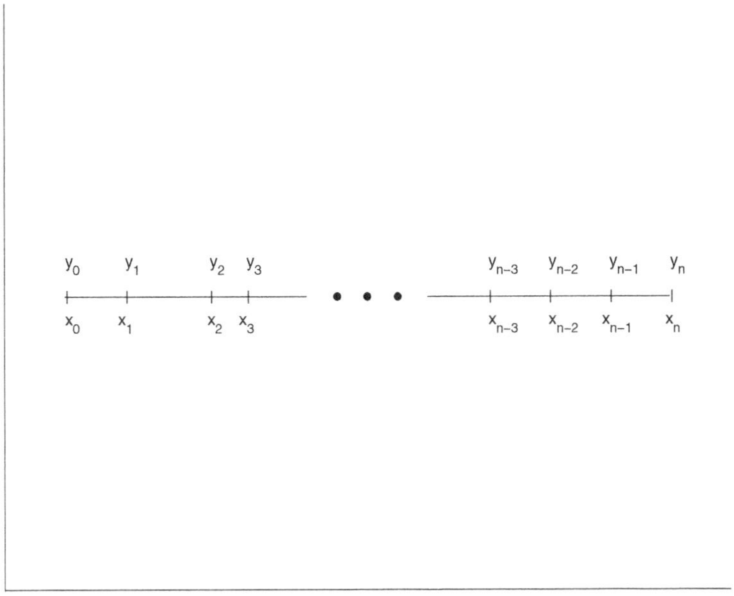

Figure 3.1 Computational grid with varying step size

This is known as the Euler algorithm for the numerical solution of differential equations.

Example 3.5.2 *Use the Euler algorithm to solve*

$$y' = x + y^2, \quad y(0) = 0 \tag{3.59}$$

on the interval $[0, 1]$ with step size of $\frac{1}{3}$ (see Fig. 2.2). Use linear interpolation to find an approximate value for $y(0.5)$.

Solution 3.5.2 *Using Equation (3.58) we have*

$$y_1 = 0 + \frac{1}{3} \cdot (0 + 0^2) = 0$$

$$y_2 = 0 + \frac{1}{3} \left(\frac{1}{3} + 0^2 \right) = \frac{1}{9}$$

$$y_3 = \frac{1}{9} + \frac{1}{3} \left(\frac{2}{3} + \left(\frac{1}{9} \right)^2 \right) = \frac{82}{243}.$$

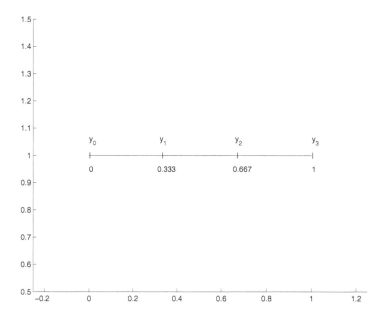

Figure 3.2 Computational grid for example 3.4.2

Table 3.2 Solution of example 2

x	y
0	0
1/3	0
2/3	1/9
1	82/243

Usually, this is presented in table form (Table 3.2).

To compute $y(0.5)$, we observe that $x = 0.5$ is the midpoint of the interval $\left[\frac{1}{3}, \frac{2}{3}\right]$. Therefore, if linear interpolation is used, then $y(0.5)$ is the average of y_1 and y_2 (i.e., $y(0.5) = \frac{1}{18}$).

It is appropriate at this point to ask about the accuracy of this scheme. It is clear that at each point in the computation we neglected terms of order h^2 or higher in the Taylor expansion. Since there are n steps, we can estimate the cumulative error to be of order nh^2. However, $n = \frac{b-a}{h}$. We deduce then that the cumulative error is order h. Thus, the error depends strongly on the step size chosen. Moreover, this implies that the accuracy of the solution usually improves as h becomes smaller. In fact, it is common practice to gauge the accuracy of the solution by applying the algorithm with step sizes h and $\frac{h}{2}$.

If the two solutions agree with each other on the common grid points up to, let's say, three digits, then we say that this is the accuracy of the numerical solution.

Remark: Observe that as h becomes smaller, the number of arithmetic operation needed to compute the solution over the whole interval increases. As a result, the round-off error in these computations increases.

Example 3.5.3 *Solve the differential equation*

$$y' = 2x - y, \quad y(0) = 1 \tag{3.60}$$

with step sizes of $\frac{1}{3}$ and $\frac{1}{6}$ and compare with the exact solution.

Solution 3.5.3 *The differential equation*

$$y' + y = 2x \tag{3.61}$$

is a first order linear inhomogeneous equation. To solve it we must find the general solution of the homogeneous equation

$$y' + y = 0 \tag{3.62}$$

and a particular solution of Equation (3.61). The general solution of Equation (3.62) is

$$y = Ce^{-x} . \tag{3.63}$$

We now try to find a particular solution in the form

$$y_p = u(x)e^{-x} \tag{3.64}$$

(variations of parameters). Substituting Equation (3.64) in Equation (3.61) yields

$$u'(x) = 2xe^x.$$

Hence,

$$u(x) = (2x - 2)e^x$$

and therefore

$$y(x) = Ce^{-x} + 2(x - 1).$$

To determine the constant C, we use the initial condition

$$y(0) = 1 = C - 2,$$

i.e., $C = 3$.

A tabulation of this solution and the numerical solution using the Euler algorithm is given in Table 3.3.

Table 3.3 Solution of $y' + y = 2x$ using the Euler method:

x	Solution of h = 1/3	Solution of h = 1/6	Exact Solution
0	1	1	1
1/6		0.833	0.873
1/3	0.667	0.750	0.816
1/2		0.736	0.820
2/3	0.667	0.780	0.874
5/6		0.872	0.970
1	0.889	1.00	1.104

We now show how this algorithm can be extended to systems of equations. To this end, we consider the system

$$\mathbf{y}' = \mathbf{F}(x, \mathbf{y}), \quad \mathbf{y}(0) = \mathbf{c} \tag{3.65}$$

where

$$\mathbf{y} = \begin{pmatrix} y^1 \\ \vdots \\ y^n \end{pmatrix}, \quad \mathbf{F} = \begin{pmatrix} f^1 \\ \vdots \\ f^n \end{pmatrix} \tag{3.66}$$

and $f^i = f^i(x, y^1, \ldots, y^n)$ on $[a, b]$.

As before, we subdivide the interval $[a, b]$ into sub-intervals of step size h and apply a first order Taylor expansion of y^i on each of the sub-intervals $[x_k, x_{k+1}]$:

$$y^i(x_{k+1}) = y^i(x_k) + hy^{i'}(x_k) = y^i(x_k) + hf^i(x_k, y_k^1, \ldots, y_k^m), i = 1, \ldots, m. \tag{3.67}$$

This can be rewritten transparently using vector notation as

$$\mathbf{y}_{k+1} = \mathbf{y}(x_{k+1}) = \mathbf{y}_k + h\mathbf{F}(x_k, \mathbf{y}_k). \tag{3.68}$$

This is the vector version of Equation (3.58).

Example 3.5.4 *Solve the following system numerically by using Euler algorithm:*

$$\begin{aligned} u' &= u + 2v \\ v' &= u - v, \quad u(0) = 0, \quad v(0) = 1 \end{aligned}$$

on $[0, 1]$ with $h = 0.25$.

Solution 3.5.4 *Applying Equation (3.68), we infer that*

$$\begin{aligned} u_{k+1} &= u(x_{k+1}) = u_k + h(u_k + 2v_k) \\ v_{k+1} &= v(x_{k+1}) = v_k + h(u_k - v_k). \end{aligned}$$

Applying this recursively, we obtain

$$\begin{aligned} u_1 &= u(0.25) = 0 + 0.25(0 + 2.1) = 0.5 \\ v_1 &= v(0.25) = 1 + 0.25(0 - 1) = 0.75 \\ u_2 &= u(0.5) = 0.5 + 0.25(0.5 + 2 \cdot 0.75) = 1 \\ v_2 &= v(0.5) = 0.75 + 0.25(0.5 - 0.75) = 0.6875 \\ u_3 &= u(0.75) = 1 + 0.25(1 + 2 \cdot 0.6875) = 1.594 \\ v_3 &= v(0.75) = 0.6875 + 0.25(1 - 0.6875) = 0.766 \\ u_4 &= u(1.0) = 1.594 + 0.25(1.594 + 2 \cdot 0.766) = 2.376 \\ v_4 &= v(1.0) = 0.766 + 0.25(1.594 - 0.766) = 0.973 \end{aligned}$$

Exercises

1. Solve the following equations numerically by using the Euler algorithm and compare with the exact solution. Graph your results.

   a. $y' = 2x + y$ on $[0, 1]$, $y(0) = 1$, $h = 0.2$

    b. $y' = y - y^2$ on $[0, 1]$, $y(0) = -1$, $h = 0.2$

    c. $y' = xy - 1$ on $[1, 2]$, $y(1) = 0.5$, $h = 0.2$

    d. $y' = (2 + y)y - x$ on $[0, 1]$, $y(0) = 0$, $h = 0.2$

    e. $y' = y \sin x$ on $[0, \pi]$, $y(0) = 1$, $h = \frac{\pi}{4}$.

2. Solve the equations in Ex. 1 with $\frac{h}{2}$ and compare with those obtained previously.

3. Solve the equations in Ex. 1 using the Taylor series expansions. Estimate the number of terms needed to obtain an accuracy of 10^{-3}.

4. Solve
$$y' = \sin x + y^2, \quad y(0) = 1$$
on $[0, 1]$ with $h = 0.25$ and use interpolation to find $y\left(\frac{2}{3}\right)$. Next, solve this equation with $h = \frac{1}{3}$ and compare the two values obtained for $y\left(\frac{2}{3}\right)$.

5. Use a computer to solve the equation in Ex. 1 with $h = 0.01$ and $h = 0.001$ and compare with the exact solution.

6. Solve the equation in Ex. 4 with $h = 0.1$ and $h = 0.01$ to find a better approximation for $y\left(\frac{2}{3}\right)$.

7. Solve the following systems numerically by using the Euler algorithm
$$A. \ u' = u + 2v, \ v' = 4u - 3v$$
$$B. \ u' = u + v, \ v' = u - v + 2x$$
with the initial conditions $u(0) = 0$, $v(0) = -1$ on $[0, 1]$ and $h = 0.2$. Can you solve these systems analytically?

Hint: Eliminate v from the second equation by using the first equation.

8. Use the "extended Euler method" where
$$y_{k+1} = y_k + hy'_k + \frac{h^2}{2} y''_k$$
to solve
$$y' = x^2 + 2y, \quad y(0) = 1, \quad \text{on } [0, 1]$$
with $h = 0.2$. Compare with the solution obtained from the Euler algorithm with the same step size and the exact solution.

9. Estimate the cumulative error in the "Extended Euler Method" in terms of h.

3.6   FINITE DIFFERENCE APPROXIMATIONS

Before discussing improvements to the Euler algorithm and the solution of boundary value problems, we must first introduce the notation of $O(h^k)$ (pronounced as "O-big" of h^k). Moreover, we shall derive finite difference approximations to the derivatives of a function $f(x)$ in terms of its values at some set of discrete points.

Definition 3.6.1 *Let $f(x)$ and $g(x)$ be two functions which are continuous at $x = 0$.*

We say that f is $O(g)$ if

$$0 < \left| \lim_{x \to 0} \frac{f(x)}{g(x)} \right| < \infty. \tag{3.69}$$

Example 3.6.1 $\sin x = O(x)$ *since*

$$\lim_{x \to 0} \frac{\sin x}{x} = 1. \tag{3.70}$$

Example 3.6.2 $f = 10^4 x^3 + 3x^2 + x$ *is $O(x^3)$ since*

$$\lim_{x \to 0} \frac{f(x)}{x^3} = 10^4. \tag{3.71}$$

Example 3.6.3 $x^4 + x^5$ *is not $O(x^3)$ since*

$$\lim_{x \to 0} \frac{x^4 + x^5}{x^3} = 0. \tag{3.72}$$

Example 3.6.4 *The remainder of the Taylor expansion of $f(x+h)$ around x*

$$R_n(h) = f(x+h) - \sum_{k=0}^{n} \frac{f^{(n)}(x) h^k}{k!} \tag{3.73}$$

is $O(h^{n+1})$ if $f^{(n+1)}(x) \neq 0$.

In fact, the remainder of the Taylor expansion is

$$R_n(h) = \sum_{k=n+1}^{\infty} \frac{f^{(k)}(x)h^k}{k!} = h^{n+1}\left\{\frac{f^{(n+1)}(x)}{(n+1)!} + h\,\frac{f^{(n+2)}(x)}{(n+2)!} + \ldots\right\} \quad (3.74)$$

$$\lim_{h\to 0}\frac{R_n(h)}{h^{n+1}} = \frac{f^{(n+1)}(x)}{(n+1)!}. \quad (3.75)$$

Therefore, we can write

$$f(x+h) = \sum_{k=0}^{n}\frac{f^{(k)}(x)h^k}{k!} + O(h^{n+1}). \quad (3.76)$$

From the definition we now infer the following:

Theorem 3.6.1 *If f is $O(h^n)$, g is $O(h^m)$, and $m \leq n$ then*

1. *$f \pm g$ is $O(h^m)$*

2. *$f \cdot g$ is $O(h^{n+m})$*

3. *f/g is $O(h^{n-m})$.*

Example 3.6.5 *The local error in the Euler algorithm (i.e., the error in each step) is $O(h^2)$ while the cumulative (global) error is $O(h)$.*

In fact, the basic approximation in the Euler method is

$$y_{k+1} = y_k + hy'_k + O(h^2). \quad (3.77)$$

Since there are n steps and $n = \frac{b-a}{h}$, the global error is $\frac{1}{h}O(h^2) = O(h)$.

The notation of O-big is used to give the user a rough idea about the behavior of the error committed by an algorithm as $h \to 0$. This helps to compare between different algorithms and determines which is superior.

We now introduce finite difference approximations.

If a symbolic representation of a function $f(\mathbf{x})$ is not known, then we consider the function to be "known numerically" on a domain D if its values on a grid of points in this domain are known and if by using these values we can calculate an acceptable approximation to the value of u at any other point in D (by interpolation). It follows, then, that in order to solve a differential equation on D, we have to introduce a grid of points on this domain and construct an algorithm to calculate the values of the unknown function u at these points. Thus, if (in one dimension) $\mathbf{x}_i, i = 1, \ldots, m$, are the grid points, then the fundamental unknowns that have to be computed are

$$f_i = f(\mathbf{x}_i) \quad i = 1, \ldots, m. \tag{3.78}$$

Since the differential equation and the initial (or boundary) conditions at hand are the only means to compute these quantities, we must develop an approximation scheme for the derivatives of f in terms of these data. To see how this can be done, consider the one-dimensional case where, to begin with, we assume that the $x_i's$ are equispaced (See Fig. 3.3):

$$x_i - x_{i-1} = h$$

(where h is constant for all values of i).

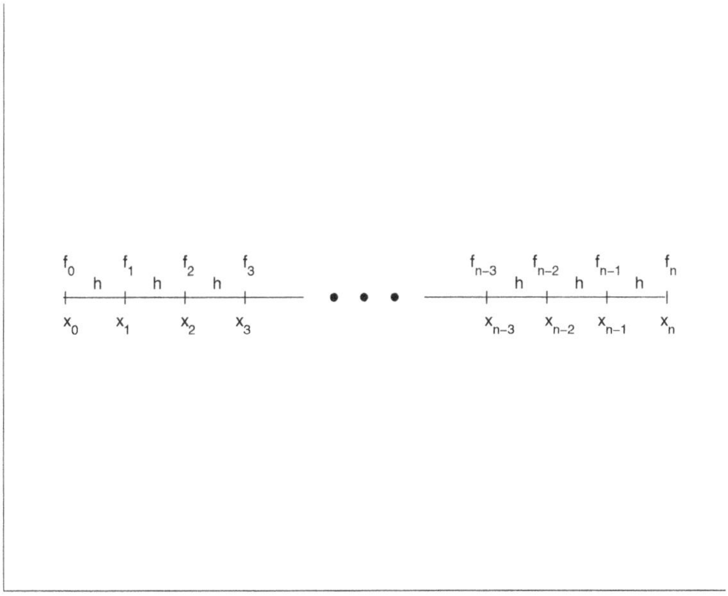

Figure 3.3 Equispaced computational grid

Thus, suppose that we are given $(x_i, f_i), i = 0, \ldots, n$, where $x_{i+1} - x_j = h = $ const. To find an approximation for $f_i' = f'(x_i)$, we use the Taylor expansion of f around x_i:

$$f_{i+1} = f(x_i + h) = f_i + h f_i' + \frac{h^2}{2} f_i'' + \frac{h^3}{3!} f_i''' + \frac{h^4}{4!} f_i^{(4)} + \ldots . \tag{3.79}$$

Hence,

$$f_i' = \frac{f_{i+1} - f_i}{h} - \frac{h}{2} f_i'' - \frac{h^2}{3!} f_i''' - \cdots \tag{3.80}$$

or

$$f_i' = \frac{f_{i+1} - f_i}{h} + O(h) . \tag{3.81}$$

This formula is called the forward difference approximation for f_i' (as the point x_{i+1} is used). Similarly, we can derive a backward difference approximation for f_i' using

$$f_{i-1} = f(x_i - h) = f_i - h f_i' + \frac{h^2}{2!} f_i'' - \frac{h^3}{3!} f_i''' + \frac{h^4}{4!} f_i^{(4)} \cdots . \tag{3.82}$$

This leads to

$$f_i' = \frac{f_i - f_{i-1}}{h} + O(h). \tag{3.83}$$

In both of these approximations, the error is $O(h)$. However, we can obtain a better approximation by using a "central difference" (i.e., use an interval where x_i is at the center). Thus, by subtracting Equation (3.82) from Equation (3.79) we obtain

$$f_i' = \frac{f_{i+1} - f_{i-1}}{2h} + \frac{2h^2}{3!} f_i''' + \frac{2}{5!} h^5 f_i^{(5)} + \cdots$$

or

$$f_i' = \frac{f_{i+1} - f_{i-1}}{2h} + O(h^2) . \tag{3.84}$$

We note that although this is a superior approximation, it cannot be used at the ends of the computational interval (i.e., at x_0 and x_n).

Similar approximation can be derived for the second order derivatives. In fact, adding Equation (3.79) and Equation (3.82) leads to

$$f_i'' = \frac{f_{i+1} - 2f_i + f_{i-1}}{h^2} + O(h^2) . \tag{3.85}$$

This is a central difference formula. A forward difference formula for f_i'' can be derived by using f_{i+2}. In fact,

$$f_{i+2} = f(x_1 + 2h) = f_i + 2h f_i' + 2h^2 f_i'' + \frac{4h^3}{3} f_i'' + \cdots . \tag{3.86}$$

Subtracting Equation (3.79) multiplied by 2 from Equation (3.86) leads to

$$f_i'' = \frac{f_{i+2} - 2f_{i+1} + f_i}{h^2} + O(h). \tag{3.87}$$

Finite difference approximation for higher order derivatives can be obtained in a similar fashion. For instance, we have the following central difference formula for the 4th derivative

$$f_i^{(4)} = \frac{f_{i+2} - 4f_{i+1} + 6f_i - 4f_{i-1} + f_{i-2}}{h^4} + O(h^2). \tag{3.88}$$

Example 3.6.6 *Suppose that the values of f at $x_0, x_0 + h$, and $x_0 + 3h$ are given. Develop a finite difference approximation for $f'(x_0)$ with $O(h^2)$ error.*

Solution 3.6.1 *From the Taylor expansion around x_0 we have*

$$f(x_0 + h) = f(x_0) + hf'(x_0) + \frac{h^2}{2!}f''(x_0) + \frac{h^3}{3!}f'''(x_0) + \dots$$

$$f(x_0 + 3h) = f(x_0)3hf'(x_0) + \frac{9h^2}{2!}f''(x_0) + \frac{27}{3!}f'''(x_0) + \dots .$$

To obtain an $O(h^2)$ approximation for $f'(x_0)$, we must find a combination of $f(x_0+h)$ and $f(x_0+3h)$ which cancels the h^2 terms in these expansions. Thus,

$$9f(x_0 + h) - f(x_0 + 3h) = 8f(x_0) + 6hf'(x_0) - 3h^3 f'''(x_0) + \dots,$$

i.e.

$$f'(x_0) = \frac{9f(x_0 + h) - f(x_0 + 3h) - 8f(x_0)}{h} + O(h^2).$$

When the grid points are not equispaced, the general technique described above can be easily adapted to derive appropriate approximations. Thus, if

$$x_i - x_{i-1} = h_i,$$

then

$$f_{i+1} = f(x_i + h_{i+1}) = f(x_i) + h_{i+1}f'(x_i) + O(h^2) \tag{3.89}$$

$$f_{i-1} = f(x_i - h_i) = f(x_i) - h_if'(x_i) + O(h^2),$$

which yields

$$f_i' = f'(x_i) = \frac{f_{i+1} - f_{i-1}}{h_i + h_{i+1}} + O(h). \tag{3.90}$$

Similarly, for the second-order derivative,

$$f_i'' = 2\frac{h_if_{i+1} + h_{i+1}f_{i-1} - (h_i + h_{i+1})f_i}{(h_ih_{i+1})(h_i + h_{i+1})} + O(h). \tag{3.91}$$

From Equations (3.90) and (3.91), we infer that in general the accuracy of our difference approximations is better with an equispaced grid. Equations (3.90) and (3.91) or their generalizations to higher dimensions are, therefore, useful near irregular boundaries where their use is mandatory.

3.6.1   Extension to Higher Dimensions

The formulas derived above in one dimension have a natural extension to higher dimensions. For example, for an equispaced grid in two dimensions with step size h, we have (see Fig. 3.4)

$$u_{i-1,j} = u(x_i - h, y_j) = u(x_i, y_j) - h\left(\frac{\partial u}{\partial x}\right)_{ij} + \frac{h^2}{2}\left(\frac{\partial^2 u}{\partial x^2}\right)_{ij} \tag{3.92}$$

$$+O(h^3)$$

$$u_{i+1,j} = u(x_i + h, y_j) = u(x_i, y_j) + h\left(\frac{\partial u}{\partial x}\right)_{ij} + \frac{h^2}{2}\left(\frac{\partial^2 u}{\partial x^2}\right)_{ij} + O(h^3).$$

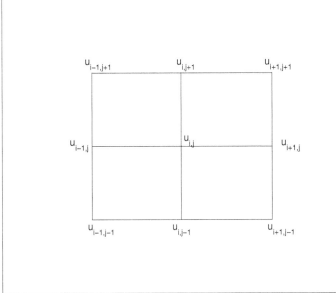

Figure 3.4   Two dimensional computational grid around $u_{i,j}$

Therefore,

$$\left(\frac{\partial u}{\partial x}\right)_{ij} = \frac{u_{i+1,j} - u_{i-1,j}}{2h} + O(h^2) \tag{3.93}$$

$$\left(\frac{\partial^2 u}{\partial x^2}\right)_{ij} = \frac{u_{i+1,j} + u_{i-1,j} - 2u_{ij}}{h^2} + O(h^2).$$

The corresponding formulas for $(\partial u/\partial y)_{ij}$ and $(\partial^2 u/\partial y^2)_{ij}$ should be obvious.

Exercises

1. Derive a backward finite difference formula for f_i''.

2. Prove that

$$f_i' = \frac{-f_{i+1} + 8f_{i+1} - 8f_{i-1} + f_{i-2}}{12h} + O(h^4) .$$

3. Prove the finite difference formula for $f_i^{(4)}$ which is given in Equation (3.88).

4. Derive an approximation for $f''(x_0)$ with $O(h^4)$ error using $f_{i+2}, f_{i+1}, f_i, f_{i-1}$ and f_{i-2} . (Use a symbolic computation package if available).

5. Derive a forward difference approximation for $f'''(x_0)$ with $O(h)$ error using $f_{i+3}, f_{i+2}, f_{i+1}$ and f_i .

6. Derive a central difference approximation for $f'''(x_0)$ with $O(h^2)$ error.

7. Derive a central difference formula for $f'(x_i)$ with $O(h^2)$ error if the grid points are not equispaced, i.e., $h_i = x_{i+1} - x_i$ are not constant.

3.7 MODIFIED EULER AND RUNGE-KUTTA METHODS

It is natural to inquire as to how one can improve the Euler method. One possibility which was discussed already is to use smaller step size h. However as h becomes smaller, the corresponding computational effort increases rapidly; furthermore, the accumulated numerical error (due to round-off errors) increases. The other options available are to use either higher order Taylor expansions (or their equivalents) or a "predictor-corrector" scheme. We illustrate both of these through the modified Euler algorithm which is a prototype for the Runge-Kutta class of algorithms.

3.7.1 Modified Euler Algorithm

To describe this algorithm we start with a geometrical argument.

The Euler algorithm to solve

$$y' = f(x, y), \quad y(a) = y_0 \tag{3.94}$$

applies a linear approximation to $y(x)$ at x_1 to compute $y(x_{i+1})$.

$$y_{i+1} = y_i + hf(x_i, y_i)$$

This linear approximation is based on the tangent to the graph of $y(x)$ at x_i whose slope is given by $y_i' = f(x_i, y_i)$ (see Fig. 3.5).

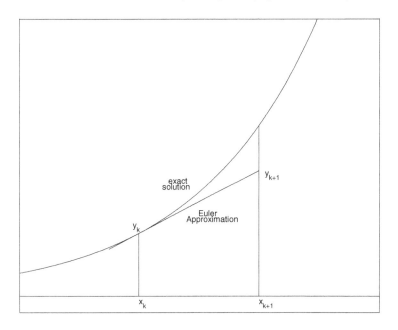

Figure 3.5 Euler's Numerical Approximation

Since the function $y(x)$ in general is not linear, it stands to reason that a better approximation for $y(x_{i+1})$ can be obtained if an "averaged slope" of $f(x)$ over $[x_i, x_{i+1}]$ is used. An appealing way to implement this will be to average the slopes at x_i and x_{i+1}, i.e.,

$$y_{i+1} = y_i + \frac{h}{2} \left[y_i' + y_{i+1}' \right] . \tag{3.95}$$

The only obstacle for the use of this formula is that

$$y_{i+1}' = f(x_{i+1} y_{i+1}). \tag{3.96}$$

This implies that in order to compute y_{i+1}' we must know y_{i+1} first!

To circumvent this difficulty we can utilize the Euler algorithm to obtain a **predictor** for the value of y_{i+1}

$$y_{i+1}^{(p)} = y_i + hf(x_i, y_i) . \tag{3.97}$$

Table 3.4 Solution of $y' + y = 2x$ with modified Euler method

x	y^p	y^c	$|y^p - y^c|$	Exact Solution
0	1	1	0	1
	0.800	0.860	0.06	0.856
0.2	0.860	0.853	0.01	
0.4	0.768	0.817	0.05	0.811
0.6	0.813	0.854	0.04	0.846
0.8	0.923	0.956	0.03	0.948
1	1.08	1.112	0.03	1.104

This value can be used then in eq. (3.95) to obtain a "corrected" value of y_{i+1}

$$y_{i+1}^{(c)} = y_i + \frac{h}{2} \left[f(x_i, y_i) + f\left(x_{i+1}, y_{i+1}^{(p)}\right) \right] .\qquad(3.98)$$

Should the difference between $y^{(p)}$ and $y^{(c)}$ be larger than some preset error tolerance ϵ, we can use $y^{(c)}$ as a new $y^{(p)}$ and iterate the process until

$$\left| y_{i+1}^{(p)} - y_{i+1}^{(c)} \right| < \epsilon .\qquad(3.99)$$

(Obviously we must set a limit on the number of these iterations to avoid the possibility of an infinite loop.)

We illustrate the implementation of this algorithm through the following example.

Example 3.7.1 *Solve*

$$y' = 2x - y, \quad y(0) = 1\qquad(3.100)$$

with $h = 0.2$ on $[0, 1]$. Use error tolerance of $\epsilon = 0.05$ and one predictor-corrector iteration.

Solution 3.7.1 *Using Equation (3.97), (3.98) we obtain Table 3.4 for the solution of Equation (3.100). To gauge the accuracy of the numerical solution we present in the last column of this table the values of the exact solution*

$$y = 3e^{-x} + (2x - 2).\qquad(3.101)$$

We demonstrate now that the modified Euler algorithm (without iterations) is equivalent to the use of second order Taylor expansions for $y(x)$ on each of the sub-intervals $[x_i, x_{i+1}]$. In fact according to this scheme

$$y_{i+1} = y_i + hy'_i + \frac{h^2}{2}y''_i + O(h^3). \qquad (3.102)$$

Replacing $y''_i = (y'_i)'$ by its forward finite difference approximation (eq. (3.79)).

$$y''_i = (y'_i) = \frac{y'_{i+1} - y'_i}{h} + O(h) \qquad (3.103)$$

leads to

$$y_{i+1} = y_i + h\left\{ y'_i + \frac{h}{2}\left[\frac{y'_{i+1} - y'_i}{h} + O(h) \right] \right\} \qquad (3.104)$$

$$+ O(h^3) = y_i + \frac{h}{2}\left\{ y'_i + y'_{i+1} \right\} + O(h^3). \qquad (3.105)$$

This also shows that the local error in the modified Euler algorithm is $O(h^3)$. Consequently, this leads to the following estimate for the cumulative (global) error

$$E = nO(h^3) = \frac{b - a}{h}O(h^3) = O(h^2). \qquad (3.106)$$

3.7.2   Runge-Kutta Methods

The basic idea that underlie, these methods is similar to the modified Euler algorithm. They use a "sampling" of $y(x)$ on $[x_i, x_{i+1}]$ to obtain an approximation for y_{i+1} which is equivalent to the use of a higher order Taylor expansion. Thus we speak of second, fourth, sixth or higher order Runge-Kutta methods according to the number of points used.

For example to determine a fourth order Runge-Kutta algorithm we try a prescription of the form,

$$y_{i+1} = y_i + a_1 k_1 + a_2 k_2 + a_3 k_3 + a_4 k_4 \qquad (3.107)$$

where

$$\begin{aligned}
k_1 &= hf(x_i, y_i) \\
k_2 &= hf(x_i + \alpha_1 h, y_i + \alpha_1 k_1) \\
k_3 &= hf(x_i + \alpha_2 h, y_i + \alpha_2 k_2) \\
k_4 &= hf(x_i + \alpha_3 h, y_i + \alpha_3 k_3).
\end{aligned} \qquad (3.108)$$

The unknowns a_i, α_i in these equations have to be determined so that Equation. (3.107) is equivalent to a fourth order Taylor expansion. It can be shown (after a long algebra) that this leads to eight equations in seven unknowns. Thus the choice of the parameters in this integration scheme is somewhat redundant. A classical solution is in the form

$$
\begin{aligned}
y_{i+1} &= y_i + \frac{1}{6}(k_1 + 2k_2 + 2k_3 + k_4) \\
k_1 &= hf(x_i, y_i), k_2 = hf\left(x_i + \frac{1}{2}h, y_i + \frac{1}{2}k_1\right) \\
k_3 &= hf\left(x_i + \frac{1}{2}h, y_i + \frac{1}{2}k_2\right), \\
k_4 &= hf\left(x_i + h, y_i + \frac{1}{2}k_3\right).
\end{aligned}
\tag{3.109}
$$

The freedom in the choice of the parameters of these algorithms (which exists for Runge-Kutta algorithms of any order) can be used to impose on the solution certain physical constraints. Thus if the differential equation under consideration describes the trajectory of a system in which energy is conserved, then the parameters of the Runge-Kutta algorithm can be chosen so that this conservation law is satisfied in each of the integration steps. (These are referred to as "symplectic algorithms".)

For more detailed treatment of the Runge-Kutta methods the reader is referred to the literature.

Exercises

Solve the equations in exercises 1-8 using the modified Euler algorithm with one predictor-corrector iteration

1. $y' = x + y^2, y(-1) = 0$ on $[-1, 1]$ with $h = 0.2$.

2. $y' = x^2y - x, y(0) = 1$ on $[0, 1]$ with $h = 0.2$.

3. $y' = xy^2, y(0) = -1$ on $[0, 1]$ with $h = 0.2$.

4. $y' = x^2 - y^2, y(0) = 0.5$ on $[0, 2]$ with $h = 0.2$.

5. $y' = x \sin y, y(0) = 0$ on $[0, 1]$ with $h = 0.2$.

6. Solve the differential equations in exercises 1-5 using the modified Euler algorithm without iterations but with $h = 0.1$.

7. Use Runge-Kutta of order four to solve exercises 1-5.

8. The following system of equations model an ecosystem of fish and sharks. Use Runge-Kutta of order four to solve these equations and plot the solution

$$u' = u - u^2 - 0.1uv \qquad u(0) = 2, \quad v(0) = 0.1$$

$$v' = -v + 0.1uv \qquad 0 \le t \le 100$$

(we are using here normalized units for the populations).

Hint: Solve these equations using MATLAB or similar.

9. Generalize the modified Euler algorithm to a system of equations and use it to solve the following:

$$u' = u + 2v + k$$

$$v' = u - 3v$$

$$u(0) = 0, \quad v(0) = 1 \text{ on } [0, 1] \text{ with } h = 0.2 \text{ and } 0.1$$

10. Solve, using different numerical methods, the following system and compare the results

$$u' = u^2 - v^2 + x$$

$$v' = u + v$$

$$u(0) = 1, v(0) = 0 \text{ on } [0, 2]. \text{ Use different step sizes.}$$

11. Solve

$$y' + 1000y = 0 \quad y(0) = 1$$

on $[0, 100]$ using different step sizes. Compare with the exact solution.

Remark 3.7.1 *Problems like this where numerical solution diverges from the correct solution are called "stiff".*

3.8 BOUNDARY VALUE PROBLEMS

The treatment of boundary value problems is totally different from the one for initial value problems which was presented in previous sections. The objective of this section is to present the basic ideas for solving boundary value problems for linear ordinary differential equations. Two prominent methods exist in the literature to solve these problems. These are the finite differences and the finite element methods. Both techniques reduce the original boundary value problem to a system of linear algebraic equations, but they are conceptually different from each other. In this section we consider only the finite difference method. The finite element method will be introduced in the chapter on variational principles. We present the application of finite difference method to boundary value problems through examples.

Example 3.8.1 *Solve the equation for the harmonic oscillator*

$$y'' + \omega^2 y = 0, \quad \omega \neq 0, \quad 0 \leq x \leq 1 \tag{3.110}$$

subject to the boundary conditions $y(0) = y(1) = 0$ with $h = \frac{1}{4}$.

Solution 3.8.1 *To solve this problem with $h = \frac{1}{4}$ we must compute*

$$y_1 = y(0.25), \quad y_2 = y(0.5), \quad y_3 = y(0.75)$$

($y_0 = y(0)$ and $y_4 = y(1)$ are given). To derive a system of linear equations for these unknowns we apply the central difference formula for y'' (Equation (3.85)) to approximate the differential equation. At $x = 0.25$ we have

$$\frac{y_2 + y_0 - 2y_1}{0.25^2} + \omega^2 y_1 = 0 .$$

Similarly at $x = 0.5$ and $x = 0.75$ we have

$$\frac{y_3 + y_1 - 2y_2}{0.25^2} + \omega^2 y_2 = 0$$

$$\frac{y_4 + y_2 - 2y_3}{0.25^2} + \omega^2 y_3 = 0$$

Since $y_0 = y_4 = 0$, these equations represent a system of three equations in the three unknowns y_1, y_2, y_3. In matrix form we can rewrite these equations as

$$\begin{pmatrix} \omega^2 - 32 & 16 & 0 \\ 16 & \omega^2 - 32 & 16 \\ 0 & 16 & \omega^2 - 32 \end{pmatrix} \begin{pmatrix} y_1 \\ y_2 \\ y_3 \end{pmatrix} = \begin{pmatrix} 0 \\ 0 \\ 0 \end{pmatrix}$$

This is a system of homogeneous equations, and a nontrivial solution (i.e., not all y_i are zero) exists only if the coefficients matrix A of this system is singular, viz. its determinant is zero. Since

$$\det A = (\omega^2 - 32)[(\omega^2 - 32)^2 - 2 \times 16^2]$$

we infer that a nontrivial solution to our problem exists only if

$$\omega = \pm 4\sqrt{2}, \quad \pm 4\sqrt{2 - \sqrt{2}}, \quad \pm 4\sqrt{2 + \sqrt{2}}, \tag{3.111}$$

e.g. for $\omega = 4\sqrt{2}$ we obtain

$$y_1 = \lambda, \quad y_2 = 0, \quad y_3 = -\lambda, \quad -\infty < \lambda < \infty$$

where λ is a parameter.

It is interesting to compare this solution with the analytic one. The general solution of Equation (3.110) is

$$y = A \cos \omega x + B \sin \omega x .$$

The first boundary condition then implies

$$y(0) = 0 = A .$$

To satisfy the second we must have

$$y(1) = B \sin \omega = 0 .$$

The solution of this equation is $B = 0$ (trivial solution for y) unless

$$\omega = n\pi, \quad n = 1, 2, \ldots, . \tag{3.112}$$

Thus the analytic solution has infinite values of ω for which a nontrivial solution $y = B \sin n\pi x$ exists.

To see the relation between the numerical and analytic solution we point out that the values of ω in Equation (3.111) are

$$\omega = \pm 3.061, \quad \pm 5.657, \quad \pm 7.391 .$$

These represent approximations to the first three values of ω in Equation (3.112). Observe that we can ignore the negative values of ω since they lead to the same solution for y as the positive values. We can expect, naturally, a better agreement between the numerical and analytic solutions as h becomes smaller.

Example 3.8.2 *Solve the following boundary value problem*

$$y'' + xy = x, \quad 0 \leq x \leq 1$$

subject to the conditions $y(0) = 0$, $y(1) = \frac{1}{16}$ *with* $h = \frac{1}{4}$.

Solution 3.8.2 *As in the previous example we use Equation (3.85) to approximate the differential equation for* $y_1 = y\left(\frac{1}{4}\right)$, $y_2 = y\left(\frac{1}{2}\right)$ *and* $y_3 = y\left(\frac{3}{4}\right)$. *This leads to*

$$\frac{y_2 + y_0 - 2y_1}{\left(\frac{1}{4}\right)^2} + 0.25y_1 = 0.25$$

$$\frac{y_3 + y_1 - 2y_2}{\left(\frac{1}{4}\right)^2} + 0.5y_2 = 0.5$$

$$\frac{y_4 + y_2 - 2y_3}{\left(\frac{1}{4}\right)^2} + 0.75y_3 = 0.75$$

where $y_0 = y(0)$, $y_4 = y(1)$.

Rewriting this in matrix form we have

$$\begin{pmatrix} -31.75 & 16 & 0 \\ 16 & -31.50 & 16 \\ 0 & 16 & -31.25 \end{pmatrix} \begin{pmatrix} y_1 \\ y_2 \\ y_3 \end{pmatrix} = \begin{pmatrix} 0.25 \\ 0.5 \\ -0.25 \end{pmatrix}.$$

Since the coefficient matrix of this system is non-singular, the system has a unique solution

$$y_1 = -0.024, \quad y_2 = -0.033, \quad y_3 = -0.009.$$

Example 3.8.3 : *Solve*

$$u'' + (1 + x^2)u' + u = x, \quad u(0) = 0, \quad u(1) = 1 \tag{3.113}$$

on $[0, 1)$ *with step size* $h = 0.25$.

Solution 3.8.3 : *Since* $h = 0.25$, *the numerical solution of this problem is equivalent to the computation of the three unknowns (see Fig. 3.6)*

$$u_1 = u(0.25), \quad u_2 = u(0.5), \quad u_3 = u(0.75)$$

(Note that $u_0 = u(0) = 0$ *and* $u_4 = u(1) = 1$.)

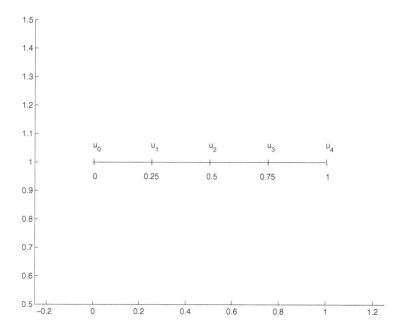

Figure 3.6 Computational grid for example 3.7.3

Using Equation (3.113) and the finite difference formulas in Equations (3.84) and (3.85), we obtain the following equations at $x = 0.25, 0.5$ and 0.75, respectively:

$$\frac{u_0 + u_2 - 2u_1}{h^2} + [1 + (0.25)^2]\frac{u_2 - u_0}{2h} + u_1 = 0.25 \qquad (3.114)$$

$$\frac{u_1 + u_3 - 2u_2}{h^2} + [1 + (0.5)^2]\frac{u_3 - u_1}{2h} + u_2 = 0.5$$

$$\frac{u_4 + u_2 - 2u_3}{h^2} + [1 + (0.75)^2]\frac{u_4 - u_2}{2h} + u_3 = 0.75$$

Using the boundary conditions and rearranging these equations lead to

$$-31u_1 + 18.125u_2 = 0.25 \qquad (3.115)$$

$$13.5u_1 - 31u_2 + 18.5u_3 = 0.5$$

$$12.875u_2 - 31u_3 = -18.375$$

The solution of this system to three decimals is

$$u_1 = 0.385 \quad u_2 = 0.672 \quad u_3 = 0.872$$

In the previous examples we showed how to solve problems where the boundary conditions are imposed on the unknown function. (These are referred to as "**Dirichlet Boundary Conditions**"). However boundary conditions can be imposed on the derivatives as well. These are called "**Neumann Boundary Conditions.**" Furthermore, boundary conditions can involve both the value of the unknown function and it derivatives at the boundary. These are called "**Robin Boundary Conditions**" or "**Boundary Conditions of the third kind.**" We demonstrate the treatment of such problems through the following example.

Example 3.8.4 *A spring-mass system is governed by the following equation*

$$\ddot{y} + 2\dot{y} + y = t \quad 0 \le t \le 1$$

subject to the boundary conditions $\dot{y}(0) = 1, \quad y(1) = 0$. *Solve this problem with* $h = 0.25$.

Solution 3.8.4 *Since* $y(0)$ *is not given, the problem involves four unknowns.*

$$y_n = y(nh), \quad n = 0, 1, 2, 3$$

$(y_4 = y(1) = 0)$ (see Fig. 3.7). Approximating the differential equation at $t_n = nh$ using Equations (3.83), (3.85) we obtain

$$\frac{y_{-1} + y_1 - 2y_0}{(0.25)^2} + 2\frac{y_1 - y_{-1}}{2 \cdot 0.25} + y_0 = 0 \tag{3.116}$$

$$\frac{y_0 + y_2 - 2y_1}{0.25)^2} + 2\frac{y_2 - y_0}{2 \cdot 0.25} = y_1 = 0.25 \tag{3.117}$$

$$\frac{y_1 + y_3 - 2y_2}{(0.25)^2} + 2\frac{y_3 - y_1}{2 \cdot 0.25} = y_2 = 0.5 \tag{3.118}$$

$$\frac{y_2 + y_4 - 2y_3}{(0.25)^2} + 2\frac{y_4 - y_2}{2 \cdot 0.25} = y_3 = 0.75 \tag{3.119}$$

where we had to add a fifth (fictitious) point $y_{-1} = y(-0.25)$. Equations (3.116), (3.119) form a system of four equations in five unknown. To add one more equation we use Equation (3.83) to approximate the boundary condition $\dot{y}(0) = 1$.

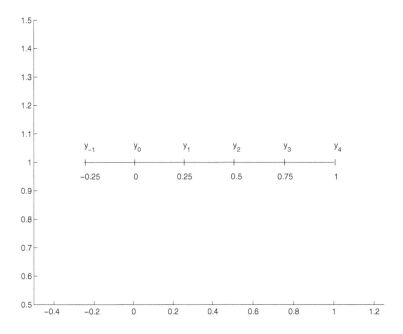

Figure 3.7 Computational grid for example 3.7.4

This leads to

$$\frac{y_1 - y_{-1}}{0.5} = 1 \ . \tag{3.120}$$

Together Equations (3.116), (3.120) form a system of five equations in five unknowns whose solution is

$$y_0 = 0.619, \quad y_1 = 0.412, \quad y_2 = -0.255, \quad y_3 = -0.123.$$

Exercises

1. Solve the boundary value problem given by Equation (3.113) with $h = 10^{-1}, 10^{-2}$. What can be said about the coefficient matrix for the resulting system of equations?

2. Compare the exact and numerical solution of

$$u'' + 2u' + u = x \quad u(0) = 1, u(1) = 3$$

with $h = 10^{-1}, 10^{-2}$.

CHAPTER 4

Stability Theory

CONTENTS

4.1 GENERAL INTRODUCTION

In the 19th and early 20th centuries most mathematical models for physical phenomena used linear differential equations which can be solved by analytic techniques. If a model turned out to be nonlinear, then some methods were used to "linearize" the model to obtain approximate solutions. With the rapid advancement of science and engineering and the advent of computers, it was found advantageous to consider sophisticated models where nonlinearities and their impact cannot be ignored. However while computers can generate "heaps" of numbers, it is not easy to extract insights about the system evolution under different conditions from this data.

Stability theory addresses the issue of model nonlinearities by making the observation that in many models an initial state of the system will evolve into a steady state (i.e., a state that is time independent). It was found that for many practical purposes the transient states of the system (i.e., those that are time dependent) are of "marginal" importance while the steady states and their properties are of utmost practical importance.

Since most mathematical models provide only an approximation to reality, a basic issue about the steady states is their "response" to "deviations" or perturbations of the system from the steady state. In other words if the system "somehow" deviates from the the steady state, will it return to it or "run away from it." As an example consider a car running smoothly on a highway then hits a pothole on the road. This will cause the body of the car to vibrate and the question arises then whether these oscillations will "damp out" and the car return to its "normal steady state", or a breakdown will occur and the car settle (after a short transient state) into a new steady state (i.e., the car is not drivable and has to be repaired) which is "far" from the original state. In other words "how stable was the original state of the car to perturbations." The answer to this fundamental question is the subject of stability theory.

From another point of view one has to remember that mathematical models contain parameters that have to be evaluated experimentally and hence, are subject to measurement errors. As a result the actual steady state of a given system might be somewhat different from the one predicted by the model equations. The question arises again as to whether these "small deviations" are "benevolent" (i.e. have little or no practical impact) or "fatal."

Historically the study of stability theory was initiated by H. Poincare and A. M. Liaponouv. Poincare was interested in the stability of the solar system and in particular whether the Earth orbit is stable. That is whether the Earth eventually "fall" into the Sun or recede from it. (This is still an open question.) A. M. Liaponouv considered the stability of various mechanical systems in particular under the influence of gravity. Today stability theory is still an important and evolving branch of science and engineering. In this chapter we give an exposition of the basic ideas of this theory.

To motivate our study and introduce some of the basic ideas we revisit the population model for the number of fish in a pool (see Chapter 2 Sec. 2.4) and assume that the size $N(t)$ of the population in this (simplified) ecological model is given by

$$\frac{dN}{dt} = (N-1)(N-3), \quad N(0) = N_0 \tag{4.1}$$

where the quadratic term in N represents the competition for resources (including food).

At a steady state $\frac{dN}{dt} = 0$ and hence Equation (4.1) has two steady (or equilibrium) states $N = 1$ and $N = 3$.

To study the stability of these steady states we note that Equation (4.1), though nonlinear, can be integrated by elementary methods and we obtain

$$N(t) = \frac{(N_0 - 3)e^{2t} - 3(N_0 - 1)}{(N_0 - 3)e^{2t} - (N_0 - 1)}. \tag{4.2}$$

From Equation (4.2) we infer that whenever $N_0 \neq 0$ the population $N(t)$ will approach the steady state $N = 1$ as $t \to \infty$ even if N_0 is very small or large. (However, N_0 must be positive). Thus, a (small) perturbation of this ecological system from $N = 1$ will cause its state to "come back" to this equilibrium while similar deviations from $N = 3$ will cause the the population to "run away" from this steady state and converge to $N = 1$ as $t \to \infty$. We conclude, therefore, that the steady state $N = 1$ is (asymptotically) stable while $N = 3$ is unstable.

To demonstrate these results graphically we plotted in Fig. 4.1 the solutions to (4.2) with different initial conditions.

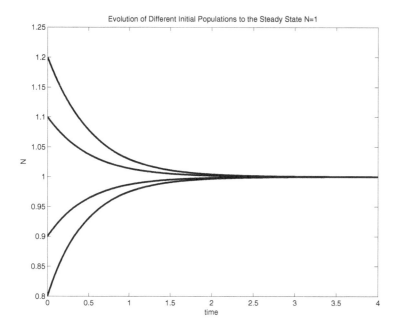

Figure 4.1 Population evolution from different initial conditions

The objective of stability theory (and phase space methods) is to derive these results about the stability of the steady states *without solving analytically or numerically the model equations under consideration.* This is a crucial feature since most current mathematical models lead to systems of nonlinear differential equations which cannot be solved analytically in closed form. Furthermore, numerical algorithms for the solution of these equations consume inordinate amounts of CPU time and in many cases do not converge or converge to the wrong solution.

The underlying idea of these techniques is to consider the model equations as an algebraic relationship between the state variables and their derivatives. Thus Equation (4.1) is viewed as an algebraic relation between N and dN/dt. To analyze the stability of the steady states we have only to remind ourselves (from elementary calculus) that $\frac{dN}{dt} > 0$ implies that N is increasing while $\frac{dN}{dt} < 0$ implies that N is decreasing with time. Thus in the "phase space" $\left(N, \frac{dN}{dt}\right)$ Equation (4.1) is described by Fig. 4.2. (Remember that N represents in this model the population size and therefore $N < 0$ is meaningless.)

We see from this figure that if $N < 3$ then $\frac{dN}{dt} < 0$ and hence N is a decreasing function. On the other hand if $N > 3$ then $\frac{dN}{dt} > 0$ and N

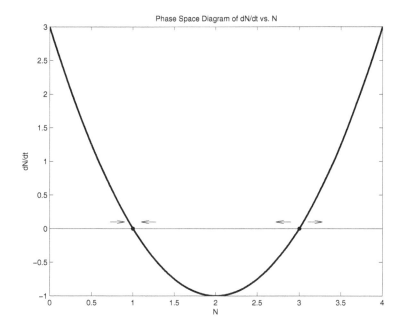

Figure 4.2 stability of the steady states N=1, N=3. Arrows indicate the evolution of perturbations from these states

will increase (these facts are indicated by arrows in the diagram). This is characteristic behavior of unstable steady state at $N = 3$. Similarly for the steady state $N = 1$, $\frac{dN}{dt} > 0$ if $N < 1$ (N is increasing with time) and $\frac{dN}{dt} < 0$ if $1 < N < 3$ (N is decreasing with time). Hence $N = 1$ is stable. We point out that these results about the steady states of this model were obtained without reference to the actual solutions of the model Equation (4.1).

Exercises

1. Solve the following equations in order to determine whether each of the steady states is stable or unstable.

   (a) $\frac{dN}{dt} = aN - bN^2$, $a, b > 0$, $N > 0$

   (b) $\frac{dN}{dt} = -aN + bN^2$, $a, b > 0$, $N > 0$

   (c) $\frac{dN}{dt} = aN + bN^2$, $a, b > 0$ $-\infty < N < \infty$

2. Use phase space techniques to deduce the stability of the steady states in Ex 1.

3. Use phase space techniques to discuss the stability of the steady states for

(a) $\frac{dN}{dt} = \alpha N + \beta N^2 + \gamma N^3$, $\quad \beta^2 - 4\alpha\gamma > 0$.

(b) $\frac{dN}{dt} = (N - \alpha)(N - \beta)(N - \gamma)$. What happens when $\alpha = \beta = \gamma$ or $\alpha = \beta$.

4. A model of two societies M, N where M exploits N was suggested by May and Noy-Meir. According to this model if M, N are the population sizes of the two populations respectively, then

$$\frac{dN}{dt} = aN(1 - N/A) - BM\frac{N^2}{1 + N^2}$$

where a, A, $B\ M$ are constants. In this equation the first term represents the natural growth of society N while the second represents the loss to this society due to the encounter between the two societies. Discuss the steady states of this model and their stability when M is small, moderate, and large.

Hint: To find the steady states plot the two terms in the model equation separately. The steady states are represented by the intersection of these two curves.

4.2 TWO SPECIES MODEL

Mathematical models which lead to systems of first order ordinary differential equations appear in many applications. It should be observed also that when the model equations contain second (or higher) order differential equations, then these equations can be rewritten as a system of first order equations. In particular when the model consists of one ordinary differential equation of second order, then this equation is equivalent to a system of two first order equations.

When the model is represented by two first order equations, then phase space methods in conjunction with graphical representation can be used to determine the stability of the steady states. In this section we demonstrate this technique through an example.

Example 1: Predator-Prey Ecological System.

Model: In this model we consider an ecological system which consists of two species. The first (Prey) consumes vegetable food and a second species (Predator) consumes vegetable and the first species for food (e.g., fish and sharks in the ocean or humans and cows). If only one of these species exists in the ecosystem, then its population will be governed by an equation similar to Equation(4.1). To model the interaction between these two species we shall assume that at each encounter between members of the two species there is a "chance" that the predator will consume the prey. Hence this interaction can be modeled as being proportional to the size of the two populations.

Denoting the size of the two populations at time t by $F(t)$, $S(t)$ respectively, the ecological system under consideration will be modeled by

$$\frac{dF}{dt} = F(a_1 - b_1 F - c_1 S) \tag{4.3}$$

$$\frac{dS}{dt} = S(a_2 - b_2 S + c_2 F) \tag{4.4}$$

where a_i, b_i, c_i, $i = 1, 2$ are non-negative constants.

4.2.1 Steady States

At the steady states of this system $\frac{dF}{dt} = \frac{dS}{dt} = 0$. Hence, these steady states of this model are given by the simultaneous solutions of

$$F(a_1 - b_1 F - c_1 S) = 0 \tag{4.5}$$

$$S(a_2 - b_2 S + c_2 F) = 0. \tag{4.6}$$

These solutions are:

1.

$$F = S = 0$$

2.

$$F = \frac{a_1}{b_1}, \ S = 0$$

3.

$$F = 0, \ S = \frac{a_2}{b_2}$$

4. The intersection of the two lines

$$a_1 - b_1 F - c_1 S = 0 \qquad (4.7)$$

$$a_2 - b_2 F + c_2 S = 0 \qquad (4.8)$$

if such an intersection exists in the first quadrant of the $F - S$ (phase) plane (obviously negative populations are meaningless).

4.2.2 Stability Analysis

To analyze the steady states with respect to their stability, using phase space techniques, we must first find the regions in the $F - S$ plane for which $\frac{dF}{dt} \lessgtr 0$, $\frac{dS}{dt} \lessgtr 0$. To accomplish this we note that $\frac{dF}{dt} > 0$ when

$$F(a_1 - b_1 F - c_1 S) > 0.$$

But $F \geq 0$ and, therefore, $\frac{dF}{dt}$ is positive if and only if

$$a_1 - b_1 F - c_1 S > 0, \ F > 0. \qquad (4.9)$$

Similarly $\frac{dF}{dt}$ is negative if and only if

$$a_1 - b_1 F - c_1 S < 0, \ F > 0. \qquad (4.10)$$

It follows then that the dividing line in the phase plane between the regions in which $\frac{dF}{dt}$ is positive and negative is given by

$$a_1 - b_1 F - c_1 S = 0. \qquad (4.11)$$

Similar analysis for $\frac{dS}{dt}$ shows that
$\frac{dS}{dt} > 0$ if and only if

$$a_2 - b_2 S + c_2 F > 0, \ S > 0. \qquad (4.12)$$

$\frac{dS}{dt} < 0$ if and only if

$$a_2 - b_2 S + c_2 F < 0, \quad S > 0. \tag{4.13}$$

Therefore the two regions in which $\frac{dS}{dt}$ is positive or negative are separated by the line

$$a_2 - b_2 F_2 + c_2 F_1 = 0. \tag{4.14}$$

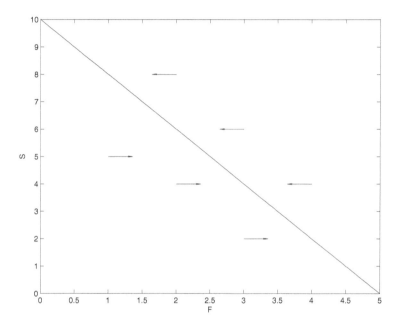

Figure 4.3 Regions in phase plane of Equations (12.1)-(12.2) where $\frac{dF}{dt}$ is positive and negative

For $a_1 = 1$, $b_1 = 0.2$, $c_1 = 0.1$, $a_2 = 0.5$, $b_2 = 0.4$ and $c_2 = 0.5$ we plot the lines given by Equations (12.9), (12.12) separately in Figs 4.3 and 4.4 respectively indicating by arrows the regions where $\frac{dF}{dt}$ and $\frac{dS}{dt}$ are positive and negative. In Fig. 4.5 we plotted these two lines together to show by arrows the response of the system to any perturbation from the steady state represented by the intersection of these two lines. Two trajectories starting from $(3.5, 4)$ and $(2.5, 3)$ are also plotted in this figure. This demonstrates that any perturbation from the steady state will decay in time and the population of the two species will return to the steady state. This implies that the two species will coexist in this ecosystem.

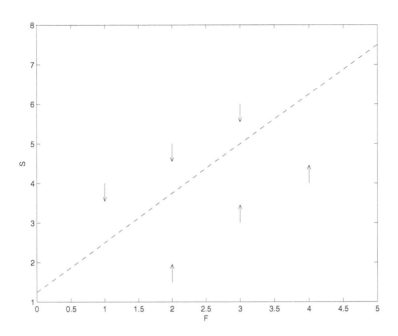

Figure 4.4 Regions in phase plane of Equations (12.1)-(12.2) where $\frac{dS}{dt}$ is positive and negative

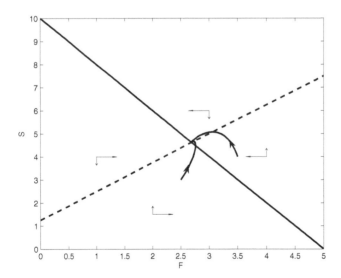

Figure 4.5 Stability of the steady state represented by the intersection of the lines $\frac{dF}{dt} = \frac{dS}{dt} = 0$

Using similar analysis we find that

1. $(0,0)$ is unstable

2. $[\frac{a_1}{b_1}, 0]$ is unstable in the S-direction and hence unstable.

3. $[0, \frac{a_2}{b_2}]$ is unstable in the F-direction and hence unstable.

4.3   BASIC CONCEPTS

In the previous section of this chapter we introduced the basic concepts of stability theory in an intuitive way. A graphical representation of these concepts is illustrated by the following examples. In the first case a ball is placed on top of a mountain and hence in unstable steady state. In the second case the ball is placed at the bottom of a "valley" (with ground friction. In this case the steady state is "asymptotically stable." That is, a perturbation from the steady state will decay in time, and the ball will return to the original steady state. Finally we consider a ball on a plane (with friction) where a perturbation from the original state will leave the ball at some distance from the original steady state but not "far away". (The distance remains bounded due to friction.) In this case we say that the steady state is "stable".

In the following we reintroduce and formalize these concepts.

Definition 4.3.1 *A system of ordinary differential equations*

$$\frac{d\mathbf{x}}{dt} = F(\mathbf{x}, t) \tag{4.15}$$

where $\mathbf{x} = (x_1(t), \ldots, x_n(t))$ *and*

$$F(\mathbf{x}, t) = \begin{pmatrix} f_1(\mathbf{x}, t) \\ \vdots \\ f_n(\mathbf{x}, t) \end{pmatrix}, \quad \mathbf{x} \in R^n \tag{4.16}$$

is called **autonomous** *if* $F(\mathbf{x}, t) = F(\mathbf{x})$, *i.e., the independent variable* t *does not appear explicitly in* F.

We observe that a non-autonomous system is equivalent to an autonomous system with an additional equation. In fact if we define

$$\mathbf{z} = \begin{bmatrix} \mathbf{x} \\ t \end{bmatrix}, \quad \mathbf{G}(z) = \begin{bmatrix} \mathbf{F}(\mathbf{x}, t) \\ 1 \end{bmatrix} \tag{4.17}$$

then

$$\frac{d\mathbf{z}}{dt} = \mathbf{G}(\mathbf{z}) \tag{4.18}$$

is equivalent to Equation (4.15)

In the following we consider only autonomous systems. Such systems are referred to as *dynamical systems.*

Definition 4.3.2 *The phase space of the system*

$$\frac{d\mathbf{x}}{dt} = \mathbf{F}(\mathbf{x}) = \begin{pmatrix} f_1(\mathbf{x}) \\ \vdots \\ f_n(\mathbf{x}) \end{pmatrix}, \quad \mathbf{x} \in R^n \tag{4.19}$$

is the space R^{2n} with coordinates $(x_1, \ldots, x_n, \dot{x}_1, \ldots \dot{x}_n)$,

where $\dot{x}_i = \frac{dx_i}{dt}$. Thus each point in phase space represents a unique state of the system. For a system of particles satisfying Newton's second law the phase space consists of all possible values of the momentum and position variables.

Example 4.3.1 *The equations of motion for a point particle of mass m under the influence of an external force $\mathbf{F} = \mathbf{F}(\mathbf{x})$ is*

$$m \frac{d^2\mathbf{x}}{dt^2} = \mathbf{F}(\mathbf{x}), \quad \mathbf{x} \in R^3 \tag{4.20}$$

which is equivalent to

$$\frac{d\mathbf{x}}{dt} = \mathbf{v}, \quad m \frac{d\mathbf{v}}{dt} = \mathbf{F}(\mathbf{x}). \tag{4.21}$$

Hence the phase space of Equation (4.20) is R^6 consisting of the points $(\mathbf{x}, \mathbf{v})$.

Definition 4.3.3 *A steady state ($\equiv$ critical point or equilibrium state) of the system (4.19) is a point $\mathbf{x}_0$ such that $\mathbf{F}(\mathbf{x}_0) = \mathbf{0}$.*

We observe that a steady state can be isolated; i.e., there exists a neighborhood of $\mathbf{x}_0$ which contains no other steady state or it might be part of a continuous set of critical points.

Example 4.3.2 *The system*

$$\frac{dx}{dt} = a_1 x + b_1 y, \quad \frac{dy}{dt} = a_2 x + b_2 y, \tag{4.22}$$

has a continuous set of steady states when $a_1 b_2 - b_1 a_2 = 0$. *In fact the steady state conditions* $\frac{dx}{dt} = \frac{dy}{dt} = 0$ *lead to*

$$a_1 x + b_1 y = 0, \quad a_2 x + b_2 y = 0, \tag{4.23}$$

and these equations are linearly dependent (that is they represent the same equation) when $a_1 b_2 - b_1 a_2 = 0$. *It follows then that any point on the line* $y = -\frac{a_1 x}{b_1}$, $b_1 \neq 0$ *is a steady state of the system. (When* $b_1 = 0$, *any point on the line* $x = 0$ *is a steady state.)*

On the other hand this analysis shows that $(0, 0)$ is an isolated steady state of the system represented by Equation(4.22) when $a_1 b_2 - b_1 a_2 \neq 0$, since then the coefficient matrix of Equation (4.23) is invertible.

In general it is rather difficult to analyze the stability of a system with a continuous set of steady states. Therefore, in the following we restrict our discussion to systems with isolated critical points.

The following is a generalization of this example.

Theorem 4.3.1 *Let* $\mathbf{F}(\mathbf{x})$ *be analytic and* $\mathbf{x}_0$ *a steady state of the system represented by Equation (4.19). A sufficient condition for* $\mathbf{x}_0$ *to be isolated is that the Jacobian matrix of* $\mathbf{F}(\mathbf{x})$ *at* $\mathbf{x}_0$

$$J(\mathbf{x}_0) = \left[\frac{\partial f_i}{\partial x_j} \right] (\mathbf{x}_0) \tag{4.24}$$

is nonsingular.

Proof Since $\mathbf{F}(\mathbf{x})$ is analytic, it has a Taylor series expansion that converges to $\mathbf{F}(\mathbf{x})$. This Taylor expansion can be used to approximate this function around the critical point $\mathbf{x}_0$.

$$\begin{aligned} \mathbf{F}(\mathbf{x}) &= \mathbf{F}(\mathbf{x}_0) + J(\mathbf{x}_0)(\mathbf{x} - \mathbf{x}_0) + O(|\mathbf{x} - \mathbf{x}_0|^2) \\ &= J(\mathbf{x}_0)(\mathbf{x} - \mathbf{x}_0) + O(|\mathbf{x} - \mathbf{x}_0|^2). \end{aligned} \tag{4.25}$$

Therefore, if another steady state exists in every neighborhood of $\mathbf{x}_0$, then in the vicinity of $\mathbf{x}_0$ all higher order terms in Equation (4.25) become negligible in comparison to the linear term, and the steady state must satisfy

$$J(\mathbf{x}_0)(\mathbf{x} - \mathbf{x}_0) = 0 . \tag{4.26}$$

But $J(\mathbf{x}_0)$ is nonsingular which implies that the only solution of this system is $\mathbf{x} = \mathbf{x}_0$. Hence $\mathbf{x}_0$ is isolated. Note, however, that this theorem sets only a sufficient but not necessary condition for $\mathbf{x}_0$ to be isolated.

Example 4.3.3 *The only critical point of the system*

$$\frac{dx}{dt} = y^2, \ \frac{dy}{dt} = x^2 \tag{4.27}$$

is $(0,0)$ *although* $J(0,0)$ *is singular.*

When a steady state is isolated, it is convenient, for conceptual and practical reasons, to translate the steady state *under consideration* to the origin by the transformation $\mathbf{w} = \mathbf{x} - \mathbf{x}_0$. The equations of the system will become,

$$\frac{d\mathbf{w}}{dt} = \mathbf{G}(\mathbf{w}) \tag{4.28}$$

where $\mathbf{G}(\mathbf{w}) = \mathbf{F}(\mathbf{w} + \mathbf{x}_0)$

Example 4.3.4 *Consider the system*

$$\frac{dx}{dt} = 4 - xy, \ \frac{dy}{dt} = 4x - y^3 \tag{4.29}$$

whose steady states are $(2, 2)$ *and* $(-2, -2)$. *To translate* $(2, 2)$ *to the origin we perform the transformation*

$$u = x - 2, \ v = y - 2 \tag{4.30}$$

and the resulting form of the system given by Equation (4.29), is

$$\begin{aligned}
\frac{du}{dt} &= -2u - 2v - uv \\
\frac{dv}{dt} &= 4u - 12v - 6v^2 - v^3
\end{aligned} \tag{4.31}$$

Similarly if we wish to consider the steady state $(-2, -2)$ *we perform the translation*

$$u = x + 2, \ v = y + 2 \tag{4.32}$$

The resulting form of the system in Equation(4.29) is

$$\frac{du}{dt} = 2u + 2v - uv$$
$$\frac{dv}{dt} = 4u - 12v + 6v^2 - v^3. \tag{4.33}$$

Definition 4.3.4 *A trajectory (≡ orbit, path) of the system given by Equation (4.19) is a solution of this system* $\mathbf{x} = \mathbf{x}(t)$ *subject to an initial condition* $\mathbf{x}(0) = \mathbf{c}$.

Sometimes such a trajectory is written as $\mathbf{x}(t, \mathbf{c})$.

Definition 4.3.5 *Let* $\mathbf{x}_0$ *be a steady state of the system, Equation (4.19). We say that*

1. $\mathbf{x}_0$ *is* **stable** *if for any given* $\epsilon > 0$ *there exists a* $\delta > 0$ *so that whenever* $| \mathbf{x}(0) - \mathbf{x}_0 | < \delta$ *then* $| \mathbf{x}(t) - \mathbf{x}_0 | < \epsilon$ *for all* $t > 0$.

2. $\mathbf{x}_0$ *is* **asymptotically stable** *if it satisfies the condition for being stable and in addition* $| \mathbf{x}(0) - \mathbf{x}_0 | < \delta$ *implies that*

$$\lim_{t \to \infty} | \mathbf{x}(t) - \mathbf{x}_0 | = 0 . \tag{4.34}$$

Thus, $\mathbf{x}_0$ is stable if whenever the initial state of the system is near the steady state then its trajectory will remain in its vicinity for all times. On the other hand if these trajectories approach $\mathbf{x}_0$ as $t \to \infty$ then we say that $\mathbf{x}_0$ is asymptotically stable.

Definition 4.3.6 *A steady state that is neither stable nor asymptotically stable is called* **unstable**.

Example 4.3.5 *The differential equation governing the motion of the linear pendulum (without friction) which was presented in Chapter 2, is*

$$\frac{d^2\theta}{dt^2} + \omega^2\theta.$$

The general solution of this equation is

$$\theta = A\cos(\omega t + \phi).$$

where A and ϕ are integration constants which are determined by the initial conditions. A small perturbation from the steady state $\theta = 0$, $\dot{\theta} = 0$ will take the system to a new (time dependent) state but the "distance" between the equilibrium state and the new state will remain bounded and does not go to 0 as time goes by. Hence the steady state is stable. On the other hand if we add friction to this system the amplitude of the oscillations will decay to zero as $t \to \infty$ and the equilibrium state of the system becomes asymptotically stable.

Definition 4.3.7 *The integral curves of the system represented by Equation (4.19) are the solutions of the system*

$$\frac{dx_1}{f_1} = \ldots = \frac{dx_n}{f_n} \ . \tag{4.35}$$

Thus the integral curves of Equation (4.19) are the trajectories of this system parametrized in terms of the $x_i's$ rather than in terms of the "extraneous" variable t.

Example 4.3.6 *Find the trajectories and integral curves of*

$$\frac{dx}{dt} = \alpha x, \ \frac{dy}{dt} = \beta y. \tag{4.36}$$

Solution 4.3.1 *The trajectories of the system are*

$$x = C_1 e^{\alpha t}, \ \ y = C_2 e^{\beta t} \tag{4.37}$$

where C_1, C_2 are arbitrary constants. The corresponding integral curves are $y^\alpha = Cx^\beta$ where C is a constant.

Example 4.3.7 *Find the integral curves of the equations of the spring-mass system (with no friction).*

Solution 4.3.2 *The equation of motion for the spring mass system for small displacements from the steady state is*

$$\frac{d^2 x}{dt^2} + kx = 0 \ . \tag{4.38}$$

This is equivalent to the system

$$\frac{dx}{dt} = v$$
$$\frac{dv}{dt} = -kx \ . \tag{4.39}$$

The integral curves of this system satisfy

$$\frac{dx}{dv} = \frac{v}{-kx} \tag{4.40}$$

and hence

$$kx^2 + v^2 = C^2$$

where C^2 is a constant. We conclude, therefore, that the integral curves of the spring-mass system are ellipses (see Fig. 4.9). It follows from this result that the steady state $(0,0)$ of the system given by Equation (4.39) is stable but not asymptotically stable. This result is obvious from a physical point of view since the system contains no friction to damp the motion.

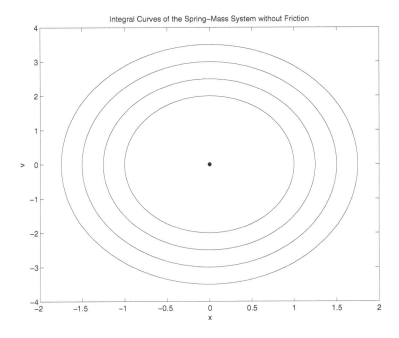

Figure 4.6 Integral curves of the spring-mass system without friction

Exercises

1. Find the form of the following systems when each of their steady states is translated to the origin.

   (a) $\frac{dx}{dt} = x(x - 2)(y - 2), \qquad \frac{dy}{dt} = (x + 2)(y - 1)^2$

(b) $\frac{dx}{dt} = \sin 2\pi x \cos 2\pi y$, $\quad \frac{dy}{dt} = (\sin \pi y)^2 - 1$

2. Find the trajectories, integral curves and stability of the critical point at the origin for the following systems

(a) $\frac{dx}{dt} = -x$, $\quad \frac{dy}{dt} = 2x - y$

(b) $\frac{dx}{dt} = -3x + y$, $\quad \frac{dy}{dt} = x - 3y$

(c) $\frac{dx}{dt} = nz - ky$, $\quad \frac{dy}{dt} = kx - mz$, $\quad \frac{dz}{dt} = my - nx$

3. Find the integral curves and stability of the critical points for the following systems.

(a) $\ddot{x} = x^3$

(b) $m\ddot{x} + b\dot{x} + kx = 0$, $\quad m, b, k > 0$. What happens if $b < 0$?

(c) $\ddot{\theta} + \nu^2 \sin \theta = 0$

(nonlinear pendulum). What happens when $|\theta|$ is small and it is possible to approximate $\sin \theta$ by θ.

4.4 LINEARIZABLE DYNAMICAL SYSTEMS

When a dynamical system is nonlinear, it is impossible, in general, to find analytical expressions for the trajectories or integral curves of the system. As a result it is, impossible to infer the stability of its steady states directly. To overcome this issue one usually attempts to approximate the equations of the dynamical system in the vicinity of a steady state by a linear system of equations. When such an approximation exists, we say that the system is **linearizable** (near the steady state under consideration). This linear approximation can be used then to deduce the stability of the steady state (with some exceptions).

Definition 4.4.1 *Let* $\mathbf{x}_0$ *be an isolated critical point of the system*

$$\frac{d\mathbf{x}}{dt} = \mathbf{F}(\mathbf{x}) = \begin{pmatrix} f_1(\mathbf{x}) \\ \vdots \\ f_n(\mathbf{x}) \end{pmatrix}. \tag{4.41}$$

We say that the system, Equation (4.41), is linearizable at $\mathbf{x}_0$ if the Jacobian of F at $\mathbf{x}_0$ is nonsingular, i.e.

$$\det J(\mathbf{x}_0) = \left| \left(\frac{\partial f_i}{\partial x_j}(\mathbf{x}_0) \right) \right| \neq 0 . \tag{4.42}$$

Example 4.4.1 *The system*

$$\frac{dx}{dt} = (x-1)(x-3)^2, \quad \frac{dy}{dt} = y \tag{4.43}$$

is linearizable at the steady state $(1,0)$ *but is not so at the steady state* $(3,0)$.

To treat systems which are linearizable at the critical point $\mathbf{x}_0$ we first translate this point to the origin and then take the Taylor expansion of each $f_i(\mathbf{x})$ around the origin

$$f_i(\mathbf{x}) = \sum_{j=1}^{n} a_{ij} x_j + O(|\mathbf{x}|^2) . \tag{4.44}$$

It follows then that the linear approximation to the original system around the origin is given by

$$\frac{d\mathbf{x}}{dt} = A\mathbf{x} \tag{4.45}$$

where $A = J(0)$ is a constant coefficient matrix.

We observe that Equation (4.42) implies that not all the coefficients $a_{i1}, \ldots, a_{in}$ in the expansion of the function $f_i(\mathbf{x})$ are zero. Therefore, there exists a neighborhood of $\mathbf{0}$ where the linear terms are dominant in the corresponding equation. This can be considered as the reason for the relationship between the stability of the linear system Equation (4.45) at the origin and the original system.

Example 4.4.2 *The system*

$$\frac{dx}{dt} = x + y^2 - 6, \quad \frac{dy}{dt} = 2y - x^2$$

has a critical point at $x = 2, y = 2$. *The determinant of the Jacobian at this point is*

$$\det J(1,1) = \left| \begin{array}{cc} \frac{\partial f_1}{\partial x} & \frac{\partial f_1}{\partial y} \\ \frac{\partial f_2}{\partial x} & \frac{\partial f_2}{\partial y} \end{array} \right| (2,2) = \left| \begin{array}{cc} 1 & 4 \\ -4 & 2 \end{array} \right| \neq 0. \tag{4.46}$$

Hence this system is linearizable at this steady state. To compute the linear approximation we first translate this steady state to the origin by the translation

$$u = x - 2, \quad v = y - 2 . \tag{4.47}$$

The system now takes the form

$$\frac{du}{dt} = u + 4v + v^2, \quad \frac{dv}{dt} = -4u + 2v - u^2, \tag{4.48}$$

and the resulting linear approximation at the origin is

$$\frac{du}{dt} = u + 4v, \quad \frac{dv}{dt} = -4u + 2v. \tag{4.49}$$

In matrix notation this takes the following form:

$$\frac{d}{dt}\begin{bmatrix} u \\ v \end{bmatrix} = \begin{bmatrix} 1 & 4 \\ -4 & 2 \end{bmatrix}\begin{bmatrix} u \\ v \end{bmatrix}. \tag{4.50}$$

When the system Equation (4.41), is linearizable at the steady state $\mathbf{x}_0$, it is possible to characterize the stability of this steady state (after the translation to the origin) in terms of the eigenvalues of the matrix A in Equation (4.45). We start with the following theorem.

Theorem 4.4.1 *The critical point at the origin of the system Equation (4.45), is*

1. *Asymptotically stable if all the eigenvalues of A have negative real parts.*

2. *Stable if all the eigenvalues of A have non-positive real parts and every eigenvalue of A which has a zero real part is a simple eigenvalue of A.*

3. *Unstable if (1) and (2) are false.*

Example 4.4.3 *The eigenvalues of the matrix A in Equation (4.50) are*

$$\lambda_{\pm} = \frac{3 \pm i\sqrt{55}}{2}.$$

Therefore this steady state at $(0,0)$ of Equation (4.49) is unstable.

Example 4.4.4 *The critical point at the origin of the system*

$$\frac{dx}{dt} = \begin{bmatrix} 1 & -2 & 4 \\ 7 & -8 & 10 \\ 2 & -2 & 1 \end{bmatrix} x \tag{4.51}$$

is asymptotically stable since the eigenvalues of A are -3, -1, -2.

The relationship between the stability of the steady states of the systems (4.41) and (4.45) at $\mathbf{x} = \mathbf{0}$ is summarized by following theorem.

Theorem 4.4.2 *If the steady state at the origin of the linearized system Equation (4.45), is asymptotically stable or unstable, then the same is true for the steady state of the original system Equation (4.41). However, if the origin is a stable steady state for the linearized system Equation (4.45), then the stability of Equation (4.41) at this point is indeterminate, i.e., the stability of the original system at this point cannot be deduced from that of the linearized system.*

Example 4.4.5 *The system*

$$\frac{d}{dt} \begin{bmatrix} x \\ y \end{bmatrix} = \begin{bmatrix} 0 & 2 \\ -2 & 0 \end{bmatrix} \begin{bmatrix} x \\ y \end{bmatrix} - \alpha \begin{bmatrix} 0 \\ x^2 y \end{bmatrix} \tag{4.52}$$

is almost linear at the critical point $\mathbf{0}$. Its linearization at this point is given by

$$\frac{d}{dt} \begin{bmatrix} x \\ y \end{bmatrix} = \begin{bmatrix} 0 & 2 \\ -2 & 0 \end{bmatrix} \begin{bmatrix} x \\ y \end{bmatrix} = Ax . \tag{4.53}$$

Since the eigenvalues of A are $\pm 2i$, it follows from theorem 4.3.1 that $(0,0)$ is a stable critical point of the system Equation (4.53). However, it can be shown that the original system Equation (4.52), is asymptotically stable if $\alpha > 0$ and unstable if $\alpha < 0$. (See exercise 6 in Section 6).

Exercises

1. Show that the solution $x = 0$ of

$$a_k x^{(k)} + a_{k-1} x^{(k-1)} + \ldots + a_1 x' + a_0 x = 0 \tag{4.54}$$

is stable if and only if all the roots of the polynomial

$$p(\lambda) = a_0\lambda^n + \ldots + a_0$$

have non-positive real parts and all roots with zero real parts are simple.

Hint: Solve Equation (4.54)

For the following exercises find the critical points at which the system is almost linear and discuss their stability.

2. $\frac{dx}{dt} = (x-1)(x-2)^2(y-1), \quad \frac{dy}{dt} = (x-1)^2(y-2)$

3. $\frac{dx}{dt} = y(1 + \cos x), \quad \frac{dy}{dt} = x(1 + \sin y)$

4. $\ddot{x} + \beta(x^2 - 1)\dot{x} + kx = 0, \quad b > 0.$ (Van der Pol eq.)

5. $\ddot{x} + \beta\dot{x} - x + 2x^3 = 0$ (Duffin's equation)

4.5 LINEARIZABLE SYSTEMS IN TWO DIMENSIONS

The previous section presented a general framework that relates, under proper conditions, the stability of a steady state of a nonlinear dynamical system to its linearization at this point. However in two dimensions it is possible carry out further classification, of these steady states in terms of their "phase portrait" (i.e., phase diagram). In the following we describe this classification.

Let the system

$$\frac{dx}{dt} = G(x,y), \quad \frac{dy}{dt} = F(x,y) \tag{4.55}$$

have an isolated critical point at the origin. If this system is linearizable and the Jacobian at $(0,0)$ is nonsingular. We can take the Taylor expansion of G, F around this point and rewrite Equation (4.55) as

$$\frac{dx}{dt} = ax + by + f_1(x,y) \tag{4.56}$$

$$\frac{dy}{dt} = cx + dy + f_2(x,y). \tag{4.57}$$

Since the Jacobian of Equation (4.55) is nonsingular at $(0,0)$, $ad - bc \neq 0$.

Moreover, f_1, f_2 are of order $| \mathbf{x} |^2$. Therefore, the linear approximation to the system given by Equation (4.55) near the origin is given by

$$\frac{dx}{dt} = ax + by, \quad \frac{dy}{dt} = cx + dy, \quad ad - bc \neq 0. \tag{4.58}$$

To solve the system Equation (4.58), we make the ansatz that the solution is of the form

$$x = A \exp(\lambda t) \quad y = B \exp(\lambda t) \tag{4.59}$$

(same λ for both x and y). Substituting Equation (4.59) in Equation (4.58) yields then

$$(a - \lambda)A + bB = 0 \tag{4.60}$$

$$cA + (d - \lambda)B = 0. \tag{4.61}$$

Equations (4.60), (4.61) form a system of linear homogeneous equations for the coefficients A, B. A nontrivial solution for these coefficients exists if and only if the determinant of the coefficients of this system vanishes, i.e.,

$$\begin{vmatrix} a - \lambda & b \\ c & d - \lambda \end{vmatrix} = \lambda^2 - (a + d)\lambda + (ad - bc) = 0. \tag{4.62}$$

It follows that a nontrivial solution exists only if λ is an eigenvalue of the coefficient matrix of the system Equation (4.58), and the corresponding solution of Equation (4.62), $\begin{bmatrix} A \\ B \end{bmatrix}$, is the eigenvector related to this eigenvalue. The stability, instability, and the phase portrait of the steady state at the origin is determined, therefore, by the eigenvalues and eigenvectors of this coefficient matrix.

In the following we provide a classification for all these possibilities.

Case 1: Equation (4.62) has two real unequal roots λ_1, λ_2 of the same sign.

The general solution of the system, Equation (4.58), is given by

$$\begin{aligned} x &= A_1 e^{\lambda_1 t} + A_2 e^{\lambda_2 t} \\ y &= B_1 e^{\lambda_1 t} + B_2 e^{\lambda_2 t}. \end{aligned} \tag{4.63}$$

If λ_1, λ_2 are positive, the steady state at the origin is unstable. If on the other hand both are negative, the steady state is asymptotically stable. (Observe

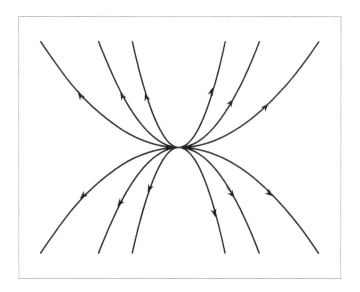

Figure 4.7 Improper node (unstable)

that these eigenvalues cannot be zero since we are assuming that the coefficient matrix of Equation (4.58) is not singular.) A typical illustration of the integral curves of the system in this case is shown in Fig. 4.7. A steady state with the phase portrait shown in this figure is called improper node (stable or unstable).

Case 2: Equation (4.62) has two real roots with opposite signs.

Under the assumption of this case Equation (4.63) still represents the solution of the system, Equation (4.58). However, now, the steady state is always unstable since one of the eigenvalues is positive. Its phase portrait is given in Fig. 4.8. Since there is a direction in which the steady state is stable, it is referred to as a "saddle point."

Case 3: Equation (4.62) has two equal roots.

Since the coefficient matrix of Equation(4.62) is nonsingular $(ad - bc \neq 0)$ it is obvious that this root must be real and different from zero. Hence the solution of the system, Equation (4.58), is given by

$$x = (A_1 + A_2t)e^{\alpha t}, \quad y = (B_1 + B_2t)e^{\alpha t} . \tag{4.64}$$

This implies that the steady state is asymptotically stable if $\alpha < 0$ and unstable if $\alpha > 0$. The phase portrait of such a steady state is referred to as improper or proper node (Figs. 4.7 or 4.9 respectively).

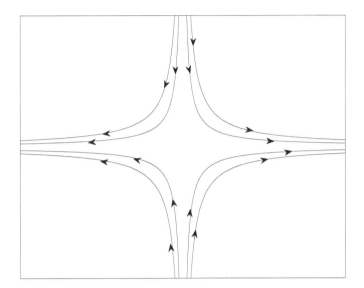

Figure 4.8 Saddle point

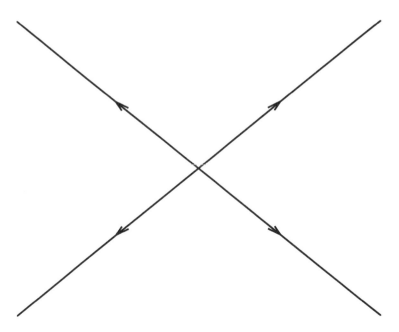

Figure 4.9 Proper node (unstable)

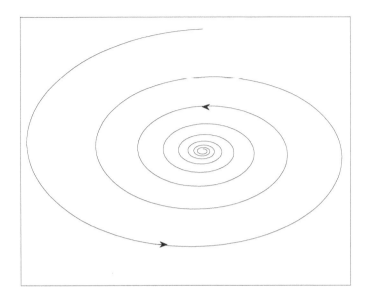

Figure 4.10 Spiral (stable)

Case 4: Equation (4.62) has complex roots with nonzero real part

$$s_\pm = \beta \pm i\omega, \quad \beta \neq 0.$$

The solution of the system, Equation (4.58), is

$$\begin{aligned}
x &= e^{\beta t}(A_1 \cos \omega t + A_2 \sin \omega t) \\
y &= e^{\beta t}(B_1 \cos \omega t + B_2 \sin \omega t) \,.
\end{aligned} \tag{4.65}$$

The critical point is unstable if $\beta > 0$ and asymptotically stable if $\beta < 0$. The phase portrait of the steady state is a spiral, regardless of the stability of the steady state, as shown in Fig. 4.10.

Case 5: Equation (4.62) has pure imaginary roots

$$s = \pm i\omega$$

The solution of the system, Equation (4.58), is represented by Equation (4.65) with $\lambda = 0$. The critical point is stable and its phase portrait as shown in Fig. 4.5. The steady state is called a "center".

As to the relationship between the phase portraits, at the origin of the system, Equation (4.55), and its linearization Equation (4.58), we have the following

Theorem 4.5.1 *The phase portraits at $(0,0)$ of the systems, Equation (4.55) and Equation (4.58) are the same except in cases 3 and 5 where a spiral is possible as an additional phase portrait.*

To motivate this result we note that the additional terms in Equation (4.56) might destroy the exact equality of the roots (case 3) or add to them a small real part when they are purely imaginary as in case 5.

It should be also noted that when the roots of Equation (4.62) are purely imaginary the stability of the steady state at $(0,0)$ of the system, Equation (4.55), cannot be deduced from those of the system, Equation (4.58). This conforms to the statement of theorem 3.4.2 in the previous section.

Exercises

For the following systems classify and draw the phase portrait for the steady state at the origin.

1. $m\ddot{x} + b\dot{x} + kx = 0 \quad m, b, k > 0$. What happens if $b < 0$

2. $\ddot{x} + b\dot{x} + kx^2 = 0, \quad k > 0$. Consider separately the case $b > 0$ and $b < 0$.

3. $\frac{dx}{dt} = \sin(x - y), \quad \frac{dy}{dt} = e^{x-y} - \cos(x - y)$

4. $\frac{dx}{dt} = x + y + 2y^2 - x^2, \quad \frac{dy}{dt} = y - 2x + y^3$

5. $\frac{dx}{dt} = 2y - x^3, \quad \frac{dy}{dt} = 2x + y^2$ (compare with the exact solution)

6. $\frac{dx}{dt} = y - 3x, \quad \frac{dy}{dt} = e^x + e^y \sin x - 1$

4.6   LIAPOUNOV METHOD

In the last two sections of this chapter we discussed dynamical systems which are linearizable near an isolated steady state. We showed that under this restriction there is a close correspondence between the stability of the steady state of the original system and its linear approximation. The question naturally arises as what techniques can be used to determine the stability of the steady state when the dynamical system is not linearizable.

Another aspect of stability theory which was not discussed so far is the size of the "**basin of stability**" of the steady state. That is if $\mathbf{x}_0$ is an asymptotically stable steady state of the system what is the size of the "maximal perturbation" to the system which will decay in time to $\mathbf{x}_0$.

To answer these questions Liapounov (direct) method is a powerful tool which has been applied successfully in many practical applications. In the following we present the essence of this technique and its application to gradient dynamical systems.

Definition 4.6.1 *Let the function $F : R^n \rightarrow R$ be defined on a domain $\Omega \in R^n$ containing the origin and $F(\mathbf{0}) = 0$.*

*1. F is said to be **positive definite** on Ω if $F(\mathbf{x}) > 0$ for all $\mathbf{x} \neq \mathbf{0}$ in Ω.*

*2. F is said to be **positive semidefinite** on Ω if $F(\mathbf{x}) \geq 0$ for all $\mathbf{x} \in \Omega$.*

In a similar fashion one can define functions which are **negative definite** and **negative semidefinite** on Ω.

Example 4.6.1 *The function*

$$F(x, y, z) = x^2 + y^2 + z^2 \tag{4.66}$$

is positive definite in R^3.

Example 4.6.2 *The function*

$$F(x, y) = \sin x^2 + y^2 \tag{4.67}$$

is positive definite on the disk $0 \leq x^2 + y^2 < \pi$.

Example 4.6.3 *The function*

$$F(x, y) = (3x - y)^2 \tag{4.68}$$

is positive semidefinite in R^2 since $F(x, y) = 0$ on the line $y = 3x$ but positive otherwise.

Example 4.6.4 *Show that the function*

$$F(x, y) = ax^2 + bxy + cy^2 \qquad (4.69)$$

is positive definite in R^2 if $c > 0$ and $4ac - b^2 > 0$.

Solution 4.6.1 *By adding and subtracting $\frac{b^2}{4c}x^2$ from the expression of $F(x, y)$ we can rewrite it as*

$$F(x, y) = \left[\frac{b}{2\sqrt{c}} x + \sqrt{c}\, y\right]^2 + \frac{4ac - b^2}{4c} x^2 \qquad (4.70)$$

which is obviously positive definite if $c > 0$ and $4ac - b^2 > 0$.

Consider now the autonomous dynamical system

$$\frac{d\mathbf{x}}{dt} = G(\mathbf{x}) = \begin{bmatrix} g_1(\mathbf{x}) \\ \vdots \\ g_n(\mathbf{x}) \end{bmatrix} \qquad (4.71)$$

which has an isolated steady state at $\mathbf{x} = \mathbf{0}$.

Theorem 4.6.1 *(Liapounov) Let $F(\mathbf{x})$ be, a positive definite function with continuous derivatives on a domain Ω containing the origin and define*

$$\dot{F}(\mathbf{x}) = D_G F(\mathbf{x}) = gradF \cdot \mathbf{G} = \sum_{i=1}^{n} \frac{\partial F}{\partial x_i} \cdot g_i . \qquad (4.72)$$

If on some domain containing the origin

1. *$\dot{F}(x)$ is negative definite then the steady state $\mathbf{x} = \mathbf{0}$ is asymptotically stable,*

2. *$\dot{F}(\mathbf{x})$ is negative semidefinite then $\mathbf{x} = \mathbf{0}$ is stable,*

3. *$\dot{F}(\mathbf{x})$ is positive definite then $\mathbf{x} = \mathbf{0}$ is unstable.*

Observe that this theorem makes no reference to the solution of the system, Equation (4.71), in contrast to the technique of linearization.

Example 4.6.5 *Discuss the stability of the steady state* $\mathbf{x} = \mathbf{0}$ *for the system*

$$\dot{x} = -2y + \alpha x(x^2 + y^2) = g_1(x, y) \tag{4.73}$$

$$\dot{y} = 2x + \alpha y(x^2 + y^2) = g_2(x, y) \tag{4.74}$$

where α *is a parameter.*

Solution 4.6.2 *Let* $F = x^2 + y^2$, *then*

$$\dot{F}(x, y) = gradF \cdot \mathbf{G} = 2\alpha(x^2 + y^2)^2. \tag{4.75}$$

Therefore, Liapounov theorem implies that if $\alpha < 0$, *then the steady state at the origin is asymptotically stable. If* $\alpha = 0$ *the state is stable while if* $\alpha > 0$ *the state is unstable.*

We infer from this example that the major difficulty inherent in this method is that one has to construct (or find) a proper positive definite function $F(\mathbf{x})$ (which is refered to as Liapounov function) that satisfies the requirements of Liapounov theorem. Several ad hoc algorithms, for particular applications, were suggested in the literature to overcome this difficulty. However, most of the important applications of this method remain limited to systems for which $F(\mathbf{x})$ can be deduced from physical arguments. These are gradient and conservative dynamical systems. (For many of these systems the function F is the Hamiltonian or a function related to it.)

Definition 4.6.2 *A dynamical system, Equation (4.71), is called a* **gradient system** *if there exists a (smooth) function* $\phi(\mathbf{x})$ *so that*

$$\mathbf{G}(x) = -grad\,\phi(\mathbf{x}) . \tag{4.76}$$

$\phi(\mathbf{x})$ *is called the* **potential function** *of the system.*

Theorem 4.6.2 *Let* $\phi(\mathbf{x})$ *be a potential function for the system, Equation (4.71), and let* $\mathbf{x}_0$ *be an isolated local minimum of* ϕ. *Then* $\mathbf{x}_0$ *is an asymptotically stable steady state of the system, Equation (4.71).*

Proof: Since ϕ is smooth and $\mathbf{x}_0$ is a local minima of ϕ, there exists a neighborhood Ω of $\mathbf{x}_0$ in which

$$F(\mathbf{x}) = \phi(\mathbf{x}) - \phi(\mathbf{x}_0) \tag{4.77}$$

is positive definite. Furthermore since $\mathbf{x}_0$ is an isolated minimum in Ω, grad $F = $ grad $\phi \neq \mathbf{0}$.

On Ω we have

$$\dot{F}(x) = \operatorname{grad} F \cdot \mathbf{G} = \operatorname{grad} F \cdot (-\operatorname{grad}\phi) = - \mid \operatorname{grad} F \mid^2 . \tag{4.78}$$

Hence $\dot{F}(\mathbf{x})$ is negative definite on some neighborhood of $\mathbf{x}_0$. It follows then from Liapounov theorem that $\mathbf{x}_0$ is asymptotically stable.

Definition 4.6.3 *A force field* $\mathbf{F}(\mathbf{x})$ *is said to be conservative if there exists a smooth function* $\phi(\mathbf{x})$ *so that*

$$\mathbf{F}(\mathbf{x}) = -grad\,\phi(\mathbf{x}).$$

In particular a mechanical dynamical system is said to be conservative if the force field acting on the system is conservative.

Example 4.6.6 *The equation of motion for a particle of mass m under the action of a force F is given by Newton's second law*

$$m\,\frac{d^2\mathbf{x}}{dt^2} = \mathbf{F}(\mathbf{x}) . \tag{4.79}$$

If F is conservative, then

$$m\,\frac{d^2\mathbf{x}}{dt^2} = -grad\,\phi . \tag{4.80}$$

Rewriting equation (4.80) as a system of first order equation we obtain

$$\frac{d\mathbf{x}}{dt} = \mathbf{v} \tag{4.81}$$

$$\frac{d\mathbf{v}}{dt} = -\frac{1}{m}grad\,\phi . \tag{4.82}$$

It follows from Equations (4.81)-(4.82) that at a steady state $(\mathbf{x}_0, \mathbf{v}_0)$ *of such a particle,*

$$\mathbf{v}_0 = \mathbf{0}, \quad grad\,\phi(\mathbf{x}_0) = 0$$

i.e., $\mathbf{x}_0$ *corresponds to a local extremum of* ϕ.

Theorem 4.6.3 *A stable steady state* $(\mathbf{x}_0, \mathbf{0})$ *of a particle in a conservative force field corresponds to an isolated local minima of* ϕ.

Proof: The total energy of the system under consideration is

$$E(\mathbf{x}, \mathbf{v}) = \frac{1}{2}m\mathbf{v}^2 + \phi(\mathbf{x}) . \tag{4.83}$$

If $\mathbf{x}_0$ is a local minima of ϕ, then

$$F(\mathbf{x}, \mathbf{v}) = E(\mathbf{x}, \mathbf{v}) - E(\mathbf{x}_0, \mathbf{0})$$

is positive definite on some domain around $(\mathbf{x}_0, \mathbf{0})$. Moreover, from, Equations (4.81)-(4.82) it follows that

$$
\begin{aligned}
\dot{F}(\mathbf{x}, \mathbf{v}) &= \operatorname{grad} F \cdot (\mathbf{v}, \frac{-1}{m}\operatorname{grad} \phi) \\
&= (\operatorname{grad} \phi, m\mathbf{v}) \cdot (\mathbf{v}, -\frac{1}{m}\operatorname{grad}\phi) = 0 .
\end{aligned}
\tag{4.84}
$$

Hence by Liapounov theorem $(\mathbf{x}_0, \mathbf{0})$ is a stable steady state.

We remark that theorem (4.6.3) is a special case of Lagrange theorem. This theorem states the following

Theorem 4.6.4 *Steady state* $(\mathbf{x}_0, \mathbf{0})$ *of a system of particles in a conservative force field is stable if* $\mathbf{x}_0$ *is an isolated local minima of* $\phi(\mathbf{x})$.

Exercises

1. Let $f(0) = 0$, $f(x) > 0$ on the interval $[0, b]$ and $f(x) < 0$ on the interval $[-b, 0]$. Show that

$$F(x, y) = \frac{1}{2}y^4 + \int_0^x f(t)dt$$

is positive definite on

$$\Omega = \{(x, y); -b < x < b, \quad -\infty < y < \infty\}.$$

What happens if y^4 is replaed by y^2 ?.

2. Show that
$$H(x, y) = ax^2 + bxy + cy^2$$
is negative definite if $c < 0$ and $4ac - b^2 > 0$.

3. For the system
$$\ddot{x} + \alpha f(x) = 0, \quad \alpha > 0$$
where $f(x)$ satisfies the assumptions of exercise 1, show that $x = 0$, $\dot{x} = 0$ is a stable steady state.

Hint: Let $F(x, \dot{x}) = \frac{1}{2}\dot{x}^2 + \alpha \int_0^x f(t)dt$.

4. Apply the results of exercise 3 to the nonlinear pendulum
$$\ddot{x} + \omega^2 \sin x = 0, \quad \omega^2 = \frac{g}{L}.$$

5. Show that the following two systems are stable or asymptotically stable at **0**

   (a) $\frac{du}{dt} = -u - 2uv^2$, $\frac{dv}{dt} = -v - vu^2$
   (b) $\frac{du}{dt} = -u^3 + 2uv^2$, $\frac{dv}{dt} = -2u^2v - 4v^3$.

   Hint: Consider a Liapounov function of the form
   $$F(x, y) = au^2 + bv^2, \quad a, b > 0.$$

6. For the system
$$\frac{dx}{dt} = -ny - xf(x, y)$$
$$\frac{dy}{dt} = nx - yf(x, y), \quad n = 1, 2, \ldots$$
show that **0** is an asymptotically stable state when $f(x, y) > 0$ in some neighborhood of the steady state and unstable if $f(x, y) < 0$ in some neighborhood of **0**.

   Hint: Note that this is a generalization of Example 4.6.5.

4.7 PERIODIC SOLUTIONS (LIMIT CYCLES)

In many important practical applications, one has to consider dynamical systems, Equation (4.71), with periodic motions. That is, some of the trajectories "allowed" by the system satisfy either exactly or approximately $\mathbf{x}(t+T) = \mathbf{x}(t)$, for some $T > 0$. In these cases the steady state of the system is represented by a one dimensional curve in R^n (usually referred to as a **limit cycle**) rather than a point. Perhaps one of the most important examples of such a system relates to the motion of Earth and the other planets around the Sun. This motion is subject to perturbations due to the gravitational field of the other planets and other celestial bodies in the system. It follows then that the answer to the question of stability of this periodic motion is of paramount importance to the very existence of life on Earth. However, periodic motion is important also in many mundane applications, e.g. waves of various types (water waves, electromagnetic waves, etc). In this section we present some elementary techniques, to determine the existence and stability of these limit cycles. However, in general the existence and stability of limit cycles for a given dynamical system is a rather difficult mathematical problem.

We start with, few examples.

Example 4.7.1 *Consider the following almost linear system with a critical point at* $(0,0)$:

$$\frac{dx}{dt} = \alpha^2 x - \omega y - x(x^2 + y^2) \tag{4.85}$$

$$\frac{dy}{dt} = \omega x + \alpha^2 y - y(x^2 + y^2) \tag{4.86}$$

where α *and* ω *are constants. The linearization of this system at the steady state* $(0,0)$ *is given by*

$$\frac{dx}{dt} = \alpha^2 x - \omega y, \quad \frac{dy}{dt} = \omega x + \alpha^2 y. \tag{4.87}$$

The eigenvalues of the coefficient matrix for this system are $\alpha^2 \pm \omega i$. *Therefore,* $(0,0)$ *is an unstable spiral point. Thus a trajectory originating from a point near* $(0,0)$ *will spiral away from this point. However, we now show that these spiral trajectories remain bounded. Moreover trajectories of the system, Equations (4.85)-(4.86) which originate from a point far away from the origin are actually directed inward.*

To prove these statements we recast Equations (4.85), (4.86) using polar coordinates

$$x = r\cos\theta \quad y = r\sin\theta \tag{4.88}$$

and note that

$$x\frac{dx}{dt} + y\frac{dy}{dt} = r\frac{dr}{dt} \tag{4.89}$$

$$y\frac{dx}{dt} - x\frac{dy}{dt} = -r^2\frac{d\theta}{dt} . \tag{4.90}$$

The system, Equation (4.85), (4.86) now takes the form

$$\frac{dr}{dt} = r(\alpha^2 - r^2), \quad \frac{d\theta}{dt} = \omega . \tag{4.91}$$

Therefore, if a trajectory starts from a point which satisfies $r > \alpha$, then $\frac{dr}{dt} < 0$ and the trajectory will spiral inward. If on the other hand a trajectory starts from a point which satisfies $r < \alpha$, then $\frac{dr}{dt} > 0$ and the trajectory will spiral outward. Moreover, for the circle $r = \alpha$ we have $\frac{dr}{dt} = 0$ which implies that

$$r = \alpha = \text{const}, \quad \theta = \omega t + t_0 \tag{4.92}$$

is a **limit cycle** of this system. To sum it up: Trajectories of this system with initial condition $r_0 > \alpha$ will spiral toward the limit cycle, and the same is true for those with $r_0 < \alpha$. It follows then that this periodic solution is, asymptotically stable limit cycle.

Example 4.7.2 *Systems with variable damping*
   We first observe that the solution of the spring-mass system

$$m\ddot{x} + b\dot{x} + kx - 0 \ m, k > 0 \tag{4.93}$$

is

$$x(t) = e^{-\alpha t}\left[C_1 e^{\beta t} + C_2 e^{\beta t}\right]$$

where

$$\alpha = \frac{b}{m}, \quad \beta = \frac{\sqrt{b^2 - 4km}}{m} .$$

Therefore, when b is positive the motion is damped and the amplitude of $x(t)$ will decrease in time. However, when $b < 0$ (negative damping), the amplitude of $x(t)$ will increase in time. Thus the sign of the frictional coefficient b determines if the motion is damped or not.

Now consider Rayleigh equation.

$$\ddot{x} - b(1 - \dot{x}^2)\dot{x} + kx = 0, \quad b, k > 0 . \tag{4.94}$$

In this equation the "effective" frictional coefficient is $-b(1 - \dot{x}^2)$. Therefore when $(\dot{x})^2 < 1$, the damping in this equation is negative but when $(\dot{x})^2 > 1$, the "damping" is positive. Therefore, the amplitude of the motion described by this equation will increase for small velocities and decrease for large ones. Therefore, it stands to reason that in between these two types of motions there will be an oscillation of constant amplitude, i.e., a periodic motion.

Here are some formal definitions and theorems regarding limit cycles.

Lemma 4.7.1 *A trajectory of a dynamical system in phase space is closed if and only if it corresponds to a periodic solution of the system.*

Proof: If the solution $\mathbf{x}(t)$ of the dynamical system is periodic, there exist $T > 0$ so that $\mathbf{x}(t + T) = \mathbf{x}(t)$ for all $t > 0$. This obviously implies that the trajectory of the system in phase space is also closed. The reverse is also obvious.

Definition 4.7.1 *A closed trajectory X of a dynamical system which has **nearby** open trajectories spiraling towards it from both the inside and outside as $t \to \infty$ is called, asymptotically stable limit cycle. Similarly if nearby trajectories neither approach nor recede from X, we say that it is (neutrally) stable.*

Example 4.7.3 *The trajectories of the linear pendulum without damping are neutrally stable.*

Definition 4.7.2 *A point y is a limit point of a trajectory $x(t)$ of a dynamical system*

$$\frac{d\mathbf{x}}{dt} = \mathbf{F}(\mathbf{x}) \tag{4.95}$$

if there exists a sequence $\{t_n\}$, $\lim\limits_{n \to \infty} t_n = \pm\infty$ so that $\lim x(t_n) = y$.

Definition 4.7.3 *A set X is called a limit set of Equation (4.95) if each of its points is a limit point of some trajectory of this system.*

Example 4.7.4 *An isolated asymptotically stable steady state of a dynamical system is obviously a limit point of all nearby trajectories.*

Example 4.7.5 *The trajectory $r = \alpha$ in Example 4.7.1 is a limit set for the system, Equations (4.85),(4.86), since each of its points is a limit point for the trajectories that spiral towards $r = \alpha$.*

Exercises

1. Use the same analysis as in example 4.7.2 to show (intuitively) that the Van der Pol equation

$$\frac{d^2u}{dt^2} + b(1 - u^2)\frac{du}{dt} + ku = 0 \tag{4.96}$$

   where b and k are positive constants admits a limit cycle.

2. Show that the system

$$\frac{du}{dt} = -v + uh(r), \quad \frac{dv}{dt} = u + vh(r) \quad r^2 = u^2 + v^2 \tag{4.97}$$

   has limit cycles which correspond to the roots of the function $h(r)$.

   Hint: Use a polar representation of the system, Equation (4.97).

3. Determine the periodic solutions and their stability for the system, Equation (4.97) if

   (a) $h(r) = r(r - 1)^2(r - 3)(r - 4)$

   (b) $h(r) - r^2 - 4$

   (c) $h(r) = \sin nr, \quad n = 1, 2, \ldots$

4. Under what conditions on $h(r)$, does the following system admit a limit cycle

$$\frac{du}{dt} = 2u + v - uh(r)$$
$$\frac{dv}{dt} = -u + 2v - vh(r) \tag{4.98}$$

   Hint: Use polar representation of the system.

5. Determine the periodic solutions and their stability for the system, Equation (4.98), if

   (a) $h(r) = r^2 - n^2, \ n = 1, 2, \ldots$

   (b) $h(r) = 2 \cos \pi r$.

6. Perform a qualitative analysis similar to example 4.7.2 to show that the steady state $x = 0$, $\dot{x} = 0$ of

$$\ddot{x} + bx^2 \dot{x} + kx = 0$$

is asymptotically stable if $b > 0$ and unstable if $b < 0$. Apply these results to the system, Equation (4.52), in Section 4.

Bibliography

[1] N.P. Bhatia, G.P. Szeg, Stability Theory of Dynamical Systems, Springer-Verlag, 2002

[2] R. Borrelli, C. Coleman, and W. Boyce, Differential Equations Laboratory Workbook: A Collection of Experiments, Explorations and Modeling Projects for the Computer Wiley, NY, 1992.

[3] W.E. Boyce and R. Diprima, Elementary Differential Equations and Boundary Value Problems, 3rd edition, J. Wiley and Sons, 1977.

[4] Clive L. Dym - Stability Theory and Its Applications to Structural Mechanics, Dover, NY.

[5] M.W. Hirsch and S. Smale, Differential Equations, Dynamical Systems and Linear Algebra, Academic Press, 1974.

[6] R. K. Miller and A. N. Michel, Ordinary Differential Equations, Academic Press, 1982.

[7] J. Palis, Jr. and W. de Melo, Geometric Theory of Dynamical Systems, Springer-Verlag, 1982.

[8] A. Pillay - Geometric Stability Theory, Oxford University press 1996.

[9] N. Rouche, P. Habets and M. Laloy, Stability Theory by Liapounov's Direct Method, Springer-Verlag, 1977.

CHAPTER 5

Bifurcations and Chaos

CONTENTS

5.1 INTRODUCTION

In the previous chapter we discussed various methods to analyze the stability of the equilibrium states of a dynamical system when the values of the system parameters are known and fixed. The objective of bifurcation theory is to investigate what happens to the type, number, and stability of the steady states as a result of a continuous change in some (or all) of the system parameters. If, as a result, the system undergoes a "sudden" change in its properties or behavior, we say that the system underwent a bifurcation.

The motivation for such analysis stems from the fact that in many real life situations the values of the model parameters are not known accurately or might actually be (very) slowly varying functions of time which we approximate by constants to simplify the model Equations.

Here are some examples of "real life" bifurcations.

1. Phase transitions.

   If we cool down water, it will remain liquid until we reach $0°C$. If we continue the cooling process below $0°C$, the water will undergo a "phase transition" and turn to ice. Thus the "liquid water system" undergoes a bifurcation at $0°C$. Similarly a bifurcation occurs when the water is heated over $100°C$ (at which point it turns to "gas").

2. Earthquakes.

   The Earth's continents "float" on the liquid core of the Earth. When these plates collide, stress starts to build up, and one plate starts to glide over the other. As these stresses build up, a point is reached where the plates break and we experience an earthquake. Thus an earthquake represents a bifurcation in the plate's structure.

3. Buckling of columns.

   If we increase the vertical load on a (vertical) column, a point will be reached where the column will buckle and break (if the "load" is a building then the building will collapse.) Under these circumstances the column undergoes a bifurcation.

4. Magnetic hysteresis.

   If we place a magnet in a weak magnetic field whose direction is opposite to its magnetization, "nothing" will happen to the magnet. However, as we increase the field intensity, there will come a point when all of a sudden the magnet will change its poles (the north magnet pole becomes the southern one and vice versa). Thus a bifurcation occurs. If we reverse now the direction of the field, another bifurcation will occur and the magnetization of the magnet will return to its original state. These changes in the magnetization of the magnet are represented by the "hysteresis loop" (see Fig. 5.5).

5.2   BIFURCATIONS OF CO-DIMENSION ONE

Definition: Let

$$\dot{\mathbf{x}} = \mathbf{F}(\mathbf{x}, \boldsymbol{\lambda}), \quad \mathbf{x} \epsilon R^n, \quad \boldsymbol{\lambda} \epsilon R^m \tag{5.1}$$

be a dynamical system with m dynamical parameters $\boldsymbol{\lambda} = (\lambda_1, \ldots \lambda_m)$.

We say that a bifurcation is of co-dimension k if this is the smallest dimension of parameter space which contains such a bifurcation (observe that there are no restrictions on n). In particular, bifurcations of co-dimension one require only one dynamical parameter.

The steady states of the dynamical system Equation (5.1) satisfy $\dot{\mathbf{x}} = 0$. Therefore, to find the bifurcation points of this system one has to study the nature of the solutions to the Equation $\mathbf{F}(\mathbf{x}, \boldsymbol{\lambda}) = 0$ as $\boldsymbol{\lambda}$ varies. If $\boldsymbol{\lambda}_0$ is a bifurcation point, then in order to study this bifurcation it is a common practice to reduce Equation (5.1) to a "local form" by the following transformation

$$\mathbf{y} = \mathbf{x} - \mathbf{x}_0, \quad \boldsymbol{\mu} = \boldsymbol{\lambda} - \boldsymbol{\lambda}_0 \tag{5.2}$$

where $\mathbf{x}_0$ is a stable steady state of the system near $\boldsymbol{\lambda}_0$.

Table 5 contains a list of all co-dimension one bifurcations and "generic" dynamical systems which contain them.

We now discuss each of these bifurcations in detail.

Table 5.1 List of Co-dimension One Bifurcations

Name	Generic Representation
Trans-critical	$\dot{x} = \lambda x - x^2$
Saddle Point	$\dot{x} = \lambda - x^2$
Pitchfork	$\dot{x} = x(\lambda - x^2)$
Sub-critical	$\dot{x} = \lambda + x - x^3$
Hopf Bifurcation	$\dot{x} = y + \lambda x - x(x^2 + y^2)$ $\dot{y} = x + \lambda y - y(x^2 + y^2)$

5.2.1 Trans-critical Bifurcation

The system

$$\dot{x} = \mu x - x^2 = x(\lambda - x) \tag{5.3}$$

has two steady states $x = 0$ and $x = \lambda$ for all possible values of the dynamical parameter λ. The linearization of the system around $x = 0$ yields

$$\dot{x} = \lambda x. \tag{5.4}$$

It is obvious then that this state is asymptotically stable when $\lambda < 0$ and unstable for $\lambda > 0$ (since $x(t) = Ce^{\lambda t}$).

The linearization of the system around the steady state $x = \lambda$ is obtained by introducing

$$y = x - \lambda \tag{5.5}$$

and hence

$$\dot{y} = -y(y + \lambda) . \tag{5.6}$$

For the linear part we then have

$$\dot{y} = -\lambda y, \tag{5.7}$$

i.e., this state is asymptotically stable for $\lambda > 0$ and unstable for $\lambda < 0$. The

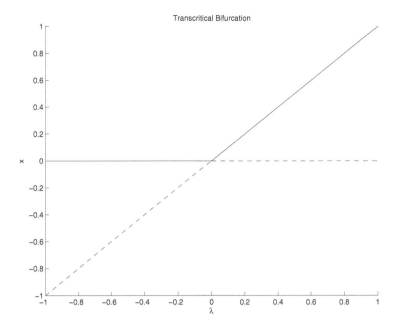

Figure 5.1 Trans-critical bifurcation

stability of these steady states is plotted on a "bifurcation diagram" where solid and dashed lines represent asymptotically stable and unstable states respectively (Fig. 5.1). From this diagram we see that for both states $\lambda = 0$ is a bifurcation point. A dynamical system in the stable steady state $x = 0$ for $\lambda < 0$ will become unstable at $\lambda = 0$ and will "transition" to the new stable state $x = \lambda$ as λ becomes positive. In this new state the system will usually exhibit different properties than in the state $x = 0$. We refer to this phenomenon as "exchange of stability" between the steady states(a phase transition is an example of this bifurcation).

5.2.2 Saddle Point Bifurcation

This type of bifurcation has two generic representations.

A.

$$\dot{x} = \lambda - x^2 \ . \tag{5.8}$$

The system has two steady states $x = \pm\sqrt{\lambda}$ for $\lambda > 0$ and none when $\lambda < 0$ (since x is assumed to be real). To investigate the stability of the

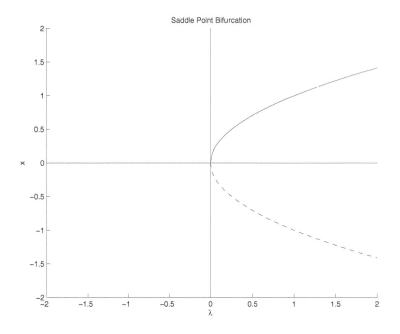

Figure 5.2   Saddle Point Bifurcation

state $x = \sqrt{\lambda}$ we introduce $y = x - \sqrt{\lambda}$. The system then becomes

$$\dot{y} = -2\sqrt{\lambda}y - y^2 \ .$$

Using linearization around $y = 0$ we infer that the state $x = \sqrt{\lambda}$ is asymptotically stable. Similarly, for $x = -\sqrt{\lambda}$ we find that the steady state is unstable. This is represented by the bifurcation diagram of Fig. 5.2.

B.

$$\dot{x} = \lambda + x^2 \tag{5.9}$$

The system has two steady states for $x = \pm\sqrt{-\lambda}$ for $\lambda < 0$ and none for $\lambda > 0$. In this case the steady state $x = \sqrt{-\lambda}$ is unstable while $x = -\sqrt{-\lambda}$ is asymptotically stable (see Fig. 5.3).

5.2.3   Pitchfork Bifurcation

The system

$$\dot{x} = \mu x - x^3 = x(\lambda - x^2) \tag{5.10}$$

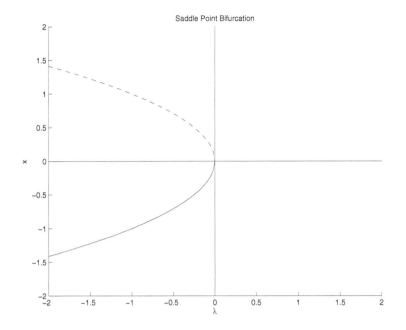

Figure 5.3 Saddle point bifurcation

has a steady state $x = 0$ for all λ and two additional steady states $x = \pm\sqrt{\lambda}$ for $\lambda > 0$. The linearization around $x = 0$ is

$$\dot{x} = \lambda x \tag{5.11}$$

which implies that $x = 0$ is asymptotically stable for $\lambda < 0$ and unstable for $\lambda > 0$. To linearize the system around the steady state $x = \sqrt{\lambda}, \lambda > 0$, we introduce $y = x - \sqrt{\lambda}$ to obtain

$$\dot{y} = -2\lambda y - 3\sqrt{\lambda} y^2 - y^3 . \tag{5.12}$$

The linearization of this Equation around $y = 0$ yields

$$\dot{y} = -2\lambda y, \quad \lambda > 0 \tag{5.13}$$

which implies that this steady state is asymptotically stable. Similarly, one can show that $x = -\sqrt{\lambda}, \lambda > 0$ is asymptotically stable.

The name of this bifurcation comes from its diagram (Fig. 5.4) which looks like a fork.

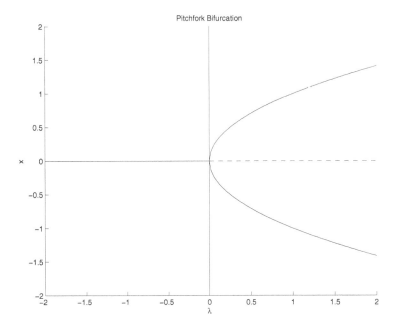

Figure 5.4 Pitchfork Bifurcation

5.2.4 Subcritical Bifurcation (Hysteresis)

To analyze the system

$$\dot{x} = \lambda + x - x^3 \tag{5.14}$$

we first note that a third order polynomial always has three roots in the complex plane. These roots can be all real or one real and two complex conjugates. The polynomial $p(x) = \lambda + x - x^3$ has three real roots on the interval $\lambda \in \left[-\frac{2}{9}\sqrt{3}, \frac{2}{9}\sqrt{3} \right]$ and only one real root outside this interval. We conclude therefore that the system has three steady states for $\lambda \in \left[-\frac{2}{9}\sqrt{3}, \frac{2}{9}\sqrt{3} \right]$ and only one outside this interval (See Fig. 5.5). To analyze the stability of these steady states we let $\lambda = 0$. At this point the three steady states are $x = 0, x = \pm 1$. For $x = 0$ the linearization of the dynamical system is $\dot{x} = x$, and we conclude that this state is unstable. For $x = 1$ we introduce $y = x - 1$ and linearize. This yields

$$\dot{y} = -2y \tag{5.15}$$

and hence the steady state is asymptotically stable. Similarly one can show that $x = -1$ is asymptotically stable. While this analysis has been carried out only for $\lambda = 0$, it is possible to show that it is valid for other values of λ in the

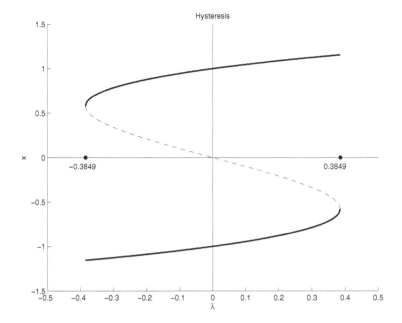

Figure 5.5 Subcritical Bifurcation

interval $\left[-\frac{2}{9}\sqrt{3}, \frac{2}{9}\sqrt{3}\right]$ (as shown in Fig. 5.5). Thus on this interval the system has one unstable state and two asymptotically stable states.

For values of λ which are outside the interval $\left[-\frac{2}{9}\sqrt{3}, \frac{2}{9}\sqrt{3}\right]$, it is straightforward to show that the corresponding steady state is asymptotically stable. For example, if $\lambda = 6$, then the corresponding steady state's $x = 2$. Applying the transformation to local form Equation (5.2) to Equation (5.14), viz.

$$\mu = \lambda - 6, \quad y = x - 2$$

we obtain

$$\dot{y} = \mu - 11y - 6y^2 - y^3.$$

The linearized form of this Equation at $\mu = 0$ is $\dot{y} = -11y$ which shows that the state $y = 0$ (viz. $x = 2$) is asymptotically stable.

If initially $\lambda < -\frac{2}{9}\sqrt{3}$, then the system must be in the lower (stable) state. When λ becomes larger than $-\frac{2}{9}\sqrt{3}$, the system acquires a second stable steady state but since the lower steady state is stable, it will remain in its original state. However, when λ becomes larger than $\frac{2}{9}\sqrt{3}$, the lower steady state ceases to exist and the system will "jump" to the upper stable steady state and will remain there for all $\lambda > \frac{2}{9}\sqrt{3}$. A similar picture unfolds when λ is decreasing.

If $\lambda > \frac{2}{9}\sqrt{3}$, the system will be in the upper steady state and will remain in this state until $\lambda = -\frac{2}{9}\sqrt{3}$ when this state ceases to exist. The actual state of the system is described therefore by the hysteresis diagram (Fig. 5.5).

5.2.5 Hopf Bifurcation

The two dimensional system

$$
\begin{aligned}
\dot{x} &= -y + \lambda x - x(x^2 + y^2) \\
\dot{y} &= x + \lambda y - y(x^2 + y^2)
\end{aligned}
\tag{5.16}
$$

has an obvious steady state $\mathbf{x} = (x, y) = \mathbf{0}$ for all λ. The existence of another attractor of dimension one can be inferred by recasting this system in polar coordinates (r, θ) using:

$$
r\frac{dr}{dt} = x\frac{dx}{dt} + y\frac{dy}{dt}
\tag{5.17}
$$

$$
-r^2\frac{d\theta}{dt} = y\frac{dx}{dt} - x\frac{dy}{dt}.
\tag{5.18}
$$

This leads to

$$
r\frac{dr}{dt} = r^2(\lambda - r^2)
\tag{5.19}
$$

$$
-r^2\frac{d\theta}{dt} = -r^2.
\tag{5.20}
$$

or

$$
\frac{dr}{dt} = r(\lambda - r^2)
\tag{5.21}
$$

$$
d\theta/dt = 1.
\tag{5.22}
$$

This implies that the system has a limit cycle for

$$
r = \sqrt{x^2 + y^2} = \sqrt{\lambda}, \quad \lambda > 0.
\tag{5.23}
$$

We discuss now the stability of these steady states.

A. $\mathbf{x} = \mathbf{0}$

The linearization of the system around this steady state is

$$
\dot{x} = -y + \lambda x, \quad \dot{y} = x + \lambda y
\tag{5.24}
$$

or in a matrix form,

$$
\dot{\mathbf{x}} = \begin{pmatrix} \lambda & -1 \\ 1 & \lambda \end{pmatrix} \mathbf{x}.
\tag{5.25}
$$

The eigenvalues of the coefficient matrix in Equaton (5.25) are

$$\mu = \lambda + \pm i. \tag{5.26}$$

The general solution of this linear system is

$$\mathbf{x} = e^{\lambda t}[c_1 \mathbf{v}_1 e^{it} + c_2 \mathbf{v}_2 e^{-it}] \tag{5.27}$$

where $\mathbf{v}_1, \mathbf{v}_2$ are the eigenvectors which correspond to the eigenvalues $\lambda \pm i$ and c_1, c_2 arbitrary constants. We conclude then that the steady state $\mathbf{x} = \mathbf{0}$ is asymptotically stable for $\lambda < 0$ and unstable for $\lambda > 0$.

B. To explore the stability of the limit cycle $r = \sqrt{\lambda}$ we observe that for $r < \sqrt{\lambda}$, $\frac{dr}{dt}$ is positive. However, when $r > \sqrt{\lambda}$, $\frac{dr}{dt}$ is negative. We conclude therefore that the limit cycle $r = \sqrt{\lambda}$ is asymptotically stable.

If we now plot the eigenvalues, Equation (5.26), in the complex plane (Fig. 5.6) we can conclude that the steady state $\mathbf{x} = \mathbf{0}$ becomes unstable as λ becomes positive and a new asymptotically stable limit cycle appears when these complex conjugate eigenvalues cross from the left to the right side of the complex plane. In fact, this is the essence of the Hopf bifurcation theorem (1950) which states that whenever the linearization of a dynamical system around a steady state has two complex conjugate eigenvalues which behave in this manner then the original steady state becomes unstable and an asymptotically stable limit cycle appears. Such a bifurcation is called "Hopf bifurcation." A complete statement of this important theorem is as follows.

Theorem:(Hopf Theorem) Assume that the system

$$\dot{\mathbf{x}} = \mathbf{F}(\mathbf{x}, \mu), \quad \mathbf{x} \in R^n, \quad \lambda \in R \tag{5.28}$$

where

$$F(\mathbf{x}) = \begin{bmatrix} f_1(\mathbf{x}) \\ \cdot \\ \cdot \\ \cdot \\ f_n(\mathbf{x}) \end{bmatrix} \tag{5.29}$$

has a steady state $(\mathbf{x}(\mu_0), \mu_0)$ and at this state

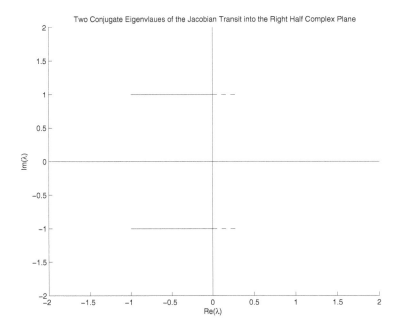

Figure 5.6 Hopf bifurcation: a steady state destabilizes as two conjugate eigenvalues move into the the right hand side of the complex plane

1. the Jacobian of the system

$$J(\mathbf{x}(\mu_0)) = \left[\frac{\partial f_j}{\partial x_i}\right](\mathbf{x}(\mu_0))$$

   has a simple pair of (pure) imaginary eigenvalues $\pm i\nu$ and all other eigenvalues of J have negative real part.

2. The two eigenvalues of $J(\mathbf{x}(\mu))$ which are imaginary at $\mu = \mu_0$ are smooth functions of μ and satisfy

$$\frac{d(Re\,\nu(\mu))}{d\mu} > 0 \ \ at \ \ \mu = \mu_0.$$

   (That is, as a function of μ these two eigenvalues move into the right half of the complex plane thus destabilizing the steady state $(\mathbf{x}(\mu_0), \mu_0)$; this is called "the transversality condition".)

Then μ_0 is a bifurcation point of the system into a limit cycle.

As an example of such a bifurcation we consider the following physical system.

Example: Brusselator Reaction.

There are several chemical reactions in which the color of the reactant in the chemical reactors changes periodically. Examples of such reactions are the Belousov-Zhabotinsky reaction, the Bray-Liebhafsky reaction, and the Brusselator reaction. These color changes are due to the fact that the concentration of some of the intermediate chemicals in the reaction oscillates in time.

A mathematical explanation for this "oscillatory behavior" is given by the following model for the Brusselator reaction. In this reaction the color of the reactants in the chemical reactor alternates periodically between blue and red. The model postulates the following sequence of (catalytic) reactions to explain this phenomenon:

$$a \to X$$
$$b + X \to Y + c$$
$$2x + Y \to 3X \tag{5.30}$$
$$X \to d.$$

Here a, b, c, d are the initial and final chemicals of these reactions. Observe that the sum of all these reactions is $a + b \to c + d$. Thus X, Y are catalysts whose presence and concentration oscillate as is indicated by the color of the reactants in the reactor.

In the following we use the same letter to denote the chemical and its concentration and assume that the rate of a reaction is proportional to the concentration of the reactants. To derive Equations for the rate of change of X, Y we observe that the production rates of X in the first and third reactions are proportional to a and X^2Y while its rate of loss in the second and fourth are bX and X. Hence

$$\frac{dx}{dt} = a - (b+1)X + X^2Y. \tag{5.31}$$

Similarly, the rate of production of Y in the first reaction is proportional to A while the loss of Y in the third reaction is proportional to X^2Y. Hence

$$\frac{dY}{dt} = bX - X^2Y . \tag{5.32}$$

Renormalizing the concentration of a to 1, we deduce that the system ,

Equations(5.31)-(5.32), has an equilibrium state $X = 1,\ \ Y = b$. Reducing Equations (5.31)-(5.32) to a local form by the transformation

$$X = 1 + U,\ \ Y = b + V \tag{5.33}$$

we obtain

$$\dot{U} = (b - 1)U + V + U^2(b + V) + 2UV \tag{5.34}$$

$$\dot{V} = -bU - V - U^2(b + V) - 2UV. \tag{5.35}$$

(Here b plays the role of the "bifurcation parameter" λ). Hence at the equilibrium state $(0, 0)$ we have

$$J(0, 0, b) = \begin{bmatrix} b - 1 & , & 1 \\ -b & , & -1 \end{bmatrix} \tag{5.36}$$

whose eigenvalues are

$$\nu = \frac{b - 2 \pm \sqrt{(b - 2)^2 - 4}}{2}. \tag{5.37}$$

We see that when $0 < b < 2$, $(0, 0)$ is a stable spiral point. At $b = 2$ the Jacobian has two pure imaginary eigenvalues which cross to the right half complex plane and all the other conditions of Hopf bifurcation theorem hold. We conclude then that $b = 2$ is a Hopf bifurcation point for the system i.e., for $b > 2$ the system has a stable limit cycle and the concentrations of X, Y will vary periodically with time.

Exercises

1. Let

$$\dot{x} = x(a - x - \lambda y)$$

$$\dot{y} = y(a - y - 2x)$$

where $a > 0$ is a constant and λ is a parameter.

(a) Interpret the dynamics of the ecosystem that is described by these Equations (as a model for interaction between species).

(b) Classify the equilibrium states and bifurcation points of this system as a function of λ .

2. Classify the bifurcation of the system

$$\dot{x} = x(\lambda + a - x)$$

where a is a constant and λ is a parameter.

3. Show that the following system undergoes a Hopf bifurcation

$$\dot{x} = \lambda x + y + x^2 y$$

$$\dot{y} = -x + \lambda y + y^3$$

4. Study the stability of the steady states and bifurcation diagram for the system

$$\dot{x} = x + 4y^2$$

$$\dot{y} = \lambda y + 2xy + 4y^3$$

5. Draw the bifurcation diagrams for the following systems

   (a) $\dot{x} = (\lambda x - x^3)(\lambda + 2x - 1)$

   (b) $\dot{x} = (\lambda x - x^3)(\lambda - x^2)$

   (c) $\dot{x} = x(1 - x)(\lambda - x)$

6. Show that the following system

$$\ddot{x} + (1 + \lambda^2 + x^2)\dot{x} + 2x - x^3 = 0$$

has a Hopf bifurcation at $\lambda = 0$. Draw the phase diagram of the system for different values of λ.

5.3   ROSSLER OSCILLATOR

There is no universal agreement about the definition of chaos. However, "an approximate" definition of a chaotic system was proposed by E. Lorenz in his 1963 seminal paper as follows:

"When the present determines the future, but the approximate present does not approximately determine the future we say that the system is chaotic."

In other words the evolution of a chaotic system is sensitive to the values of the initial conditions, and a small variation in these conditions leads (in the long run) to a completely different trajectory of the system. One usually refers to this phenomenon as "sensitivity to the initial conditions."

It was proposed that one possible route for a deterministic system to become chaotic is when it undergoes a cascade of bifurcations. The Rossler Oscillator (or attractor) was discovered in an attempt to find a "simple" system which exhibits such behavior. The system consists of the following three Equations:

$$\dot{x} = -y - z, \quad \dot{y} = x + ay, \quad \dot{z} = b + (x - c)z \tag{5.38}$$

where a, b, c are non-negative constants (or parameters). Originally this system was proposed as a theoretical model; however, since then it found applications in the modeling of some chemical reactions.

We observe that although this system is nonlinear, the nonlinearity is due to a single term xz in the third Equation. Therefore, if we let $b = 0$ and consider the system in the plane $z = 0$, we obtain a linear system

$$\dot{x} = -y, \quad \dot{y} = x + ay. \tag{5.39}$$

When $0 < a < 2$ the solution of the system is

$$x(t) = e^{\frac{at}{2}}\left[(-\alpha D_1 - \frac{aD_2}{2})\sin(\alpha t) + (-\frac{aD_1}{2}D_1 + D_2\alpha)\cos(\alpha t)\right]$$

$$y(t) = e^{\frac{at}{2}}[D_1 \cos((\alpha t) + D_2 \sin(\alpha t)]$$

where $\alpha = \sqrt{4 - a^2}$. This implies that the steady state $(0,0)$ is unstable spiral. Similarly when $a > 2$, $(0,0)$ is unstable focus.

The three dimensional system, Equation (10.89), with $c^2 - 4ab > 0$ has two steady states

$$\mathbf{x_1} = \left(\frac{c - \beta}{2}, -\frac{c - \beta}{2a}, \frac{c - \beta}{2a}\right)$$

$$\mathbf{x_2} = \left(\frac{c + \beta}{2}, -\frac{c + \beta}{2a}, \frac{c + \beta}{2a}\right)$$

where $\beta = \sqrt{c^2 - 4ab}$. To determine the stability of these two steady states, we compute the Jacobian of the system

$$J(a, b, c) = \begin{pmatrix} 0 & -1 & -1 \\ 1 & a & 0 \\ z & 0 & x - c \end{pmatrix}.$$ (5.40)

The characteristic polynomial $p(\lambda)$ of the Jacobian is

$$p(\lambda) = \lambda^3 - (a + x - c)\lambda^2 + [(a(x - c) + 1 + z]\lambda - (x - c) - az.$$

For $a = 0.1$, $b = 0.1$, $c = 4$, we have

$$\mathbf{x}_1 = (0.0025, -0.0250, 0.0250), \quad \mathbf{x}_2 = (3.997, -39.975, 39.97).$$

Thus the two steady states are well separated from each other. The corresponding eigenvalues of J at these points are

$$\lambda_\pm = 0.047 \pm 0.999i, \quad \lambda_3 = -3.992$$ (5.41)

$$\lambda_\pm = -2.973 \times 10^{-7} \pm 6.401i, \quad \lambda_3 = 0.097.$$ (5.42)

It follows then that both steady states are unstable. We now observe that eigenvalues with negative real parts attract in the direction of the corresponding eigenvector while eigenvalues with positive real parts repulse in that direction. For a (general) trajectory that starts near $\mathbf{x}_1$ the directions along the eigenvectors which correspond to $\lambda_\pm$ will cause the trajectory to spiral away from $\mathbf{x}_1$. However, as the trajectory moves away from $\mathbf{x}_1$ and the $x - y$ plane, the direction along the third eigenvector will become dominant (since the real part of $\lambda_\pm$ is small) and the trajectory will curve back to $\mathbf{x}_1$.

To illustrate the bifurcation cascade which the Rossler oscillator undergoes as the parameters in Equation(10.89) change, we let $a = b = 0.1$ and allow c to change. To illustrate the results we plot below the trajectories which start near $\mathbf{x}_1$ with $c = 4$, 6, 8.2 and 15. For $c = 4$ we obtain a trajectory which, after a transient state, settles to a one cycle trajectory. For $c = 6$ the system undergoes a bifurcation and we obtain a two cycle trajectory. For $c = 8.2$ it is a four cycle trajectory. Finally for $c = 15$ the trajectory is fully chaotic.

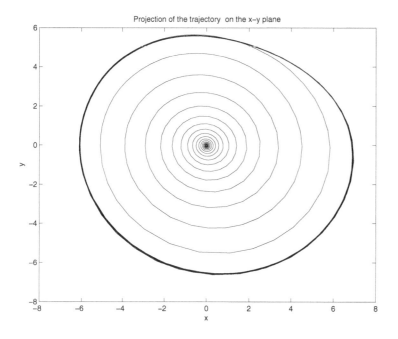

Figure 5.7   Trajectory of the Rossler oscillator starting near $\mathbf{x}_1$ with $c = 4$

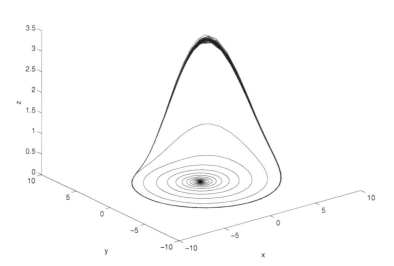

Figure 5.8   Same trajectory in 3D

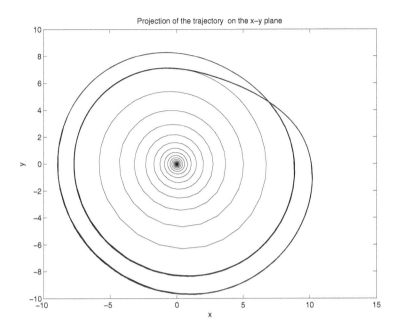

Figure 5.9 Trajectory of the Rossler oscillator starting near $\mathbf{x}_1$ with $c = 6$

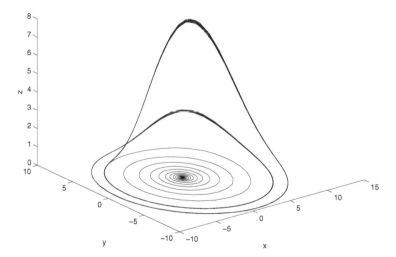

Figure 5.10 Same trajectory in 3D

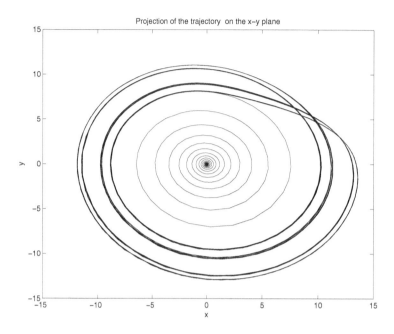

Figure 5.11   Trajectory of the Rossler oscillator starting near $\mathbf{x}_1$ with $c = 8.2$

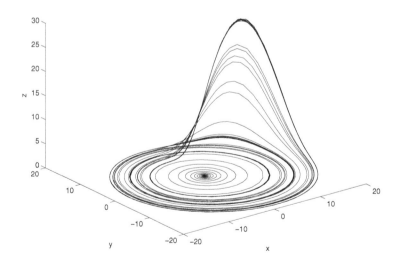

Figure 5.12   Same trajectory in 3D

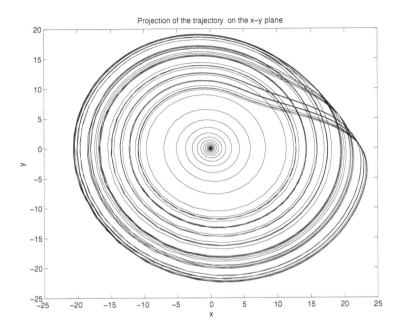

Figure 5.13 Trajectory of the Rossler oscillator starting near $\mathbf{x}_1$ with $c = 15$

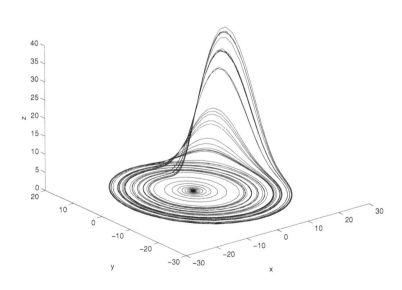

Figure 5.14 Same rajectory in 3D

5.4 LORENZ EQUATIONS

Atmospheric flow in general is governed by a set of complicated nonlinear partial differential Equations. In 1963 Lorenz "projected" these Equations (see appendix) to obtain a highly simplified set of three ordinary differential Equations

$$
\begin{aligned}
\dot{x} &= \sigma(y - x) \\
\dot{y} &= rx - y - xz \\
\dot{z} &= -bz + xy.
\end{aligned}
\tag{5.43}
$$

In these Equations σ, b are positive constants and r is a (dynamical) parameter. These Equations, which are called "Lorenz Equations," since their discovery have played a major role in the theory of nonlinear dynamical systems. In fact, they led to the establishment of new areas of research such as "chaos theory," "strange attractors," and other related topics.

In this section we address the steady states of this system, their stability, and bifurcation as r changes.

To obtain the steady states of Equation (5.43) we must solve the algebraic system

$$
\sigma(x - y) = 0
\tag{5.44}
$$

$$
rx - y - xz = 0
\tag{5.45}
$$

$$
-bz + xy = 0.
\tag{5.46}
$$

Since $\sigma \neq 0$ we obtain $x = y$ from Equation (5.44). Substituting for y in Equation (5.45) we have

$$
x(r - 1 - z) = 0 .
\tag{5.47}
$$

Thus either

$$
\mathbf{x} = (x, y, z) = \mathbf{0}
\tag{5.48}
$$

or

$$
x = y, \quad z = r - 1.
\tag{5.49}
$$

From Equation (5.46) we deduce then that two other steady states exist for $r \geq 1$

$$
\mathbf{x} = \left(\pm\sqrt{b(r - 1)}, \pm\sqrt{b(r - 1)}, r - 1 \right).
\tag{5.50}
$$

(Observe that for $r < 1$ we obtain complex values for x, y which are not admissible.) **A.** Stability of the steady state $\mathbf{x} = \mathbf{0}$

The linearization of the system, Equation (5.43), around this steady state yields

$$\dot{\mathbf{x}} = \begin{pmatrix} -\sigma & \sigma & 0 \\ r & -1 & 0 \\ 0 & 0 & -b \end{pmatrix} \mathbf{x}. \tag{5.51}$$

The characteristic polynomial of the coefficient matrix in this system is

$$p(\lambda) = (\lambda + b)[\lambda^2 + (1 + \sigma)\lambda - \sigma(r - 1)]. \tag{5.52}$$

Hence the eigenvalues are

$$\begin{aligned} \lambda_1 &= -b \\ \lambda_\pm &= -\frac{1 + \sigma}{2} \pm \sqrt{\left(\frac{1 + \sigma}{2}\right)^2 + \sigma(r - 1)}. \end{aligned} \tag{5.53}$$

We infer that if $r < 1$, all the eigenvalues are negative and therefore the steady state is asymptotically stable. However, if $r > 1$, one of the eigenvalues is positive and therefore $\mathbf{x} = \mathbf{0}$ is unstable.

B. The steady states

$$\mathbf{x} = \left(\pm\sqrt{b(r - 1)}, \pm\sqrt{b(r - 1)}, r - 1\right), r \geq 1.$$

Using Taylor expansion around these points one can show that in both cases the linearization of Equation (5.43) around these steady states is given by

$$\dot{\mathbf{x}} = \begin{pmatrix} -\sigma & \sigma & 0 \\ 1 & -1 & -\alpha \\ \alpha & \alpha & -b \end{pmatrix} \mathbf{x}, \quad \alpha = \pm\sqrt{b(r - 1)}. \tag{5.54}$$

The characteristic polynomial of the coefficient matrix is

$$p(\lambda) = \lambda^3 + (1 - \sigma + b)\lambda^2 + b(\sigma + r)\lambda + 2\sigma b(r - 1). \tag{5.55}$$

For $r = 1$ the roots of this polynomial are

$$\lambda_1 = 0, \quad \lambda_2 = -(1 + \sigma), \quad \lambda_3 = -b. \tag{5.56}$$

Thus we conclude that at this point the "new" states that appear for $r \geq 1$ are stable. For $0 < r - 1 \ll 1$ the first root of Equation (5.55) can be approximated by

$$\lambda_1 \approx \frac{2\sigma(1 - r)}{1 + \sigma} < 0, \tag{5.57}$$

i.e., the states are asymptotically stable.

To summarize for $r < 1$ the steady state $\mathbf{x} = \mathbf{0}$ is asymptotically stable, but it becomes unstable for $r > 1$. At $r = 1$ two new steady states appear, and they are asymptotically stable. It follows then that $r = 1$ is a pitchfork bifurcation point for the Lorenz system , Equation (5.43).

To demonstrate the sensitivity to initial conditions of Lorenz Equations (5.43) we simulated these Equations with $\sigma = 10$, $r = 28$ and $b = 8/3$ with two sets of initial conditions $(x, y, z) = (5, 5, 5)$ and $(x, y, z) = (5, 5.001; 5)$. (The values assigned to σ, r and b are those used by Lorenz in his 1963 paper).

Although these initial conditions are close to each other, the difference between the solutions for the y variable over the time interval $[0, 100]$ spikes to more than 30 (see Fig. 5.15). The corresponding relative difference is over 100%.

Figure 5.15 Difference between the y-solutions to Lorenz Equations

5.5 NERVE MODELS

A cell in the nervous system of a living organism is called a **neuron**. A simplified picture of a neuron is that it is composed of a "core" (referred to as "soma"), dendrites which act as receptors for signals from other neurons, and a "long extension" which is called an axon. The axon ends with synapses through which signals are transmitted to other neurons. Under proper stimulation the electric potential in the axon membrane rises above a certain threshold and the axon "fires" electrical impulses to neighboring cells to communicate with them through the synapses. The basic model for this firing process is due to

Hodgkin-Huxley. It consists of four ODEs and depends heavily on the modeling of the electro-chemical properties of sodium and potassium ions in the axon. To simplify this model conceptually FitzHugh-Nagumo introduced a model which consists of two ODEs.

$$\dot{v} = v - \frac{1}{3}v^3 - w + I_{ext} \tag{5.58}$$

$$\dot{w} = \epsilon(\beta v - \gamma w + \delta) \tag{5.59}$$

where v is the (axon) membrane potential, w is a recovery variable that provides negative feedback (that is provides damping to the amplitude of the membrane potential), and I_{ext} is a control parameter which represents the current due to the external stimulus. The parameters $\epsilon, \beta, \gamma, \delta$ are constants. These parameters can change the equilibrium (i.e. rest) state and dynamics of the model. In the FitzHugh-Nagumo model the values of these constants are usually set to $\epsilon = 0.08$, $\beta = 1$, $\gamma = 0.8$ and $\delta = 0.7$. The cubic term for v allows for regenerative self-excitation of the membrane potential. Observe that this model neglects the spatial dependence of the signal; i.e. v, w depend only on time. Such a model is usually referred to as "**space clamped**."

At equilibrium $\dot{v} = 0$ and $\dot{w} = 0$. Hence the equilibrium state is given by the intersection of the curves

$$w = v - \frac{1}{3}v^3 + I_{ext}, \quad w = \frac{\beta v + \delta}{\gamma}$$

in the phase plane (v,w). This equilibrium is unstable if I_{ext} is strong enough. In this case the model exhibits periodic spiking activity.

Numerical simulations of this model show that weak stimuli (small amplitude pulses of I_{ext}) result in small-amplitude closed trajectories in phase space around the equilibrium point. Biologically this corresponds to subthreshold responses (no firing). On the other hand strong stimuli result in large-amplitude trajectories in phase space that correspond to a firing spike. Thus the model does not have a well-defined firing threshold.

A more general form of this model replaces Equation (12.30) by

$$\dot{v} = f(v) - w + I_{ext}. \tag{5.60}$$

The Equations of this model are used also to model the propagation of waves in excitable media, such as heart tissue or nerve fiber. Thus if we let $v = v(x,t)$, $w = w(x,t)$ and add a diffusion term Dv_{xx} to (12.32) we obtain the reaction-diffusion system

$$v_t = f(v) - w + I + Dv_{xx} \tag{5.61}$$

$$w_t = \epsilon(\beta v - \gamma w + \delta) \tag{5.62}$$

where $f(v) = v(a-v)(v-1)$. In this model a represents the threshold for excitation, and ϵ the excitability of the medium.

To gain some insight into the general properties of this model we let $D = 1$, $\delta = 0$, and $I = 0$. With this simplification the model can be rewritten as

$$v_t = v_{xx} + v(a-v)(v-1) - w \tag{5.63}$$

$$w_t = bv - dw. \tag{5.64}$$

We assume that $0 < a < 1$, and $b, d \geq 0$.

A traveling wave solution $v = f(x-ct)$, $w = g(x-ct)$ of Equations(12.35) and (12.36) will have to satisfy

$$-cf' - f'' = f(a-f)(f-1) - g, \quad -cg' = bf - dg. \tag{5.65}$$

where primes denote differentiation with respect to $z = x - ct$. We can rewrite these Equations as a system of first order Equations as follows:

$$f' = h, \quad h' = -ch - f(a-f)(f-1) + g, \quad cg' = -bf + dg. \tag{5.66}$$

If $d \neq 0$ the equilibrium points of this system are at

$$f = \alpha_i, \quad h = 0, \quad g = \frac{b\alpha_i}{d}, \quad i = 1, 2, 3$$

where α_i are the roots of the third order Equation

$$x\left[(a-x)(x-1) - \frac{b}{d}\right] = 0.$$

The root $\alpha_1 = 0$ will be the unique real root (and hence the only equilibrium point of the system) if

$$(1-a)^2 < \frac{4b}{d}.$$

On the other hand, if $d = 0$ then the only equilibrium state of the system is $(0,0,0,)$.

The Jacobian of the system, Equation (9.69), at $(0,0,0)$ is

$$J(0,0,0) = \begin{pmatrix} 0 & 1 & 0 \\ a & -c & 1 \\ -b & 0 & d \end{pmatrix} \tag{5.67}$$

and the stability of this equilibrium point will be determined by the eigenvalues of this matrix.

In the special case $b = 0$ the system, Equation (9.69), decouples, and we can solve for g

$$g = A \exp(\frac{dz}{c}).$$

Therefore, if we are looking only for bounded solutions of the system , we must let $A = 0$, i.e. $g = 0$. The remaining system for (f,h) has three equilibrium points $(0,0), (1,0)$, and $(a,0)$. The first two equilibrium points are saddle-points while (a,0) is a center if $c = 0$ and unstable spiral or node if $c \neq 0$.

5.6 MISCELLANEOUS TOPICS

5.6.1 Dimension

We were all taught that a point has 0-dimension, a line is an entity of one dimension, a plane is two-dimensional, and that "real space" has three dimensions.

We now try to formalize these intuitive concepts.

Consider the unit interval $[0,1]$ in R and consider measuring sticks of length $\epsilon = \frac{1}{2^n}, n = 1, 2, \ldots$. The number $M(\epsilon)$ of such sticks that is needed to cover the interval $[0,1]$ is 2^n. Hence

$$D = \lim_{\epsilon \to 0} \frac{\ln M(\epsilon)}{\ln\left(\frac{1}{\epsilon}\right)} = 1. \tag{5.68}$$

(Here we dropped the requirement $\epsilon = \frac{1}{2^n}$). Thus the expression, Equation (5.68), yields a number that corresponds to our intuitive notions about the dimension of a line.

Now consider the unit square $[0,1] \times [0,1]$ in the plane. Instead of a measuring stick we need here a "measuring square" with sides of $\frac{1}{2^n}$. When $n = 1$

we need 4 such squares to cover $[0,1] \times [0,1]$. For $n = 2$ we need 16 squares, and for $\epsilon = \frac{1}{2^n}$ we need 2^{2n} squares. Hence

$$D = \lim_{\epsilon \to \infty} \frac{\ln M(\epsilon)}{\ln \left(\frac{1}{\epsilon} \right)} = 2. \tag{5.69}$$

In three dimensions we consider the unit cube with "measuring cubes" whose sides are of length $\epsilon = \frac{1}{2^n}$. With $n = 1$ we need 8 such cubes to cover the unit cube. With $n = 2$ we need 64 cubes, etc. Hence

$$D = \lim \frac{\ln M(\epsilon)}{\ln \left(\frac{1}{\epsilon} \right)} = 3. \tag{5.70}$$

Thus it seems appropriate to define the dimension of a metric space as

$$D = \lim_{\epsilon \to 0} \frac{\ln M(\epsilon)}{\ln \left(\frac{1}{\epsilon} \right)} \tag{5.71}$$

where $M(\epsilon)$ is the number of "cubes" with side ϵ that are needed to cover the unit cube in such a space. The dimension D defined by Equation (5.71) is called the "Hausdorff dimension" of the space.

We now apply this definition to an "exotic" set and compute its dimension.

Example 5.6.1 *Cantor Set*

To define the Cantor set we consider the interval $I = [0,1]$. We now define

$$C_1 = I - \left(\frac{1}{3}, \frac{2}{3} \right) = \left[0, \frac{1}{3} \right] \cup \left[\frac{2}{3}, 1 \right], \tag{5.72}$$

i.e. C_1 is the subset that is obtained from the unit interval by removing the middle (open) $\frac{1}{3}$ of I. Now define C_2 as

$$C_2 = C_1 - \left(\frac{1}{9}, \frac{2}{9} \right) - \left(\frac{7}{9}, \frac{8}{9} \right), \tag{5.73}$$

i.e. we remove again the middle $\frac{1}{3}$ of each of the subintervals of C_1.

Similarly we define

$$C_3 = C_2 - \left(\frac{1}{27}, \frac{2}{27} \right) - \left(\frac{7}{27}, \frac{8}{27} \right) - \left(\frac{19}{27}, \frac{20}{27} \right) - \left(\frac{25}{27}, \frac{26}{27} \right) \tag{5.74}$$

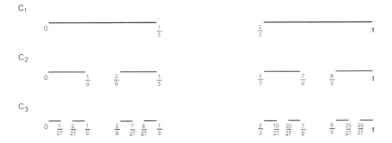

Figure 5.16 Cantor set

and so on. The Cantor set is now defined as the set remaining at the "end" of such interval divisions and subtractions, i.e. (See Fig. 5.16)

$$C = \bigcap_{i=1}^{\infty} C_n. \tag{5.75}$$

To show that C is not empty (in fact it is uncountable) we use the number representation with base 3. That is, a number $0 \le x \le 1$ is represented by

$$x = .a_1 a_2 \ldots \quad , 0 \le a_i \le 2. \tag{5.76}$$

(That is, the a_i are integers which can take the values $0, 1, 2$) where the a_is are determined by the relation

$$x = \sum_{n=1}^{\infty} \frac{a_n}{3^n}. \tag{5.77}$$

Remark 5.6.1 *The decimal and binary representation of a number $0 \le x \le 1$ is a short hand notation for*

$$x = \sum_{n=1}^{\infty} \frac{\alpha_n}{10^n}, \quad x = \sum_{n=1}^{\infty} \frac{\beta_n}{2^n} \tag{5.78}$$

respectively where $\alpha_i \epsilon \{0, 1 \ldots 9\}, \beta_i \epsilon \{0, 1\}$.

It is now clear from its construction that the Cantor set consists of all numbers in base 3 whose representation contains only the digits $\{0, 2\}$; i.e. $a_i \ne 1$ for all i.

What is the Hausdorff dimension of the Cantor set?

To answer this question we consider a yard stick of length $\epsilon = \frac{1}{3}$. It is then obvious that $M\left(\frac{1}{3}\right) = 2$. Similarly for $\epsilon = \frac{1}{9}$ we have $M\left(\frac{1}{3^2}\right) = 2^2$. In general we then have $M\left(\frac{1}{3^n}\right) = 2^n$ and hence

$$D = \lim \frac{\ln M(\epsilon)}{\ln\left(\frac{1}{\epsilon}\right)} = \frac{\ln 2}{\ln 3}. \tag{5.79}$$

Thus the Cantor set is a set with non-integer Hausdorff dimension.

Example 5.6.2 *Sierpinski Triangle*

To construct this set we start with an equilateral triangle S_0 with sides of length one. In the first iteration we split it into four equal-sided triangles (as shown in Fig. 5.17) and remove the one in the center. This yields S_1. By iterating this procedure on each of the three remaining triangles we obtain S_2 and so on. The Sierpinski triangle is the limit set that remains after all these divisions and removals.

$$S = \lim_{n \to \infty} S_n.$$

To find the Hausdorff dimension of S we note that for S_0 we have one triangle and need a measuring stick with length $\epsilon = 1$. For S_1 we have three triangles and $\epsilon = \frac{1}{2}$. For S_2 we have nine triangle and $\epsilon = \frac{1}{4}$. Hence

$$D = \lim_{\epsilon \to 0} \frac{\ln M(\epsilon)}{\ln\left(\frac{1}{\epsilon}\right)} \equiv \lim_{n \to \infty} \frac{\ln 3^n}{\ln 2^n} = \frac{\ln 3}{\ln 2}.$$

Thus the dimension of this "triangle" is greater than one but less than two. To compute the area of S we observe that S_0 has area $\frac{\sqrt{3}}{4}$. S_1 has an area of $\frac{\sqrt{3}}{4} \cdot \frac{3}{4}$ (since $\frac{1}{4}$th of the area has been removed). S_2 has an area of $\left(\frac{\sqrt{3}}{4} \cdot \frac{3}{4}\right) \cdot \frac{3}{4}$ (since we removed once again $\frac{1}{4}$ of the area of S_1). In general we then have

$$\text{area of } S_n = \frac{\sqrt{3}}{4}\left(\frac{3}{4}\right)^n$$

and hence

$$\text{area of } S = \lim_{n \to \infty} (\text{area of } S_n) = 0.$$

S_0

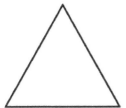 S_1

 S_2

Figure 5.17 First two iterations toward the creation of the Sierpinski Triangle

5.6.2 Liapunov Exponents

Let

$$\dot{\mathbf{x}} = \mathbf{F}(\mathbf{x}), \quad \mathbf{x} \epsilon R^n \tag{5.80}$$

be a dynamical system, and let $\mathbf{x}(t)$ be the trajectory of the solution with initial condition $\mathbf{x}(0) = \mathbf{x}_0$. Now consider another solution $\mathbf{y}(t)$ of this system with initial condition $\mathbf{y}(0) = \mathbf{x}_0 + \delta \mathbf{x}_0$.

Definition:

$$\alpha = \lim_{t \to \infty} \frac{1}{t} \ln \frac{\| \, \mathbf{y}(t) - \mathbf{x}(t) \, \|}{\| \, \delta \mathbf{x}_0 \, \|} \tag{5.81}$$

is called a Liapunov exponent of the dynamical system Equation (5.80). It is obvious from the definition that different $\delta \mathbf{x}_0$ might lead to different Liapunov exponents.

Intuitively, Liapunov exponents tell us how the solution of system (5.80) "reacts" to small changes in the initial conditions. If the Liapunov exponents of the system are all negative, then small changes in the initial conditions cause small changes in the trajectory, and one can make predictions about the evolution of the system even if the initial conditions are not known exactly. However if at least one Liapunov exponent of the system is positive, then (random) small changes in the initial conditions might lead to diverging trajectories and hence no long term predictions about the evolution of the system can be made. Thus such a system is sensitive to the values of the initial conditions and hence is called "chaotic."

This issue (of sensitivity to the initial conditions) is important from a practical point of view since in most cases the initial conditions are known only with a certain margin of error. If all the Liapunov exponents of the system are negative such errors are not significant as the approximate trajectory of the system (using inexact initial condition) remain close to the true one. However, if at least one of the Liapunov exponents is positive even minute errors in the initial condition will lead (in the long run) to large predictive errors regarding the behavior of the system.

Example 5.6.3 *The weather system of the Earth is chaotic. As a result one cannot make long term predictions about its behavior.*

Example 5.6.4 *The Lorenz system is chaotic (see exercises).*

Exercises

1. Carry out in detail the calculations needed to establish the stability of the various steady states for the Lorenz model Equations.

2. Simulate Lorenz Equations with

$$\sigma = 10, \quad r = 28, \quad b = 8/3$$

(These were the original parameters used by Lorenz in his 1963 paper) subject to the initial conditions $x = (6, 6, 6)$, $x = (6.01, 6, 6)$, $\mathbf{x} = (6.001, 6, 6)$ over the time interval $[0, 250]$.

(a) Plot the solutions of $x(t)$ on the same graph and compare (do the same for the solutions of $y(t)$ and $z(t)$).

(b) Plot the difference between the solutions of $x(t)$ on the same graph and compare (do the same for the solutions of $y(t)$ and $z(t)$).

(c) Plot x vs. y and x vs. z. Explain what you see.

(d) Compute an approximation for one of the Lyaponuv exponents of the system at $x = (6, 6, 6)$ based on the simulations of this problem.

3. Simulate the Lorenz system for other values of r's e.g., $1, 5, 10, 20$ and make similar plots to those in Ex. 2b. ANALYZE your results.

5.7 APPENDIX A: DERIVATION OF LORENZ EQUATIONS

Basic Model

Atmospheric convection in two dimensions is described by the following system of partial differential Equations:

$$\frac{\partial}{\partial t}\nabla^2\psi = -\frac{\partial(\psi, \nabla^2\psi)}{\partial(x, z)} + \nu\nabla^4\psi + \alpha g\frac{\partial\theta}{\partial x} \tag{A. 1}$$

$$\frac{\partial\theta}{\partial t} = -\frac{\partial(\psi, \theta)}{\partial(x, z)} + \frac{\Delta T}{H}\frac{\partial\psi}{\partial x} + \kappa\nabla^2\theta. \tag{A. 2}$$

Subject to the stress free boundary conditions:

$$\psi = 0, \ \nabla^2\psi = 0, \ \theta = 0, \ z = 0, 1. \tag{A. 3}$$

In this system ψ is the stream function and θ is the potential temperature

$$\theta = T(p/p_0)^{R/c_p}$$

where T is the temperature in Kelvins, p is the pressure, and p_0 is the pressure at sea level. The constants α, ν, κ, g denote respectively the coefficients of thermal expansion, the kinematic viscosity, the thermal conductivity, and the acceleration of gravity. H is the fluid layer thickness, and ΔT is the temperature difference between the upper and lower surface of the fluid (which is assumed to be held constant). We also have

$$\frac{\partial(f, g)}{\partial(x, z)} = \begin{vmatrix} \frac{\partial f}{\partial x} & \frac{\partial f}{\partial z} \\ \frac{\partial g}{\partial x} & \frac{\partial g}{\partial z} \end{vmatrix}. \tag{A. 4}$$

In his famous 1963 paper Lorenz introduced and studied a model in which the solution to Equations (A. 1)-(A. 2) is approximated by a three Fourier modes

$$\psi = \frac{\kappa(1 + a^2)\sqrt{2}}{a}X(t)\sin\left(\frac{\pi ax}{H}\right)\sin\left(\frac{\pi z}{H}\right) \tag{A. 5}$$

$$\theta = \frac{R_c}{\pi R_a}\left\{\sqrt{2}Y(t)\cos\left(\frac{\pi ax}{H}\right)\sin\left(\frac{\pi z}{H}\right) - Z(t)\sin\left(\frac{2\pi z}{H}\right)\right\} \tag{A. 6}$$

where a is a parameter and

$$R_a = \frac{g\alpha H^3\Delta T}{\kappa\nu} \tag{A. 7}$$

$$R_c = \frac{\pi^4(1+a^2)^3}{a^2} \tag{A. 8}$$

are the Rayleigh number and the critical Rayleigh numbers for the flow. This led to the following three coupled Equations for X, Y, Z.

$$\dot{X} = -\sigma X + \sigma Y \tag{A. 9}$$
$$\dot{Y} = XZ + rX - Y$$
$$\dot{Z} = XY - bZ$$

where

$$\sigma = \frac{\nu}{\kappa}, \ b = \frac{4}{(1+a^2)}, \ r = \frac{R_a}{R_c}. \tag{A. 10}$$

These Equations are usually referred to as the 'Lorenz model.' Since its appearance this model and its implications have been studied in great detail in hundreds of publications with special attention to its bifurcations as a function of the parameters σ, r, and b. It has been recognized ,however, that (as expected) the approximation of the solution to Equations (A. 1) and (A. 2) which is provided by (A. 4)-(A. 10) becomes 'poor' as r increases. This has led several authors to develop and study models with a larger number of modes [4,21].

Higher Order Models

In an attempt to relate the leading Lyaponuv exponent to the turbulence strength in the flow we consider a truncated expansion of Equations (A. 1)-(A. 2) with 3 and 6 Fourier components. For the six components model we let

$$\psi = \frac{\sqrt{2}(1+a^2)\kappa}{a} \left\{ \left[X_1(t) \sin\left(\frac{\pi a x}{H}\right) + X_2(t) \sin\left(\frac{2\pi a x}{H}\right) \right] \right. \tag{A. 11}$$
$$\left. \sin\left(\frac{\pi z}{H}\right) + X_3(t) \sin\left(\frac{\pi a x}{H}\right) \sin\left(\frac{2\pi z}{H}\right) \right\}$$

$$\theta = \frac{\sqrt{2}\Delta T R_c}{\pi R_a} \left[Y_1(t) \cos\left(\frac{\pi a x}{H}\right) + Y_2(t) \cos\left(\frac{2\pi a x}{H}\right) \right]$$
$$\sin\left(\frac{\pi z}{H}\right) - \frac{\Delta T R_a}{\pi R_c} Z(t) \sin\left(\frac{2\pi z}{H}\right) \tag{A. 12}$$

which leads after some lengthy algebra to the following six Equations for $X_1, X_2, X_3, Y_1, Y_2, Z$

$$\dot{X}_1 = \sigma Y_1 - \sigma X_1 + \frac{9}{4}\frac{(a^2 - 1)\sqrt{2}}{1 + a^2}X_2X_3 \tag{A. 13}$$

$$\dot{X}_2 = \frac{2\sigma(1 + a^2)}{1 + 4a^2}Y_2 - \frac{\sigma(1 + 4a^2)}{1 + a^2}X_2 + \frac{9\sqrt{2}}{4(1 + 4a^2)}X_1X_3 \tag{A. 14}$$

$$\dot{X}_3 = -\frac{\sigma(4 + a^2)}{1 + a^2}X_3 - \frac{9a^2\sqrt{2}}{4(4 + a^2)}X_1X_2 \tag{A. 15}$$

$$\dot{Y}_1 = \frac{3}{4}\sqrt{2}X_3Y_2 - X_1Z - Y_1 + rX_1 \tag{A. 16}$$

$$\dot{Y}_2 = -\frac{3}{4}\sqrt{2}X_3Y_1 - \frac{(1 + 4a^2)}{1 + a^2}Y_2 - 2X_2Z + 2rX_2 \tag{A. 17}$$

$$\dot{Z} = X_1Y_1 + 2X_2Y_2 - bZ \tag{A. 18}$$

Perturbations

CONTENTS

6.1 INTRODUCTION

When we model a complex system, we create at first a "prototype model." This model takes into account the "major features" of the system at hand and ignores other aspects which might have "small impact" on the system evolution. In other words the prototype model reduces the system to its "bare bones." In many cases these prototype models lead to equations with closed form solutions, viz. solutions in terms of analytical formulas. However, when refinements of the prototype model are needed, new (nonlinear) terms have to be added, and in many cases the resulting model equations have no closed form solutions. At this stage one realizes that the additional terms in the refined model might be "small" when compared to original terms in the prototype model and attempts to take advantage of this fact. The essence of perturbation theory and techniques is to develop methods that yield at least approximate solutions for these refined models. As a first step in this approach one has to rewrite the model equations in non-dimensional form. That is, the equations have to be rewritten in a form that is independent of the physical units that are being used. In this form one can see immediately which terms are small when compared to other terms in the equations and thereby treat them as a "perturbation."

It might be argued that with wide availability of fast computers and appropriate software one can solve these complicated (nonlinear) equations (of the refined model) numerically. However, in general, it is not straightforward to gain insights about the evolution of the system from the numerical solution. Furthermore, the numerical solution of a complex system of nonlinear differential equations is always a challenging problem. Algorithms to solve such a system might not converge, and even if they converge, they might yield incorrect solutions.

In the following we present some of the basic techniques of perturbations theory, but for a more comprehensive treatment we refer the reader to books on this topic.

6.2 MODEL EQUATIONS IN NON-DIMENSIONAL FORM

When we model a physical system, the variables and the value of the coefficients that appear in the model equations depend on the physical units being used. For example the length unit might be centimeter, meter, kilometer, light year (i.e. the distance that light travels in one Earth year) and so on. It is obvious then that the use of different units will impact the value of the parameters and coefficients that appear in the model equation. In order to overcome this issue one attempts to find "characteristic values" for the variables that appear in the equations and combinations of these values so that the equations are expressed in terms of "pure numbers" (independent of the physical units). We illustrate this process by a few examples.

Example 6.2.1 *Consider Newton's second law*

$$m\frac{d^2\mathbf{x}}{dt^2} = \mathbf{F}. \tag{6.1}$$

To rewrite this law in non-dimensional form let us assume that the characteristic values of mass #, length, and time in this equation are M, L, and T respectively. The force term $\mathbf{F}$ in this equation has a dimension of $\frac{mass \cdot length}{time^2}$. Therefore, we define the following dimensionless quantities,

$$\bar{m} = \frac{m}{M}, \quad \bar{\mathbf{x}} = \frac{\mathbf{x}}{L}, \quad \bar{t} = \frac{t}{T}, \quad \bar{\mathbf{F}} = \frac{\mathbf{F}T^2}{ML}. \tag{6.2}$$

Then the left hand side of Equation (12.1) becomes

$$m\frac{d^2\mathbf{x}}{dt^2} = M\bar{m}\frac{d^2(L\bar{\mathbf{x}})}{d(T^2\bar{t}^2)} = \frac{ML}{T^2}\bar{m}\frac{d^2\bar{\mathbf{x}}}{d\bar{t}^2}.$$

At the same time the right hand side of Equation (12.1) yields

$$\mathbf{F} = \frac{ML}{T^2}\bar{\mathbf{F}}.$$

Therefore, Equation (12.1) takes the following form

$$\bar{m}\frac{d^2\bar{\mathbf{x}}}{d\bar{t}^2} = \bar{\mathbf{F}}. \tag{6.3}$$

Although Equations (12.1), (12.3) look "similar," the quantities in (12.3) are dimensionless (i.e. independent of the physical units).

To see why this non-dimensionalization process is important we the discuss the following example.

Example 6.2.2 *Suppose we consider the following modification of Equation (12.1)*

$$m\frac{d^2\mathbf{x}}{dt^2} = \mathbf{F} + \alpha\left(\frac{d\mathbf{x}}{dt}\right)^2. \tag{6.4}$$

The question naturally arises as to the impact of the additional term on the solution. Is it "small" or "large" compared to the other terms of the equation? To answer this question we introduce the same non-dimensional quantities as in (12.2), and the additional term in Equation (12.4) becomes

$$\left(\frac{d\mathbf{x}}{dt}\right)^2 = \left(\frac{d(L\bar{\mathbf{x}})}{d(T\bar{t})}\right)^2 = \frac{L^2}{T^2}\left(\frac{d\bar{\mathbf{x}}}{d\bar{t}}\right)^2.$$

Therefore, Equation (12.4) leads to

$$\frac{ML}{T^2}\bar{m}\frac{d^2\bar{\mathbf{x}}}{d\bar{t}^2} = \frac{ML}{T^2}\bar{\mathbf{F}} + \alpha\frac{L^2}{T^2}\left(\frac{d\bar{\mathbf{x}}}{d\bar{t}}\right)^2. \tag{6.5}$$

Hence

$$\bar{m}\frac{d^2\bar{\mathbf{x}}}{d\bar{t}^2} = \bar{\mathbf{F}} + \frac{\alpha L}{M}\left(\frac{d\bar{\mathbf{x}}}{d\bar{t}}\right)^2. \tag{6.6}$$

It follows then that the size of the additional term in Equation (12.4) (as compared to other terms in this equation) is not determined by α alone but by the ratio $\bar{\alpha} = \frac{\alpha L}{M}$. If $\bar{\alpha} \ll 1$, we can treat the additional term in Equation (12.4) as a small perturbation. On the other hand if $\bar{\alpha} \approx 1$, such a treatment will be inappropriate.

Remark: *Another way to look on this issue is to realize that α is not a dimensionless number since each of the terms in Equation (12.4) must have the same dimension. Therefore α must have the dimension of $\frac{mass}{length}$. Hence the dimensionless form of this constant is $\frac{\alpha L}{M}$.*

Additional examples of the non-dimensionalization process are presented in Chapters 9 and 10.

6.3 REGULAR PERTURBATIONS

To illustrate how the regular perturbation procedure is used to obtain an approximate solution of an equation with a term which is "small" compared to other terms, we consider the following example.

Example 6.3.1 *Find an approximate solution to the following equation,*

$$\frac{dy}{dt} + ky + \epsilon y^3 = 0, \ \ y(0) = 2. \tag{6.7}$$

Here we assume that the equation is written already in non-dimensional form and $\epsilon \ll 1$.

Solution 6.3.1 *If $\epsilon = 0$ the solution of Equation (12.23) is*

$$y(t) = 2e^{-kt}. \tag{6.8}$$

To find approximate solution of (12.23) when $\epsilon \neq 0$ we write the desired solution $y(t, \epsilon)$ in the form of a power series around $\epsilon = 0$ viz.

$$y(t, \epsilon) = y_0(t) + \epsilon y_1(t) + \epsilon^2 y_2(t) + \dots \tag{6.9}$$

Substituting this expansion in Equation (12.23) and collecting terms with the same power of ϵ yields,

$$\left[\frac{dy_0}{dt} + ky_0\right] + \epsilon\left[\frac{dy_1}{dt} + ky_1 + y_0^3\right] + \epsilon^2\left[\frac{dy_2}{dt} + ky_2 + 3y_0^2 y_1\right] + \dots = 0. \tag{6.10}$$

$$y_0(0) + \epsilon y_1(0) + \epsilon^2 y_2(0) + \dots = 2. \tag{6.11}$$

Since ϵ is a parameter, terms with different powers of ϵ in Equation (12.26) must each equal to zero. Using Equation (12.27) we obtain

$$y_0(0) = 2, \ \ y_i(0) = 0, \ \ i \neq 0.$$

Thus we obtain the following system of differential equations

$$\frac{dy_0}{dt} + ky_0 = 0, \ \ y_0(0) = 2, \tag{6.12}$$

$$\frac{dy_1}{dt} + ky_1 + y_0^3 = 0, \ \ y_1(0) = 0, \tag{6.13}$$

$$\frac{dy_2}{dt} + ky_2 + 3y_0^2y_1, \quad y_2(0) = 0, \tag{6.14}$$

and so on. Observe that although the original Equation (12.23) is nonlinear the new system of equations is actually linear and can be solved recursively. In fact the solution for y_0 is given by Equation (12.24). Substituting this solution in Equation (9.21) we obtain

$$y_1(t) = \frac{4}{k}(e^{-2kt} - 1)e^{-kt}.$$

Similarly we can substitute the solutions for y_0 and y_1 in Equation (9.22) and solve for y_2 etc.

Hence to order ϵ the (approximate) solution of Equation(12.23) is

$$y_a(t) = 2e^{-kt} + \frac{4\epsilon}{k}(e^{-2kt} - 1)e^{-kt}. \tag{6.15}$$

It is interesting to note that although, Equation (12.23) is nonlinear, it has an exact closed form solution

$$y_e(t) = \frac{2\sqrt{k}}{\sqrt{-4\epsilon + e^{2kt}(4\epsilon + k)}}. \tag{6.16}$$

In Fig 6.1 we plotted the difference $y_a - y_e$ on the time interval $[0, 5]$ for $k = 1$ and $\epsilon = 0.1$.

Exercises

1. Use regular perturbations to find an approximate solution to order ϵ of the following equation

$$\frac{dy}{dx} + \epsilon y + 3y^2 = 0, \quad y(0) = 1.$$

2. The equation for the orbit of the planet Mercury around the Sun (in general relativity) can be reduced to the following equation

$$\frac{d^2u}{d\theta^2} + u = a(1 + \epsilon u^2)$$

where $\epsilon \ll 1$. Use first order perturbations to examine the impact of the term $a\epsilon u^2$ on the period of the orbit.

Hint: First solve this equation with $\epsilon = 0$.

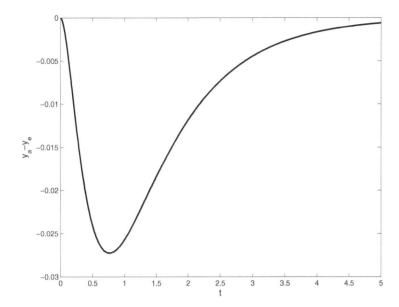

Figure 6.1 Difference between the approximate and exact solutions of Equation (12.23)

3. Use regular perturbations to find an approximate solution to order ϵ of the following initial value problem

$$\frac{d^2y}{dx^2} + 4y - \epsilon y^2 = 0, \quad y(0) = 1 + \epsilon, \quad \frac{dy}{dx}(0) = 0.$$

6.4   SINGULAR PERTURBATIONS

In general, when the perturbation term is small, its impact on the solution of the corresponding equation without the perturbation term will be small (as we showed in the previous section). However, there are cases where the perturbation yields terms in the solution which are unbounded in time. These are referred to as "**Singular Perturbations**" and we demonstrate their treatment through the following example.

Example 6.4.1 *Find an approximate solution to the following perturbed spring-mass problem (with no friction)*

$$\frac{d^2y}{dt^2} + k^2y + \epsilon y^3 = 0, \quad y(0) = 1, \quad \frac{dy}{dt}(0) = 0 \tag{6.17}$$

Solution 6.4.1 *Using the regular perturbation expansion in Equation (12.25) and substituting it in Equation (12.30) we obtain the following system of equations*

$$\frac{d^2 y_0}{dt^2} + k^2 y_0 = 0, \quad y_0(0) = 1, \quad \frac{dy_0}{dt}(0) = 0, \tag{6.18}$$

$$\frac{d^2 y_1}{dt^2} + k^2 y_1 + y_0^3 = 0, \quad y_1(0) = 0, \quad \frac{dy_1}{dt}(0) = 0, \tag{6.19}$$

$$\frac{d^2 y_2}{dt^2} + k^2 y_2 + 3y_0^2 y_1 = 0, \quad y_0(0) = 0, \quad \frac{dy_2}{dt}(0) = 0, \tag{6.20}$$

and so on. The solution for y_0 is

$$y_0 = \cos(kt), \tag{6.21}$$

and the equation for y_1 becomes

$$\frac{d^2 y_1}{dt^2} + k^2 y_1 + \cos^3(kt) = 0, \quad y_1(0) = 0, \quad \frac{dy_1}{dt}(0) = 0. \tag{6.22}$$

However

$$\cos^3(kt) = \cos(kt)(1 - \sin^2(kt))$$

which implies that Equation (12.35) has a resonance term since $\cos(kt)$ is one of the solutions of the homogeneous equation. In fact the solution of Equation (12.35) is

$$y_1 = \frac{1}{4k^2} \left[\cos(kt) + \frac{1}{8}(\cos(3kt) - 9\cos(kt) - 12kt\sin(kt)) \right]. \tag{6.23}$$

Following the steps in the previous section, the approximate solution to order ϵ of Equation (12.30) should be,

$$y(t) = \cos(kt) + \tag{6.24}$$
$$\epsilon \frac{1}{4k^2} \left[\cos(kt) + \frac{1}{8}(\cos(3kt) - 9\cos(kt) - 12kt\sin(kt)) \right].$$

*However even if $\epsilon \ll 1$ the last term in this equation will dominate as t increases. This term is referred to as **secular term**. Furthermore such unbounded terms exist also in the solution for y_2 etc.*

To overcome this problem (i.e. remove the secular term from the solution) it is customary to use a perturbation expansion for both the dependent and independent variables, viz.

$$t = s(1 + \epsilon a_1 + \epsilon^2 a_2 + \ldots) \tag{6.25}$$

$$y(s) = y_0(s) + \epsilon y_1(s) + \epsilon^2 y_2(s) + \ldots \tag{6.26}$$

where a_i, $i = 1, 2, \ldots$ are constants whose value will be determined so that the secular term(s) in the expansion disappear. (In more general expansions the constants a_i are replaced by functions of s.)

Using these formulas we have

$$\frac{dy}{dt} = \frac{dy}{ds}\frac{ds}{dt} = \frac{dy}{ds}(1 + \epsilon a_1 + \epsilon^2 a_2 + \ldots)^{-1}.$$

The Taylor expansion of

$$f(\epsilon) = (1 + \epsilon a_1 + \epsilon^2 a_2 + \ldots)^{-1}$$

around $\epsilon = 0$ is

$$f(\epsilon) = (1 - \epsilon a_1 + \ldots).$$

Therefore,

$$\frac{dy}{dt} = \frac{dy}{ds}(1 - \epsilon a_1 + \ldots).$$

Similarly for the second order derivative we obtain

$$\frac{d^2y}{dt^2} = \frac{d}{dt}\left(\frac{dy}{ds}\frac{ds}{dt}\right) = \frac{d}{dt}\left[\frac{dy}{ds}(1 - \epsilon a_1 + \ldots)\right] = \tag{6.27}$$

$$\frac{d}{ds}[\frac{dy}{ds}(1 - \epsilon a_1 + \ldots)]\frac{ds}{dt} =$$

$$\frac{d^2y}{ds^2}(1 - \epsilon a_1 + \ldots)^2 = \frac{d^2y}{ds^2}(1 - 2\epsilon a_1 + \ldots).$$

Substituting this result in Equation (12.30) and using Equation (9.70) we have

$$\left(\frac{d^2y_0(s)}{ds^2} + \epsilon\frac{d^2y_1(s)}{ds^2} + \ldots\right)(1 - 2\epsilon a_1 + \ldots) + \tag{6.28}$$

$$k^2(y_0(s) + \epsilon y_1(s) + \ldots) + \epsilon(y_0(s) + \epsilon y_1(s) + \ldots)^3 = 0$$

and for the initial conditions (since $t = 0$ corresponds to $s = 0$)

$$y_0(0) + \epsilon y_1(0) + \ldots = 1 \tag{6.29}$$

$$\frac{dy}{dt}(0) = \frac{dy}{ds}(0)(1 - \epsilon a_1 + \dots) = \quad (6.30)$$

$$\left(\frac{dy_0}{ds}(0) + \epsilon \frac{dy_1}{ds}(0) + \dots\right)(1 - \epsilon a_1 + \dots) = 0.$$

Hence to order ϵ we obtain the following two equations:

$$\frac{d^2 y_0(s)}{ds^2} + k^2 y_0(s) = 0, \quad y_0(0) = 1, \quad \frac{dy_0}{ds}(0) = 0. \quad (6.31)$$

$$\frac{d^2 y_1(s)}{ds^2} + k^2 y_1(s) + y_0(s)^3 - 2a_1 \frac{d^2 y_0(s)}{ds^2} = 0, \quad y_1(0) = 0, \quad \frac{dy_1}{ds}(0) = 0. \quad (6.32)$$

The solutions of these equations are

$$y_0(s) = \cos(ks) \quad (6.33)$$

$$y_1(s) = \frac{s}{32k^2}(-32a_1 k^3 - 12k)\sin(ks) + \frac{1}{32k^2}(\cos(3ks) - \cos(ks)). \quad (6.34)$$

We see then that the secular term will disappear if we let

$$a_1 = -\frac{3}{8k^2}$$

and the approximate solution of Equation (12.30) to first order in ϵ is

$$y(s) = \cos(ks) + \frac{\epsilon}{32k^2}(\cos(3ks) - \cos(ks))$$

$$t = s\left(1 - \frac{3\epsilon}{8k^2}\right).$$

Exercises

Use singular perturbation expansion to first order in ϵ to find an approximate solution of the following equations. Compare the solution with the numerical solution.

1.

$$\frac{d^2 y}{dt^2} + \epsilon\left(\frac{dy}{dt}\right)^2 + k^2 y = 0, \quad y(0) = 1, \quad \frac{dy}{dt}(0) = 0.$$

6.5   BOUNDARY LAYERS

When a viscous fluid flows over a plate (e.g. airplane wing), a thin layer is formed over the plate where the flow velocity undergoes a very rapid change. This thin layer is referred to as a "**Boundary Layer**". Mathematically these phenomena are usually modeled by differential equations where $0 < \epsilon \ll 1$ multiplies the highest derivative in the equations. To see how an approximate solution for these problems can be obtained using perturbation theory we consider the following prototype problem.

Example 6.5.1 *Consider the following boundary value problem on the interval* $[0, 1]$;

$$\epsilon \frac{d^2 y}{dx^2} + \frac{dy}{dx} + ky = 0, \quad y(0) = 0, \quad y(1) = 1, \quad k > 0, \quad 0 < \epsilon << 1. \quad (6.35)$$

Solution 6.5.1 *The exact solution of this linear problem is*

$$y(x) = A \exp\left(-\frac{x}{2\epsilon}\right) \sinh(\alpha x)$$

where

$$\alpha = \frac{\sqrt{1 - 4k\epsilon}}{2\epsilon}, \quad A = \frac{exp(\frac{1}{2\epsilon})}{\sinh(\alpha)}.$$

A plot of this solution with $k = 1$ and $\epsilon = 0.1$ is shown in Fig 6.2. From this figure we see that the solution rises rapidly from 0 to 2.3 over the "boundary layer" $[0, 0.15]$ and then decays smoothly over the rest of the interval.

To see the challenge that these types of problems present (from perturbation theory perspective) we attempt, at first, to obtain an approximate solution via regular perturbations. Thus, if we write the perturbation expansion

$$y(x) = y_0(x) + \epsilon y_1(x) + \epsilon^2 y_2(x) + \dots \quad (6.36)$$

and substitute in Equation (6.35) we obtain the following system of equations:

$$\frac{dy_0}{dx} + ky_0 = 0, \quad y_0(0) = 0, \quad y_0(1) = 1 \quad (6.37)$$

$$\frac{dy_1}{dx} + ky_1 = -\frac{d^2 y_0}{dx^2}, \quad y_1(0) = y_1(1) = 0 \quad (6.38)$$

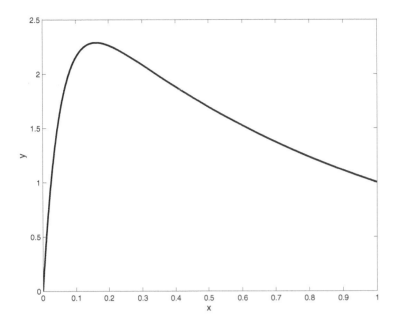

Figure 6.2 Exact solution of (6.35)

$$\frac{dy_2}{dx} + ky_2 = -\frac{d^2y_1}{dx^2}, \quad y_2(0) = y_2(1) = 0 \tag{6.39}$$

etc. Thus although Equation (6.35) is a second order equation with two boundary conditions, the perturbation expansion yields a system of first order equations with two boundary conditions for which no solution exists in general. In fact the general solution of Equation (6.37) is

$$y_0 = Ce^{-kx},$$

but since $y_0(0) = 0$, it follows that $C = 0$ and, therefore, $y_0(1) = 1$ cannot be satisfied.

To see the origin of this peculiar behavior and how it can be ameliorated we note that the second order derivative of the exact solution of Equation (6.35) has a factor of

$$\frac{exp(-\frac{x}{2\epsilon})}{\epsilon}.$$

This observation shows that the term $\epsilon\frac{d^2y}{dx^2}$ is small compared to $\frac{dy}{dx}$ and y for large x (e.g., $x > \frac{2\epsilon}{b}$. However, in a small interval around 0 it is of the same order as these two terms.

In view of this observation it is natural to divide the solution of our problem into two.

1. A solution in an inner region $0 \geq x \leq x_0$ where $x_0 > 0$ is small.

2. A solution in an exterior region $x_0 < x$.

Observe that, at the present, the exact value of x_0 remains unspecified.

Denoting the exterior solution by v we note that in this region v, and its first and second order derivatives have the same order of magnitude, and hence $\epsilon \frac{d^2 v}{dx^2}$ is negligible as compared to the other terms in Equation (6.35). Therefore, in the exterior region Equation (6.35) can be approximated correctly by a regular perturbation expansion,

$$\frac{dv_0}{dx} + kv_0 = 0, \quad v_0(1) = 1, \tag{6.40}$$

$$\frac{dv_1}{dx} + kv_1 = -\frac{d^2 v_0}{dx^2}, \quad v_1(1) = 0, \tag{6.41}$$

etc.

Observe that the point $x = 0$ is outside the "exterior region." Therefore, we do not have to take into account the boundary condition on the solution at this point.

From Equations (6.40) and (6.41) we obtain

$$v_0 = \exp[-k(x-1)] \tag{6.42}$$

$$v_1 = -k^2(x-1)\exp[-k(x-1)], \tag{6.43}$$

i.e., to first order in ϵ the outer solution is given by

$$v = [1 - \epsilon k^2(x-1)]\exp[-k(x-1)]. \tag{6.44}$$

To obtain a similar perturbation expansion in the inner region we shall perform a stretching transformation on this region so that the derivatives of the solution in this region have the "same magnitude." This can be accomplished by a transformation of the form

$$z = \epsilon^a x \tag{6.45}$$

where a is a parameter to be determined. Substituting this transformation in Equation (6.35) and denoting the inner solution by u it follows that

$$\epsilon^{1+2a}\frac{d^2u}{dz^2} + \epsilon^a\frac{du}{dz} + ku = 0, \quad u(0) = 0 \tag{6.46}$$

From this equation it follows that the derivatives of u will have the same power of ϵ if $1 + 2a = a$ i.e., $a = -1$. (Observe that due to this stretching transformation $x = \epsilon$ will correspond to $z = 1$.) Hence in the inner region Equation (6.35) takes the form

$$\frac{d^2u}{dz^2} + \frac{du}{dz} + \epsilon ku = 0, \quad u(0) = 0. \tag{6.47}$$

Applying regular perturbation expansion to Equation (6.47) yields

$$\frac{d^2u_0}{dz^2} + \frac{du_0}{dz} = 0, \quad u_0(0) = 0 \tag{6.48}$$

$$\frac{d^2u_1}{dz^2} + \frac{du_1}{dz} = -ku_0, \quad u_1(0) = 0 \tag{6.49}$$

and so on.

The solution of these equations to first order in ϵ is given by

$$u_0 = c_1(1 - e^{-z}) \tag{6.50}$$

$$u_1 = k[c_2 - c_1z - (c_2 + c_1z)e^{-z}]. \tag{6.51}$$

Now that we have perturbation solutions for the inner and outer regions a "matching principal" must be formulated so that the two solutions blend smoothly at the "edges" of the two regions. This will help to determine the constants c_1, c_2 which appear in Equations (6.50) and (6.51). To accomplish this task several "principles" were formulated in the past. The first due to Prandtl stipulates that

$$\lim_{x \to 0} v(x, \epsilon) = \lim_{z \to \infty} u(z, \epsilon). \tag{6.52}$$

Applying this principle to the zeroth order perturbation solutions (6.42) and (6.51) of our problem we find that we must satisfy

$$\lim_{x \to 0} Ae^{k(1-x)} = \lim_{z \to \infty} c_1(1 - e^{-z}). \tag{6.53}$$

Hence $c_1 = Ae^k$. However, it is easy to verify that this principle fails to determine the constant c_2 in the first order perturbation expansion.

To overcome this problem M. Van Dyke formulated the following generalized matching principle for higher order perturbation expansions. To apply this matching principle to the perturbation expansions in the inner and outer regions up to order ϵ^m,

$$v(x) = v_0(x) + \epsilon v_1(x) + \ldots + \epsilon^m v_m(x)$$

$$u(z) = u_0(z) + \epsilon u_1(z) + \ldots + \epsilon^m u_m(z),$$

one must express $v(x)$ in terms of z and $u(z)$ in terms of x, then expand the resulting expressions in powers of ϵ up to ϵ^m and finally, match the coefficients of ϵ^k, $k = 0, \ldots, m$ in the two expressions to determine the redundant constants.

As an example we implement this principle to determine the constants c_1, c_2 in Equations (6.50),(6.51) by matching the inner and outer solutions to first order in ϵ.

Rewriting the outer solution Equation (6.44) in terms of z and expanding in ϵ we obtain

$$v(z) = Ae^k[e^{-\epsilon kz} + \epsilon k^2(1 - \epsilon z)e^{-\epsilon kz}] \tag{6.54}$$
$$= Ae^k[1 - \epsilon kz + \epsilon k^2 + O(\epsilon^2)].$$

Similarly for the inner solution we have

$$u(x) = c_1(1 - e^{-x/\epsilon}) + \epsilon k[c_2 - c_1 x/\epsilon - \tag{6.55}$$
$$(c_2 + c_1 x/\epsilon)e^{-x/\epsilon}] = c_1(1 - kx) + \epsilon k c_2 + O(\epsilon^2).$$

(Note that the expression $e^{-x/\epsilon}$ for any fixed $x > 0$ converge to 0 faster than any powers of ϵ and, therefore, can be neglected). It follows then,

$$c_1(1 - kx) + \epsilon k c_2 = Ae^k[(1 - kx) + \epsilon k^2]. \tag{6.56}$$

This yields

$$c_1 = Ae^k, \quad c_2 = Ake^k. \tag{6.57}$$

Exercises

For the following two differential equations use first order perturbations to find inner, outer, and matched solutions. Solve these equations numerically. Compare the numerical and analytic solutions on the same plot.

1.
$$\epsilon \frac{d^2 y}{dx^2} + (1 + ax)\frac{dy}{dx} + k^2 y = 0, \quad y(0) = 0 \;\; y(1) = 1.$$

2.
$$\epsilon \frac{d^2 y}{dx^2} + (1 + \epsilon)\frac{dy}{dx} + y^2 = 0, \quad y(0) = 0 \;\; y(1) = 1.$$

CHAPTER 7

Modeling with Partial Differential Equations

CONTENTS

In this chapter, we illustrate this modeling process by considering various systems that are modeled in terms of partial differential equations. In particular, we concentrate on the heat, wave, and potential equations that are important in many scientific and engineering applications.

7.1 THE HEAT (OR DIFFUSION) EQUATION

Objective

Build a model that describes the temperature distribution in a metal as a function of the position and time.

Discussion

As stated, this problem does not specify if the metal composition of the body is homogeneous, nor is any information given about its shape. Because such a general problem might require a complex model, we will first attempt to build a prototype model and then compound it. To begin, we consider the heat conduction problem in a rod of length L made of homogeneous metal with a constant cross section A that is completely insulated along its lateral edges (see Fig. 7.1). (All these assumptions are, naturally, mathematical idealizations.)

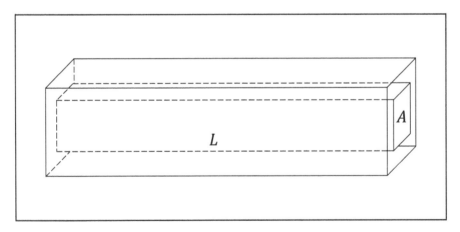

Figure 7.1 Perfectly insulated rod with constant cross section

Background

To build an acceptable model for this problem, an understanding of the concept of the flux and basic laws of thermodynamics (and physics) is necessary. We present here a short review of these pertinent ideas in one spatial dimension.

The Flux. Consider a flow of a certain physical quantity (such as mass, energy, heat, etc.). The flux $\mathbf{q}(\mathbf{x}, t)$ of this flow is defined as a vector in the direction of the flow [at $(\mathbf{x}, t)$] whose length is given by the amount of the quantity crossing a unit area (at $\mathbf{x}$) normal to the flow in unit time; that is,

$$|\mathbf{q}(\mathbf{x}, t)| = \lim_{\substack{\Delta S \to 0 \\ \Delta t \to 0}} \frac{\text{Quantity passing through } \Delta S \text{ in time } [t, t + \Delta t]}{\Delta S \Delta t} \qquad (7.1)$$

where ΔS is a (small) surface area at $\mathbf{x}$ that is normal to the flow.

Thus, the approximate amount of the physical quantity passing through a surface ΔS in time Δt is given by

$$Q(\mathbf{x}, t, \Delta S, \Delta t) \cong |\mathbf{q}(\mathbf{x}, t)| \Delta S \Delta t. \qquad (7.2)$$

If ΔS is not normal to the flow, then it must be replaced by its projection in the direction normal to the flow.

Example 7.1.1 *Consider water flowing in a river with velocity $\mathbf{v}(\mathbf{x}, t)$. To evaluate the flux of this flow at $(\mathbf{x}, t)$, we consider a small surface element ΔS normal to $\mathbf{v}(\mathbf{x}, t)$. The amount of water flowing through ΔS in time $[t, t + \Delta t]$ is given by the quantity present at t in a tube of base ΔS and height $|\mathbf{v}| \Delta t$; that is*

$$Q(\mathbf{x}, t, \Delta S, \Delta t) \cong \rho |\mathbf{v}| \Delta S \Delta t \qquad (7.3)$$

where ρ is the mass density of the water. Hence,

$$\mathbf{q}(\mathbf{x}, t) = \rho \mathbf{v} \qquad (7.4)$$

since the direction of the flow is given by $\mathbf{v}$ (see Fig. 7.2).

Basic Laws of Thermodynamics. A change ΔQ in the amount of heat in a body of mass m is accompanied by a change Δu in its (equilibrium) temperature. The relationship between these changes is given by

$$\Delta Q = cm\Delta u \qquad (7.5)$$

where c is the specific heat of the material of which the body is made, that is, the amount of heat required to raise the temperature of a body of unit mass (made of the same material) by 1 degree.

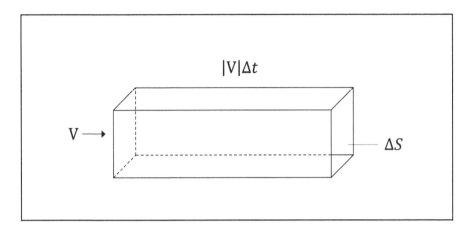

Figure 7.2 Flux-All the fluid in the tube will pass through S in time Δt

In the following discussion, we assume that Q and u are normalized so that $Q = cmu$.

Remark About Units. In Equation (7.3) (as in any equation that relates physical quantities), a consistent set of units must be used. Thus, if the MKS system of units is used, then Q (energy) is expressed in joules, mass in kilograms, u in degrees Kelvin (or Celsius), and c in joules/(kg.degK). In this book we consistently use the MKS units unless otherwise noted.

Fourier Law of Heat Conduction. Heat is transported by diffusion in the direction opposite to the temperature gradient and at a rate proportional to it. Thus, the heat flux $\mathbf{q}(\mathbf{x}, t)$ is related to the temperature gradient by

$$\mathbf{q}(\mathbf{x}, t) = -\kappa \operatorname{grad} u(\mathbf{x}, t) = -\kappa \left(\frac{\partial u(\mathbf{x}, t)}{\partial x}, \frac{\partial u(\mathbf{x}, t)}{\partial y}, \frac{\partial u(\mathbf{x}.t)}{\partial z} \right) \qquad (7.6)$$

where κ is the thermal conductivity of the material. [From Equation (7.6) we infer that its units are joules/(m.sec. degK).]

Remember that the gradient of a function gives the direction in which the function increases most rapidly while in the direction opposite to it the function decreases most rapidly. Thus, a restatement of the Fourier law is that heat flows in the direction in which the temperature decreases most rapidly (and this is the reason for the minus sign in equation (7.6). Note that by convention any constant that appears in a physical law is assumed to be positive).

Principle of Energy Conservation. Because heat is a form of energy, it must be conserved. Hence, the rate of change in the amount of heat in a body must equal the rate at which heat is flowing in less the rate at which it is flowing out (we assume that no heat is generated by the body).

Approximations and Idealizations

1. Since we assumed that the material of the rod that we are considering is homogeneous, it follows that c, k, and ρ (the material density kg/m^3) are independent of the position **x**. However, for the purpose of constructing a prototype model we further assume that they are also independent of the temperature u.

2. We assume that the length of the rod remains constant in spite of the changes in its temperature.

3. We assume that the rod is perfectly insulated along its lateral surface (idealization). Hence, heat can flow only in the horizontal direction, since a vertical flow will lead to heat accumulation along the edges, which is forbidden by the Fourier law of conduction. Therefore, we infer that the temperature at every point on a vertical cross section of the rod must be the same. Thus, the temperature u depends only on x and t; that is, $u = u(x, t)$.

4. For definiteness, we assume that heat flows in the rod from left to right, which requires the left side to be warmer than the right.

Modeling

A possible approach to model the system under consideration is to use the atomic and crystal structure of the material of the rod and build a model for the heat conduction using these microscopic variables. However, because this approach will lead to a very complex set of equations, it is not useful in our context.

Thus, we present in the following two methods to derive the **macroscopic** heat equation. We use the term *macroscopic* since we are using macroscopic variables such as u, c, k, and so on to model the system.

Infinitesimal Approach

In this method, we consider an infinitesimal element of the rod between x and $x + \Delta x$ and write the equation for the energy conservation in it.

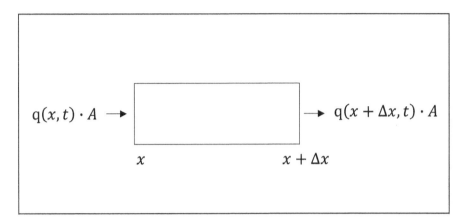

Figure 7.3   Heat flux through infinitesimal part of the rod.

Thus, since the volume of the element is $A\Delta x$, its mass Δm is given by $\rho A\Delta x$ (see Fig. 7.3). The amount of heat in this element at time t is

$$Q(x, t, \Delta x) = c\Delta m u(x, t) \tag{7.7}$$

or

$$Q(x, t, \Delta x) = c\rho A u(x, t)\Delta x. \tag{7.8}$$

The rate of change in Q is therefore given by

$$\frac{dQ}{dt} = c\rho A\frac{\partial u}{\partial t}\Delta x. \tag{7.9}$$

By the principle of energy conservation, this rate of change must equal the rate at which heat is flowing in less the rate at which it is flowing out. Hence,

$$\frac{dQ}{dt} = q(x, t) \cdot A - q(x + \Delta x, t) \cdot A. \tag{7.10}$$

Replacing $\frac{dQ}{dt}$ by

$$c\rho A\frac{\partial u}{\partial t}\Delta x \tag{7.11}$$

we have

$$c\rho A\frac{\partial u}{\partial t} = -A\frac{q(x + \Delta x, t) - q(x, t)}{\Delta x}. \tag{7.12}$$

Letting $\Delta x \to 0$, we obtain

$$c\rho \frac{\partial u}{\partial t} = -\frac{\partial q}{\partial x}. \tag{7.13}$$

From (7.6), the Fourier law of heat conduction in one dimension yields $q = -\kappa(\partial u/\partial x)$ (since u is a function of x and t only!) and, therefore,

$$c\rho \frac{\partial u}{\partial t} = \kappa \frac{\partial^2 u}{\partial x^2} \tag{7.14}$$

or

$$\frac{1}{k} \frac{\partial u}{\partial t} = \frac{\partial^2 u}{\partial x^2} \tag{7.15}$$

where $k^{-1} = c\rho/\kappa$ is called the **thermal diffusivity**. Equation (7.15) is called the **heat** (or diffusion) **equation in one (space) dimension.**

Integral Approach

In this method, we consider a *finite* section of the rod between a and b and use the principle of energy conservation to write an equation for the heat balance in this segment.

Since the amount of heat in an infinitesimal section of the rod between x and $x + \Delta x$ is given by Equation (7.8), the total amount of heat in the section $[a, b]$ is given by the integral of the expression

$$Q(t, a, b) = \int_a^b c\rho A u(x, t)dx. \tag{7.16}$$

The rate of change in this quantity is therefore given by

$$\frac{dQ}{dt} = \int_a^b c\rho A \frac{\partial u}{\partial t} dx. \tag{7.17}$$

By the principle of energy conservation, dQ/dt must equal the rate at which heat enters the section less the rate at which it leaves it. Thus,

$$\frac{dQ}{dt} = Aq(a, t) - Aq(b, t). \tag{7.18}$$

By the fundamental theorem of calculus, this equation can be rewritten as

$$\frac{dQ}{dt} = -\int_a^b A \frac{\partial q}{\partial x} dx \tag{7.19}$$

from which it follows that

$$\int_a^b c\rho A \frac{\partial u}{\partial t} dx = -\int_a^b A \frac{\partial q}{\partial x} dx. \tag{7.20}$$

By the Fourier law of heat conduction,

$$\int_a^b \left(c\rho A \frac{\partial u}{\partial t} - \kappa A \frac{\partial^2 u}{\partial x^2} \right) dx = 0. \tag{7.21}$$

But since a and b are arbitrary, Equation (17.21) implies that the integrand in this equation must also be zero. Hence,

$$c\rho \frac{\partial u}{\partial t} - \kappa \frac{\partial^2 u}{\partial x^2} = 0 \tag{7.22}$$

which is the same equation we derived using the infinitesimal approach.

Remark 7.1.1 *Although the infinitesimal and integral approaches must always (if applied correctly) yield the same result, from a conceptual modeling point of view one might be superior to the other in a given context.*

To illustrate the process of **model compounding**, we now present the derivation of the heat equation in two dimensions.

Example 7.1.2 *Derive the heat equation for a thin homogeneous plate with constant cross section (height) h. Assume that the plate is perfectly insulated on the top and bottom.*

Solution 7.1.1 *Since the plate is thin and perfectly insulated,*

$$u = u(x, y, t). \tag{7.23}$$

To derive a model for u, we use the infinitesimal approach and consider a small rectangular element that is located at a point (x, y) in the plate (see Fig. 7.4).

The amount of heat Q in this element at time t is given (approximately) by

$$Q(x, y, t, \Delta x, \Delta y) = c\rho h \Delta x \Delta y u(x, y, t). \tag{7.24}$$

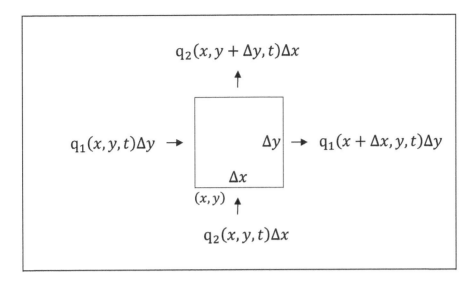

Figure 7.4 Flux in an infinitesimal section of the plate

Hence, the rate of change in Q is

$$\frac{dQ}{dt} = c\rho h \Delta x \Delta y \frac{\partial u}{\partial t}(x, y, t). \tag{7.25}$$

This rate of change must equal the rate at which heat flows into the element minus the rate at which it flows out. To compute these rates we decompose $\mathbf{q}$ into

$$\mathbf{q} = q_1 \mathbf{i} + q_2 \mathbf{j} \tag{7.26}$$

and observe that $q_1 \mathbf{i}$ is parallel to the boundary represented by the line between (x, y) and $(x + \Delta x, y)$. Hence, $q_1 \mathbf{i}$ does not contribute to the flux through this boundary. Similar considerations apply to other boundaries. It follows then that

$$\frac{dQ}{dt} = q_2(x, y, t) \Delta x h + q_1(x, y) \Delta y h - q_2(x, y + \Delta y, t)$$

$$\Delta x h - q_1(x + \Delta x, y, t) \Delta y h. \tag{7.27}$$

Equating Equation (7.25) with Equation (7.27) and dividing by $\Delta x \Delta y$ yields

$$c\rho \frac{\partial u}{\partial t} = -\frac{q_1(x + \Delta x, y, t) - q_1(x, y, t)}{\Delta x} - \frac{q_2(x, y + \Delta y, t) - q_2(x, y, t)}{\Delta y}. \tag{7.28}$$

Letting $\Delta x, \Delta y \to 0$, we obtain

$$c\rho \frac{\partial u}{\partial t} = -\left(\frac{\partial q_1}{\partial x} + \frac{\partial q_2}{\partial y}\right) = -\operatorname{div} \mathbf{q}. \tag{7.29}$$

To obtain an equation containing u only, we apply Equation (7.6) (in two dimensions). This leads to

$$\frac{1}{k}\frac{\partial u}{\partial t} = \nabla^2 u \tag{7.30}$$

where

$$\nabla^2 u = \frac{\partial^2 u}{\partial x^2} + \frac{\partial^2 u}{\partial y^2}. \tag{7.31}$$

∇^2 is the Laplace operator and $k = c\rho/\kappa$. Equation (7.30) is the **heat conduction equation in two dimensions**.

7.1.1 Burger's Equation

Burger's equation was introduced as a one-dimensional model for turbulence and has since found applications in the study of shock wave, wave transmission, traffic flow, etc. The equation is of the form

$$\frac{\partial u}{\partial t} + u\frac{\partial u}{\partial x} = \nu\frac{\partial^2 u}{\partial x^2}$$

where u is "the amplitude of the disturbance" and ν is a constant. We now show that this equation can be transformed into the heat equation.

To begin with, we remind the reader that a vector field $\mathbf{F} = (f_1(\mathbf{x}), f_2(\mathbf{x}), f_3(\mathbf{x}))$ is conservative if $curl\mathbf{F} = 0$. When this happens, there is a (potential) function $\phi(\mathbf{x})$ so that $grad\phi(\mathbf{x}) = \mathbf{F}$. This result follows from the fact that for any function $g(\mathbf{x})$, $curl(gradg(\mathbf{x})) = 0$. Similarly, a vector field $\mathbf{F}$ is called *solenoidal* if $div\mathbf{F} = 0$. Since for any vector field $\mathbf{V}$ we have $div(curl\mathbf{V}) = 0$, it is possible to prove that if a vector field is solenoidal, then there exists $\mathbf{V}$ so that $curl\mathbf{V} = \mathbf{F}$. This vector field $\mathbf{V}$ is called the "vector potential" of $\mathbf{F}$. In the two dimensional case where $\mathbf{F} = (f_1(x, y), f_2(x, y))$ and $div\mathbf{F} = 0$, i.e.,

$$\frac{\partial f_1}{\partial x} + \frac{\partial f_2}{\partial y} = 0,$$

we conclude that there exists a function ψ so that

$$f_1 = \frac{\partial \psi}{\partial y}, \quad f_2 = -\frac{\partial \psi}{\partial x}.$$

Going back to Burger's equation, we first write it in divergence form as

$$\frac{\partial u}{\partial t} + \frac{\partial}{\partial x}\left(\frac{1}{2}u^2 - \nu\frac{\partial u}{\partial x}\right) = 0.$$

We conclude, therefore, that there exists a function ψ so that

$$u = \frac{\partial \psi}{\partial x}, \quad \left(\frac{1}{2}u^2 - \nu\frac{\partial u}{\partial x}\right) = -\frac{\partial \psi}{\partial t}.$$

Using the first part of this equation to substitute for u in the second part, we obtain the following equation for ψ

$$\frac{\partial \psi}{\partial t} = \nu\frac{\partial^2 \psi}{\partial x^2} - \frac{1}{2}\left(\frac{\partial \psi}{\partial x}\right)^2. \tag{7.32}$$

We now introduce a new function η as

$$\psi = -2\nu\ln\eta.$$

(This is called the "Hopf-Cole transformation.") We then have

$$u = \frac{\partial \psi}{\partial x} = -\frac{2\nu\frac{\partial \eta}{\partial x}}{\eta}, \quad \frac{\partial \psi}{\partial t} = -\frac{2\nu\frac{\partial \eta}{\partial t}}{\eta} \tag{7.33}$$

and

$$\frac{\partial^2 \psi}{\partial x^2} = 2\nu\left[\left(\frac{\frac{\partial \eta}{\partial x}}{\eta}\right)^2 - \frac{\frac{\partial^2 \eta}{\partial x^2}}{\eta}\right].$$

Substituting these relations in Equation (7.32), we obtain that

$$\frac{\partial \eta}{\partial t} = \nu\frac{\partial^2 \eta}{\partial x^2}.$$

That is, η satisfies the heat equation. From this solution, we can recover u (that is the solution of the Burger equation) from Equation (7.33).

7.1.2 Similarity Solutions

Consider the heat equation in one dimension

$$\frac{1}{k}\frac{\partial u}{\partial t} = \frac{\partial^2 u}{\partial x^2}.$$

It is easy to see that the equation remains invariant (unchanged) under the transformations

$$\bar{x} = \alpha x, \quad \bar{t} = \alpha^2 t, \quad \bar{u} = \beta u + \gamma$$

where α, β and γ are constants. This three-parameter group of transformations contains the following one-parameter subgroup

$$\bar{x} = \alpha x, \quad \bar{t} = \alpha^2 t, \quad \bar{u} = \alpha^{2n} u$$

where m is a positive number. A similarity solution is a solution that is invariant under such a one-parameter group. In fact, if we define

$$\eta = x^2/t \quad u = t^n f(\eta)$$

then

$$\frac{\partial u}{\partial t} = n t^{n-1} f(\eta) + t^n f'(\eta)\left(-\frac{x^2}{t^2}\right),$$

$$\frac{\partial u}{\partial x} = 2t^{n-1} x f'(\eta),$$

and

$$\frac{\partial^2 u}{\partial x^2} = 2t^{n-1} f'(\eta) + 4x^2 t^{n-2} f''.$$

Substituting these results in the heat equation, we obtain

$$4k\eta f'' + (\eta + 2k) f' - nf = 0$$

which is a hypergeometric equation (the standard form of this equation is obtained after the substitution $\chi = -\eta$).

The PDE has been reduced to an ODE. The same transformations and reductions are true in higher dimensions if $u = u(r,t)$ where $r = |\mathbf{x}|$.

As another example for the application of similarity transformations, we consider a nonlinear heat equation where the diffusion coefficient depends on u

$$\frac{\partial u}{\partial t} = \frac{\partial}{\partial x}\left(D(u)\frac{\partial u}{\partial x}\right). \tag{7.34}$$

To find a similarity solution, we attempt to introduce a new variable $\eta = x^\alpha t^\beta$ where α and β are constants to be determined later and attempt to find a solution of (7.34) which is dependent only on η, viz. $u = u(\eta)$. For such a function, we have (primes denote differentiation with respect to η)

$$\frac{\partial u}{\partial t} = \beta x^\alpha t^{\beta-1} u',$$

$$\frac{\partial u}{\partial x} = \alpha x^{\alpha-1} t^{\beta} u',$$

and

$$\frac{\partial}{\partial x}\left(D(u)\frac{\partial u}{\partial x}\right) = \alpha(\alpha-1)x^{\alpha-2}t^{\beta}D(u)u' + \alpha^2 x^{2\alpha-2}t^{2\beta}\left[D(u)u'\right]'.$$

Substituting these results in Equation (7.34) and dividing by $x^{\alpha-2}t^{\beta}$, we obtain

$$\beta\frac{x^2}{t}u'(\eta) = \alpha(\alpha-1)D(u)u' + \alpha^2\eta\left[D(u)u'\right]'.$$

We see that if we choose $\alpha = 1$ and $\beta = -1/2$, i.e.,

$$\eta = \frac{x}{t^{1/2}}$$

then the explicit dependence of this equation on x and t separately disappears. The equation simplifies considerably and we have

$$\left[D(u)u'\right]' + \frac{1}{2}\eta u' = 0; \tag{7.35}$$

i.e., the partial differential equation has been reduced to an ODE. This transformation is due to Boltzmann.

7.1.3   Stephan Problem(s)

"Stephan problems" refer to a class of interface models between two phases of material (e.g., solid and liquid) where the boundary of the interface is moving in time.

For concreteness, we consider a solid and liquid in a slender pipe with a constant cross-section A which is perfectly insulated. The solid phase is on the right side and the liquid phase on the left. We denote the position of the interface between these phases by $p(t)$. Both the solid and liquid phases are assumed to satisfy the heat equation with proper coefficients

$$\frac{1}{k_S}\frac{\partial u_S}{\partial t} = \frac{\partial^2 u_S}{\partial x^2}, \quad \frac{1}{k_L}\frac{\partial u_L}{\partial t} = \frac{\partial^2 u_L}{\partial x^2} \tag{7.36}$$

where the subscripts S and L denote the solid and liquid phase respectively.

At the interface, we also have

$$u_S(p(t), t) = u_L(p(t), t);$$

i.e., the temperature varies continuously across the interface.

To model the movement of the interface between the two phases, we invoke the law of energy conservation. The heat flux across the boundary $p(t)$ is

$$Aq_S(p(t), t) - Aq_L(p(t), t).$$

This amount of heat is used to melt some of the solid in the tube. Denoting the latent heat of the solid phase by L, we have

$$Aq_S(p(t), t) - Aq_L(p(t), t) = L\rho A\dot{p}(t). \tag{7.37}$$

This is referred to as the "Stephan condition." Since Fourier law of heat conduction holds for the two phases, (7.37) becomes

$$-k_S \frac{\partial u_S(p(t), t)}{\partial x} + k_L \frac{\partial u_L(p(t), t)}{\partial x} = L\rho \dot{p}(t). \tag{7.38}$$

This equation is used to determine the position of the interface in time.

Exercises

1. Generalize the prototype model to a situation in which heat is being generated in the rod at a rate of $r(x, t)$ per unit volume.

   Hint. The rate at which heat is being generated in any infinitesimal slice $[x, x + \Delta x]$ is given by $r(x, t)A\Delta x$. Use this to modify Equation (1.6) or (1.8).

2. Generalize the prototype model to the case in which c, k, and ρ are functions of x (non-homogeneous rod).

3. Repeat Exercise 2 when c, k, and ρ are functions of the temperature u.

4. Generalize the prototype model to the case $A = A(x)$ where $A(x)$ is a function that varies slowly with x so we can still assume approximately that the temperature u is a function of x and t.

5. Newton's law of cooling states that for a non-insulated rod, the rate of the heat loss per unit length is proportional to $u - T_0$ where u is the rod

temperature and T_0 is the temperature of the surroundings. Show that the heat equation in one dimension now takes the form

$$c\rho\frac{\partial u}{\partial t} = k\frac{\partial^2 u}{\partial x^2} - a(u - T_0) \tag{7.39}$$

where a is constant.

Note. In this case we must assume that the rod is "thin" in order to be able to justify the approximation that temperature u is a function of x and t.

6. A rod is made of a material that undergoes a chemical reaction as a result of which the specific heat and conductivity change with respect to time [i.e., $c = c(t)$ and $k = k(t)$]. Derive the corresponding heat equation in one dimension.

7. Derive the heat conduction equation for a three-dimensional body by considering an infinitesimal volume.

8. If the temperature inside a homogeneous and isotropic sphere is given to be a function of the radial distance r only, show that the heat conduction equation is given by

$$\frac{1}{k}\frac{\partial u}{\partial t} = \left(\frac{\partial^2 u}{\partial r^2} + \frac{2}{r}\frac{\partial u}{\partial r}\right).$$

Hint. Use infinitesimal spherical shells.

9. Derive the heat conduction equation for an isotropic homogeneous and laterally insulated circular plate in which $u = u(r, t)$.

Hint. Consider infinitesimal rings.

10. Show that if $u(x, t)$ is a solution of the heat equation, then $\partial u/\partial x$ and $\partial u/\partial t$ are also solutions of this equation.

11. Show that Burger's equation

$$\frac{\partial u}{\partial t} + u\frac{\partial u}{\partial x} = \nu\frac{\partial^2 u}{\partial x^2}$$

can be transformed into the heat equation by introducing

$$w = -2\nu\frac{\partial}{\partial x}(\ln u).$$

7.2   MODELING WAVE PHENOMENA

Objective

Construct a prototype mathematical model for the transverse vibrations of a string with fixed ends (see Fig. 7.5).

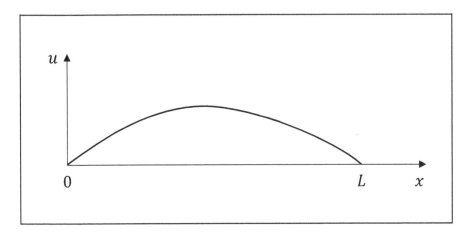

Figure 7.5   Flexed string of length L

Background

In general, an in-depth treatment of wave phenomena requires considering the elastic properties of matter and leads to a complicated set of equations. To overcome this difficulty, we make the following simplifying approximations and idealizations so that a prototype model can be constructed by applying only Newton's second law $\mathbf{F} = \mathbf{ma}$ (i.e., the force equals the mass multiplied by the acceleration) to the system under study.

Approximations and Idealizations

1. The string is rigidly attached at its end points.

2. The string vibrates in one plane.

3. No external forces act on the string (prototype model).

4. The string does not suffer from damping forces (prototype model).

5. The string is homogeneous. In particular, this implies that the linear density ρ (i.e., the mass per unit length) of the string is constant.

6. The deflection u of the string from equilibrium and its slope are always small. Consequently, we are able to assume a point on the string moves only in the vertical direction.

7. The tension force in the string is always tangential to it. This is usually expressed by saying that the string is assumed to be *perfectly flexible.*

Modeling

Consider a small segment of the string between x and $x + \Delta x$ as shown in Fig. 7.6. Before we can apply Newton's Second Law to the motion of this segment, we must make the following observations:

1. By approximation 6, the segment is not moving in the horizontal direction. Therefore if we denote by $\mathbf{T}(x, t)$ the tension in the string, the horizontal components of $\mathbf{T}(x, t)$ at x and $x + \Delta x$ must be equal (see Fig 7.6).

$$T(x) \cos \alpha = T(x + \Delta x) \cos \beta = R. \tag{7.40}$$

2. The mass of the segment $[x, x + \Delta x]$ is given by $\rho \Delta s$. Where Δs is the length of this segment in the deformed state of the string. However, since we are considering only small deflections, $|u| \ll 1$, it follows that $\Delta s \approx \Delta x$.

3. The acceleration of the segment in the vertical direction is given by $(\partial^2 u / \partial t^2)$.

4. The sum of the vertical forces acting on the segment (see Fig. 7.6)is

$$T(x + \Delta x) \sin \beta - T(x) \sin \alpha.$$

Combining all these observations and approximations, we infer from Newton's second law that

$$\rho \Delta x \frac{\partial^2 u}{\partial t^2} = T(x + \Delta x) \sin \beta - T(x) \sin \alpha. \tag{7.41}$$

Using (7.40) to elliminate T form this equation yields

$$\rho \Delta x \frac{\partial^2 u}{\partial t^2} = R(tan\beta - tan\alpha) \tag{7.42}$$

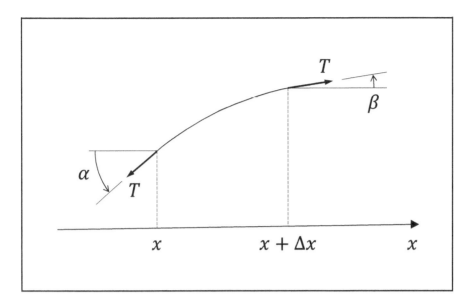

Figure 7.6 Balance of forces on a small section of the string

However since T is always tangential to the string

$$\tan \alpha = \frac{\partial u}{\partial x}(x, t) \tag{7.43}$$

$$\tan \beta = \frac{\partial u}{\partial x}(x + \Delta x, t). \tag{7.44}$$

Substituting Equations (7.43) and (7.44) in Equation (7.42) and dividing by Δx, we obtain

$$\rho \frac{\partial^2 u}{\partial t^2} = R \frac{\frac{\partial u}{\partial x}(x + \Delta x, t) - \frac{\partial u}{\partial x}(x, t)}{\Delta x}. \tag{7.45}$$

Letting $\Delta x \to 0$, we finally obtain

$$\frac{1}{c^2} \frac{\partial^2 u}{\partial t^2} = \frac{\partial^2 u}{\partial x^2} \tag{7.46}$$

where $c^2 = R/\rho$.

Equation (7.46) is called the **wave equation in one dimension**.

Remark 7.2.1 *From Fig. 7.6, we can infer that the sum of the horizontal forces acting on the string segment is $T(\cos \beta - \cos \alpha) \neq 0$. Hence, the segment must have an acceleration in the horizontal direction, which contradicts approximation 6b. However, since α and β are small, we can argue that $T(\cos \beta - \cos \alpha)$ is negligible.*

Compounding

Example 7.2.1 *Derive a model equation for very small vibrations of a vertically suspended chain whose length is L and whose mass density per unit length ρ is constant.*

Approximations

1. Since the amplitude u of the vibrations is small, we assume that a point on the chain does not change its x-coordinate (see Fig 7.7).

2. The tension $T(x, t)$ in the chain cannot be assumed to be constant in the context of this problem. In fact, in the equilibrium (vertical) position of the chain,

$$T(x) = \rho g(L - x). \tag{7.47}$$

   In the following model, we assume that Equation (7.47) gives an acceptable approximation for the tension in the vibrating chain when $|u| \ll 1$ and $\left|\frac{\partial u}{\partial t}\right| \ll 1$.

3. Other approximations and idealizations of the prototype model remain intact.

Modeling

To construct a mathematical model, we once again consider a small section of chain between $[x, x + \Delta x]$. Applying Newton's second law in the horizontal direction to such a section (see Fig. 7.8), we obtain

$$\rho \Delta x \frac{\partial^2 u}{\partial t^2} = T(x + \Delta x) \sin \beta - T(x) \sin \alpha. \tag{7.48}$$

But since α and β are small, we can once again use the approximations given by Equations (7.43) and (7.44). Hence,

$$\rho \frac{\partial^2 u}{\partial t^2} = \frac{1}{\Delta x} \left[T(x + \Delta x) \frac{\partial u}{\partial x}(x + \Delta x, t) - T(x) \frac{\partial u}{\partial x}(x, t) \right]. \tag{7.49}$$

Letting $\Delta x \to 0$, it follows then that

$$\rho \frac{\partial^2 u}{\partial t^2} = \frac{\partial}{\partial x} \left[T(x) \frac{\partial u}{\partial x}(x, t) \right]. \tag{7.50}$$

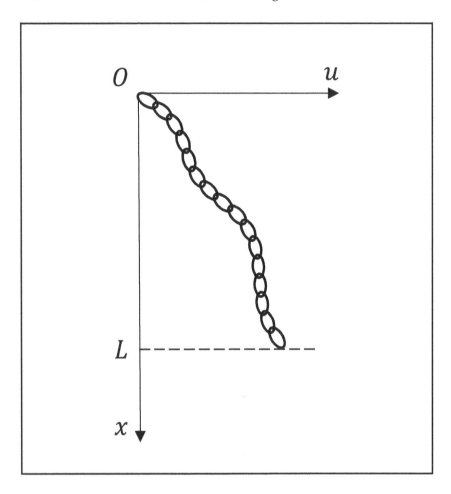

Figure 7.7 Vibrating chain

Substituting Equation (7.47) for $T(x)$, we finally obtain

$$\frac{\partial^2 u}{\partial t^2} = g \frac{\partial}{\partial x} \left[(L - x) \frac{\partial u}{\partial x} \right]. \tag{7.51}$$

Other examples of physical systems describing wave phenomena will be discussed in Sections 4 and 5.

7.2.1 Nonlinear Wave Equations

Nonlinear wave equations have many important applications. We give here a brief overview of this topic.

To begin with, we consider the nonlinear wave equation

$$u_{tt} + \frac{F''(u)}{F'(u)} u_t^2 = \lambda \left[u_{xx} + \frac{F''(u)}{F'(u)} u_x^2 \right]$$

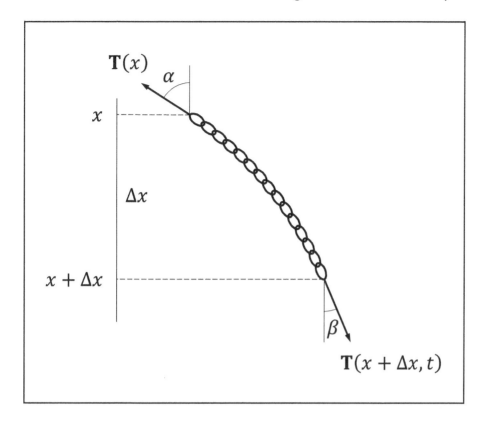

Figure 7.8 Balance of forces on a small section of the chain

where $F(u)$ is a smooth invertible function of u. This equation can be linearized by the transformation $\psi = F(u)$. In fact,

$$\psi_t = F'(u)u_t, \quad \psi_{tt} = F'(u)\left[u_{tt} + \frac{F''(u)}{F'(u)}u_t^2\right]$$

with similar expressions for the derivatives with respect to x. We infer then that

$$\psi_{tt} = \lambda\psi_{xx}.$$

An important wave equation that models long water waves of low amplitude is the Korteweg-de Vries (KdV for short) equation

$$u_t - 6uu_x + u_{xxx} = 0. \tag{7.52}$$

This equation admits "traveling wave solutions" of the the form $u = f(x - ct)$ (or $u = f(x + ct)$) in which case this equation reduces to

$$-cf' - 6ff' + f''' = 0 \tag{7.53}$$

where primes denote differentiation with respect to $z = x - ct$. This equation can be integrated once, and we obtain

$$-(c + 3f)f + f'' = C_1. \tag{7.54}$$

Multiplying this equation by f' and integrating again leads to

$$-(c + 2f)f^2 + (f')^2 = 2C_1 f + C_2 \tag{7.55}$$

where C_1, C_2 are integration constants. For the special case that $C_1 = C_2 = 0$ (which can be justified by imposing proper boundary conditions on f and f'), we obtain a simple first order equation whose solution is

$$f = -\frac{c}{2} sech^2 \left[\frac{\sqrt{c}}{2}(x - ct) \right].$$

Such a solution is called a soliton as it exhibits some special properties. One of these special properties is that when two solitons traveling in opposite directions collide, they come out of the collision with unchanged shape.

The KdV equation admits also n-soliton solutions. A two soliton solution is given by

$$f = -12 \frac{3 + 4cosh(2x - 8t) + cosh(4x - 64t)}{[3cosh(x - 28t) + cosh(3x - 36t)]^2}.$$

The Sine-Gordon (SG) equation is obtained from the linear wave equation by replacing the right hand side of this equation by some elementary function, e.g.,

$$u_{xx} - u_{tt} = \sin u.$$

This equation models superconducting transmission lines ("Josephson Junctions") and the propagation of crystal defects (and many other phenomena). The nonlinear term on the right hand side of this equation models the quantum tunneling effect of electron-pairs through the insulating material of a superconducting transmission line.

Looking again for traveling wave solutions, $u = f(x - ct)$ reduces the SG equation to

$$(1 - c^2)f'' = \sin f.$$

We can integrate this equation with respect to $z = x - ct$ by multiplying it by f' to obtain

$$(1 - c^2)(f')^2 = C_1 - 2\cos f.$$

For the special case $C_1 = 2$, we can solve this differential equation to obtain a soliton solution in the form

$$u(x,t) = 4 \arctan \left[exp \left(\pm \frac{x - ct}{\sqrt{(1 - c^2)}} \right) \right].$$

This solution with the plus (minus) sign is usually referred to as a "kink" (anti-kink) solution since u increases (decreases) monotonically by 2π as z varies from $-\infty$ to ∞.

Another wave equation which models water waves in deep ocean and the propagation of nonlinear pulses in fiber optics is the Nonlinear Schrodinger equation (NLS)

$$iu_t \pm u_{xx} + 2|u|^2 u = 0. \tag{7.56}$$

The equation with the '+' sign is referred to as the focusing NSL equation while the one with '−' sign is called the defocusing NLS equation.

To derive this equation, we consider a one dimensional superposition of waves (a "wave packet") in the form

$$f(x,t) = \frac{1}{2\pi} \int_{-\infty}^{\infty} F(k) e^{i(kx - \omega(k)t)} dk \tag{7.57}$$

where $F(k)$ is the Fourier transform of $f(x,0)$. Observe that the different waves in this superposition might have different speeds $c(k) = \omega(k)/k$. The relation $\omega = \omega(k)$ is referred to as the "dispersion relation for the wave" and $c(k)$ as the phase velocity, $v_p(k)$, of the individual waves in the packet.

We now consider such a superposition under the assumption that $F(k)$ is concentrated around a certain value of $k = k_0$ (that is $F(k) \approx 0$ outside a small interval around k_0). Such a superposition of waves is called a "wave packet." For such a packet, we can expand the dispersion relation around k_0

$$\omega = \omega_0 + a_1(k - k_0) + a_2(k - k_0)^2 + O[(k - k_0)^3].$$

Substituting this approximation in (7.57) we obtain

$$f(x,t) = e^{i(k_0 x - \omega_0 t)} \psi(x,t) \tag{7.58}$$

where

$$\psi(x,t) = \frac{1}{2\pi} \int_{-\infty}^{\infty} F(k) e^{i[(k-k_0)x - a_1(k-k_0)t - a_2(k-k_0)^2 t]} dk. \tag{7.59}$$

In Equation (7.58), $e^{i(k_0 x - \omega_0 t)}$ is called the "carrier wave" while $\psi(x, t)$ is the "envelope wave."

Letting $\nu = k - k_0$ and differentiating Equation (7.59) with respect to t, we obtain

$$a_2 \frac{\partial^2 \psi}{\partial x^2} + i \left(\frac{\partial \psi}{\partial t} + a_1 \frac{\partial \psi}{\partial x} \right) = 0. \tag{7.60}$$

By adding a "small nonlinear term $a|\psi|^2 \psi$" to this equation to account for the transmission medium nonlinearities, we obtain the "Nonlinear Schrodinger equation"

$$a_2 \frac{\partial^2 \psi}{\partial x^2} + i \left(\frac{\partial \psi}{\partial t} + a_1 \frac{\partial \psi}{\partial x} \right) + a|\psi|^2 \psi = 0.$$

This equation can be transferred to the standard form of the NLS equation (7.56) by a proper transformation of the variables. The NLS equation admits solitons as solutions.

7.2.2 Riemann Invariants

Consider the nonlinear wave equation with $c = c(\psi_x)$, i.e.,

$$c(\psi_x)^2 \frac{\partial^2 \psi}{\partial x^2} - \frac{\partial^2 \psi}{\partial t^2} = 0. \tag{7.61}$$

We can rewrite this equation as a system of two equations by introducing $\psi_t = u$ and $\psi_x = v$. In fact, since $u_x = \psi_{xt} = v_t$, we have

$$v_t - u_x = 0. \tag{7.62}$$

Furthermore, from Equation (7.61) we have

$$c(v)^2 v_x - u_t = 0. \tag{7.63}$$

Multiplying Equation (7.62) by $c(v)$ and adding to Equation (7.63) we have

$$c(v)[v_t + c(v)v_x] - [u_t + c(v)u_x] = 0. \tag{7.64}$$

Similarly, by multiplying Equation (7.62) by $c(v)$ and subtracting from Equation (7.63), we obtain

$$c(v)[v_t - c(v)v_x] - [u_t - c(v)u_x] = 0. \tag{7.65}$$

On the line where $c(v) = \frac{dx}{dt}$, Equation (7.64) becomes

$$c(v)[v_t + c(v)v_x] - [u_t + c(v)u_x] = c(v)\frac{dv}{dt} + \frac{du}{dt} = 0.$$

Letting $dG = c(v)dv$, we rewrite this equation as

$$\frac{d}{dt}[G(v) + u] = 0.$$

Similarly on the line $c(v) = -\frac{dx}{dt}$, Equation (7.65) becomes

$$c(v)[v_t - c(v)v_x] - [u_t - c(v)u_x] = c(v)\frac{dv}{dt} - \frac{du}{dt} = 0$$

which leads to

$$\frac{d}{dt}[-G(v) + u] = 0.$$

It follows then that the quantities

$$r = G(v) + u, \quad s = -G(v) + u$$

are conserved on the lines where $c(v) = \pm\frac{dx}{dt}$ respectively. These quantities are called the Riemann invariants of the Equation (7.61). The lines where $c(v) = \pm\frac{dx}{dt}$ are called the "characteristic curves" of Equation (7.61).

Exercises

1. Derive a model equation for the vibration of the string if a vertical external force $F(x, t)$ per unit length is acting on it. Especially consider the case of the gravitational force where $F = \rho g$.

2. Derive a model equation for the vibration of the string when its motion is subject to an elastic restraint and a damping force. (The restraint can be thought of as a force of ku per unit of length acting to return the string to its equilibrium position. The damping force is given by $b(\partial u/\partial t)$ per unit length acting to oppose its motion.)

3. Generalize the model equation for the vibration of the string to the case where $\rho = \rho(x)$.

4. **Longitudinal elastic waves:** Consider an elastic homogeneous rod with a constant cross section placed along the x-axis. If we apply longitudinal stresses to the rod, then a particle whose rest position is x will find itself at $x + u(x, t)$, where we are assuming that the displacement $u = u(x, t)$. It is known that the local stress, T (force/unit area), in an elastic bar satisfies $T = Eu_x$ where E is called the *elastic modulus*. Derive a model equation for $u(x, t)$ by considering a small section of the rod and the stress difference between its ends (see Fig. 7.9).

Hint. Let ρ be the mass density of the bar and A its cross section. The mass of the small section is $\rho A \Delta x$, and its acceleration is $\partial^2 u / \partial t^2$. The stress difference across the section is $EA[u_x(x + \Delta x, t) - u_x(x, t)]$. Note that $c = \sqrt{E/\rho}$ is called the *speed of sound in the medium*.

5. Derive a model equation for the vibrations of a membrane whose edges are fixed. Make similar assumptions as for the vibrating string.

Hint. Consider a small rectangular area and apply the same analysis as for the string in the x and y directions.

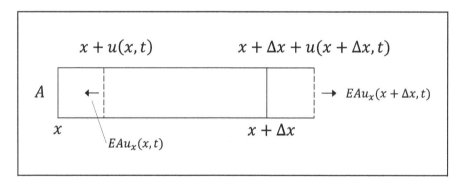

Figure 7.9 Acoustic vibrations in a metal rod

6. The pressure p and the mass flow rate u of a fluid flowing in a long pipe are related approximately by the equations

$$\frac{\partial p}{\partial t} = c\frac{\partial u}{\partial x}$$

$$\frac{\partial u}{\partial t} = \frac{\partial p}{\partial x}$$

where c is the compressibility of the fluid. Show that both p nd u satisfy the wave equation in one dimension.

7. Show that

$$u(x,t) = f(x - ct) + g(x + ct)$$

is a solution of the wave equation in one dimension when f and g are any "smooth functions."

8. Derive a model equation for the small vibrations of a string that is rotating around the x-axis at a constant angular velocity ω. Assume that at each moment all the points of the string are one plane.

9. A mass m is attached to the end of a suspended chain of length L and linear density ρ. Derive a model equation for the small vibrations of this system.

10. The results of Exercise 4 can be applied to the air vibrations in an organ pipe. However, since it is not "easy" to follow the position of air molecules, it is more natural to derive an equation for the pressure in the pipe. If it is known that the pressure is proportional to $\partial u/\partial x$, show that it must satisfy the wave equation.

7.3   SHALLOW WATER WAVES

Objective

Derive a prototype model equation that describes the phenomena of slow waves in shallow water (i.e., waves in a pool or on the seashore).

Background

An in-depth treatment of wave phenomena in fluids requires a knowledge of fluid mechanics. We simplify the derivation of this problem by applying Newton's second law and the following elementary facts.

1. The **hydrostatic pressure** p (force/unit area) in a fluid at a point of depth D below its surface is given by

$$p = \rho g D \tag{7.66}$$

where ρ is the mass density of the fluid and g is the acceleration of gravity.

2. The **principle of mass conservation** states that the rate of change of mass in a given volume equals the rate at which mass is entering the volume less the rate at which it is leaving it.

Approximations and Idealizations

1. Water is incompressible and hence has a constant density ρ independent of the pressure.

2. The motion of the fluid under consideration is two dimensional; that is, each fluid particle is constrained to move in two dimensions, x and y (see Fig. 7.10).

3. Because we are considering only slow waves, water acceleration is small and we can approximate the pressure at a point (x, y) beneath the water surface by the hydrostatic pressure

$$p(x, y, t) = \rho g[h(x, t) - y] \qquad (7.67)$$

where $h(x, t)$ is the wave height.

4. The force per unit volume in the fluid is given by the negative of the pressure gradient. In particular, the force per unit volume in the x direction is

$$F = -p_x = -\rho g h_x(x, t) \qquad (7.68)$$

where $p_x = \partial p / \partial x$, and so on.

5. Since the force F is independent of y, it is reasonable to assume that the x component of a particle velocity u is a function of x and t only; that is, $u = u(x, t)$. (We assume also that the y component of the velocity is 0.)

Modeling

Consider a small volume ΔV of water. Its mass is $\rho \Delta, V$ and its acceleration

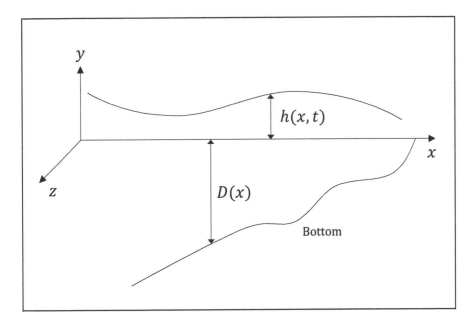

Figure 7.10 Waves over topography

is du/dt. The force acting on it in the x direction is $-\rho g h_x(x,t)\Delta V$. Hence, by Newton's second law,

$$\rho \Delta V \frac{du}{dt} = -\rho g h_x(x,t)\Delta V. \qquad (7.69)$$

(This is Lagrange's picture where the coordinate system is moving with the particle.) However, for an inertial observer (Euler picture) $u = u[x(t),t]$, therefore,

$$\frac{du}{dt} = \frac{\partial u}{\partial t}\frac{dt}{dt} + \frac{\partial u}{\partial x}\frac{dx}{dt} = \frac{\partial u}{\partial t} + \frac{\partial u}{\partial x}u. \qquad (7.70)$$

Thus, we infer from Equation (7.69) that

$$u_t + u_x u = -g h_x. \qquad (7.71)$$

Because Equation (7.71) contains two unknown quantities, we need another independent equation that relates h and u in order to be able to solve our model. Such an equation can be obtained by using the principle of mass conservation.

To apply this principle, consider the column of water between x and $x + \Delta x$ and the horizontal length Δz in the z direction (see Fig. 7.11).

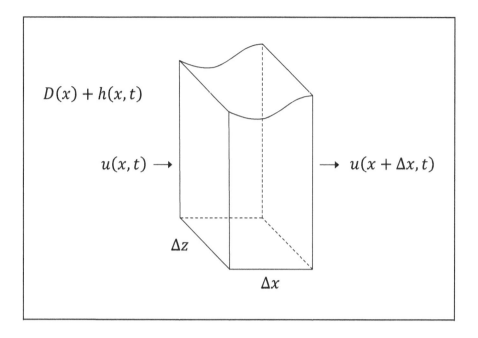

Figure 7.11 Fluid flux in a small column

The difference between the amounts of water in this volume at $t + \Delta t$ and t is

$$\Delta z \cdot \Delta x \cdot \rho[h(x, t + \Delta t) - h(x, t)]. \tag{7.72}$$

This quantity must equal the difference between the amount of water entering this volume in time Δt less the amount leaving it. Hence,

$$\Delta z \cdot \Delta t \cdot \rho\{[D(x) \quad + \quad h(x,t)]u(x,t) - [D(x + \Delta x) + h(x + \Delta x, t)]u(x + \Delta x, t)\}$$
$$= \quad \Delta z \cdot \Delta x \cdot \rho[h(x, t + \Delta t) - h(x, t)]. \tag{7.73}$$

Dividing Equation (7.73) by Δx and Δt and letting them approach zero, we obtain in the limit

$$h_t = -[u(h + D)]_x. \tag{7.74}$$

Equations (7.71) and (7.74) constitute a system of coupled partial differential equations describing shallow water waves.

We observe that Equation (7.74) is in conservative form; i.e., if we define

$$f_1 = h, \quad f_2 = u(h + D)$$

then

$$\frac{\partial f_1}{\partial t} + \frac{\partial f_2}{\partial x} = 0.$$

However, Equation (7.71) is not in this form. To re-express this equation in conservative form, we multiply Equation (7.71) by $(h+D)$ and Equation (7.74) by u and sum. Combining the derivatives and observing that $D = D(x)$, we obtain

$$[(h + D)u]_t + [(h + D)u^2]_x + g(h + D)h_x = 0.$$

Adding and subtracting $gh(h + D)_x$ in this equation we obtain

$$[(h + D)u]_t + [(h + D)u^2 + g(h + D)h]_x = gh(h + D)_x. \qquad (7.75)$$

Multiplying this equation by ρ, we can interpret it as a representation of the law of momentum conservation where the right side of the equation represents the impulse due to gravity and non-constant bottom. The terms in the expression $[(h + D)u^2 + g(h + D)h]$ represent the kinetic and potential energy of the fluid respectively.

7.3.1 Tsunamis

Waves on the oceans are generated by various causes. First, there are waves generated by winds blowing on the ocean surface. Then, there are waves due to tidal forces exerted by the gravitational interaction with the Sun and Moon systems. Finally, there are those that are generated as a result of earthquakes or other natural catastrophes. This last category of waves is referred to as tsunamis.

Tsunamis are characterized by long wavelength λ and period ω. Typically, the wavelength can range from few to hundreds of kilometers. Since the ocean depth D on Earth is less than $11 km$, it follows that for these waves we always have $\lambda/D \gg 1$. Accordingly, tsunamis can be treated as shallow water waves. As the wave approaches, land D decreases and λ/D becomes very large. Thus, decreasing depth leads to a higher concentration of the energy.

In the deep ocean, we can approximate the horizontal and vertical displacements of a particle at (x, y) due to the wave by

$$X = A\sin(kx + \omega(k)t), \quad h = -B\cos(kx + \omega(k)t)$$

where A is a constant and $B = Aky$. The factor ky has been added in this model of the tsunami wave for several reasons. To begin with, $k = 2\pi/\lambda$, and it follows that at the ocean surface the vertical amplitude of this wave is $2\pi AD/\lambda$. However, $D/\lambda \ll 1$, which implies that (as expected) the vertical amplitude of the wave is much smaller than the horizontal amplitude. In addition, the vertical displacement is 0 at the ocean bottom. Furthermore, this form of the wave satisfies the continuity equation $div\,\mathbf{u} = 0$, $\mathbf{u} = (u, v)$. In fact, the velocity components of the wave are

$$u = A\omega \cos(kx + \omega(k)t), \quad v = Aky\omega \sin(kx + \omega(k)t)$$

and therefore

$$u_x + v_y = 0.$$

Energy conservation implies then that $\omega^2 = (gD)k^2$. (Such a relation that expresses the frequency as a function of the wave number is called "a dispersion relation.") The group velocity is therefore

$$v_g = |\frac{d\omega}{dk}| = \sqrt{gD}.$$

Observe that for this dispersion relation, the phase velocity $v_p = |\frac{\omega}{k}|$ is equal also to $\sqrt{gD}$. From this expression, we infer that in the deep ocean with $D \approx 4km$, we have $v_g \approx 200m/sec$. Thus, a tsunami can propagate very fast in the open ocean. However, as the water depth decreases, the tsunami slows down. At the same time the tsunami's energy flux, which is the sum of its kinetic and potential energies, remains almost constant. Consequently, as the tsunami's speed diminishes as it travels into shallower water, its height grows. Because of this effect, a tsunami, imperceptible at sea, may grow to be several meters or more in height near the coast.

Thus, assuming no dissipation due to friction or turbulence, the tsunami energy flux is proportional to $B^2 v_g$ or $B^2\sqrt{D}$. It follows then that constant energy flow requires that the tsunami amplitude B is proportional to $D^{-1/4}$. Suppose now that at a depth of $4km$ the tsunami's vertical amplitude is $1m$ (a "normal wave"), then at depth of $2m$ (near the shore) the wave height will be approximately $6.7m$.

Exercises

1. If D is a constant and u, h, and their derivatives are small (so that nonlinear terms in these quantities can be neglected), show that $h(x, t)$ must satisfy the wave equation

$$h_{tt} = (gD)h_{xx}.$$

2. Under the same conditions as in the previous exercise, what is the equation that is satisfied by $u(x, t)$?

3. Repeat Exercises 1 and 2 when D is a linear function of x; that is,

$$D = ax + b \quad a > 0, b < 0.$$

4. Modify Equation (7.71) to include the action of a damping force (in the x direction) that is proportional to u.

7.4   UNIFORM TRANSMISSION LINE

Objective

Derive a prototype model equation for the voltage and current in a long, uniform, two-wire transmission line.

Background

(For a detailed introduction to electric circuits, see chapter 1.) The most common forms of transmission lines are coaxial and two-wire types. The coaxial transmission line consists of two concentric circular cylinders of metal. The two-wire types consist of two parallel wires, one of which is used as ground. We consider here the two-wire line.

The passage of an electric current through a cable always involves a leakage, which leads to a loss of electric energy. For short distances, this loss can usually be ignored. However, over long distances, which are found in transmission lines, these losses must be taken into account.

Modeling

Since the transmission line is uniform, we assume that the resistance (R), capacitance (C), inductance (L), and leakage (G) *per unit length* of the transmission line are constant.

To derive the required model equations, we consider a small section of the line between x and $x + \Delta x$ and apply Kirchoff's laws to a circuit that is equivalent to it (see Fig. 7.12). Thus, $R\Delta x, L\Delta x, C\Delta x$, and $G\Delta x$ are, respectively, the resistance, inductance, capacitance, and conductance of the section. The quantity $G\Delta x$ (where G is expressed in mhos or siemens) is a "virtual" resistance so that the power lost through it to the ground is equal to that due to leakage.

Applying Kirchoff's second law between A and D, (see Fig. 7.12) we obtain

$$e(x,t) - e(x + \Delta x, t) = R\Delta x i(x,t) + L\Delta x \frac{\partial i(x,t)}{\partial t}. \tag{7.76}$$

Similarly, applying Kirchoff's first law at the node B, we have

$$i(x,t) - i(x + \Delta x, t) = C\Delta x \frac{\partial e(x + \Delta x, t)}{\partial t} + G\Delta x e(x + \Delta x, t). \tag{7.77}$$

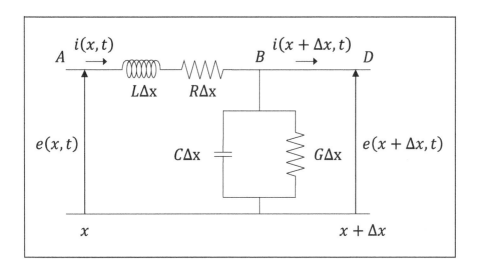

Figure 7.12 Equivalent circuit for a small section of the transmission line

Dividing these equations by Δx and letting Δx go to zero, we arrive at two differential equations:

$$\frac{\partial e}{\partial x}(x,t) = -Ri(x,t) - L\frac{\partial i(x,t)}{\partial t} \tag{7.78}$$

$$\frac{\partial i}{\partial x}(x,t) = -Ge(x,t) - C\frac{\partial e(x,t)}{\partial t}. \tag{7.79}$$

Finally, we can find an equation for $e(x,t)$ only by differentiating Equations (7.78) and (7.79) with respect to x and t, respectively, to eliminate $i(x,t)$. We obtain

$$e_{xx} = LCe_{tt} + (LG + RC)e_t + RGe. \tag{7.80}$$

Similarly, we can show that $i(x,t)$ satisfies

$$i_{xx} = LCi_{tt} + (LG + RC)i_t + RGi. \tag{7.81}$$

Equations (7.80) and (7.81) are known as the **telegraph equations**.

Special Cases

High-Frequency Limit. To qualitatively analyze this limit, consider the case where

$$e(x,t) = A(x)\sin(\omega t + \phi_1), \quad i(x,t) = B(x)\sin(\omega t + \phi_2) \tag{7.82}$$

and $\omega \gg 1$. Under these assumptions, the second term in the right-hand side of Equation (7.78), whose "effective coefficient" is $L\omega$ (the impedance), is much larger than the first term, whose effective coefficient is R. Hence, Equation (7.78) can be approximated in this limit by

$$e_x = -Li_t. \tag{7.83}$$

Similarly, Equation (7.79) reduces to

$$i_x = Ce_t. \tag{7.84}$$

Combining Equations (7.83) and (7.84) we obtain the wave equation

$$e_{xx} = LCe_{tt}, \quad i_{xx} = LCi_{tt}. \tag{7.85}$$

Low-Frequency Limit. In this case, i and e change very slowly with time. Therefore, from a similar qualitative analysis as performed for the high frequency limit, with $\omega \ll 1$, we obtain that the effective coefficient of the second term $(L\omega)$ is much smaller than R. Consequently, we can approximate Equations (7.78) and (7.79) by

$$e_x = -Ri, \quad i_x = -Ge \tag{7.86}$$

and therefore,

$$e_{xx} = RGe, \quad i_{xx} = RGi. \tag{7.87}$$

These are ordinary differential equations for i and e.

Submarine Cable. In the 19th century, telecommunication signals between the United States and Europe were transmitted by cables that were laid down on the ocean floor. For these cables, $G \cong 0$ because of their extreme insulation. Moreover, the signal frequency ω is low. Under these circumstances, we infer from equation (7.82) that the impedance $L\omega$ is much smaller than R. Hence, equation (7.78) can be approximated by

$$e_x = -Ri. \tag{7.88}$$

Furthermore, since we have assumed $G = 0$, Equation (7.79) simplifies to

$$i_x = -C\frac{\partial e}{\partial t}. \tag{7.89}$$

Combining these two equations, we can approximate Equations (7.78) and (7.79) by

$$e_{xx} = RCe_t, \quad i_{xx} = RCi_t \tag{7.90}$$

which show that e and i satisfy the one-dimensional diffusion equation.

Exercises

1. Give explicit derivations of equations (7.80) and (7.81).

2. Compare equations (7.80) and (7.81) with the model equation for a spring mass system with friction and identify the physical meaning of each term in these equations.

3. Explain in detail the approximations that lead to equations (7.83), (7.84), (7.86), (7.89), and (7.90).

4. Derive a differential equation for the voltage $e(t)$ in the circuit in Fig. 7.13 if the current $i(t)$ is known.

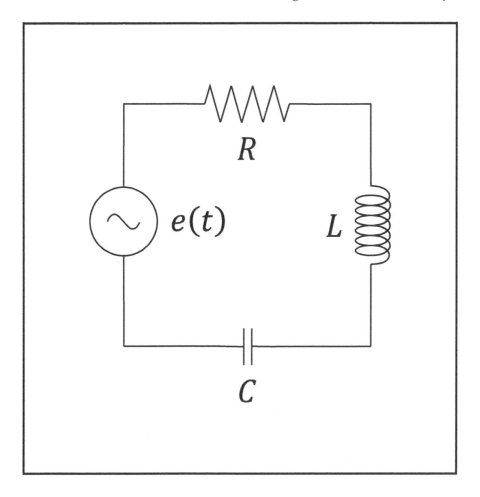

Figure 7.13 LRC circuit

7.5 THE POTENTIAL (OR LAPLACE) EQUATION

Objective

Derive model equations to compute the gravitational field of a material body.

Remark 7.5.1 *Although we restrict our discussion to the gravitational field, the static electric field would warrant similar treatment.*

Background

Newton's law of gravitation states that a point mass M attracts another point mass m by a force

$$\mathbf{F} = -\frac{GMm}{r^2}\mathbf{e}_r \tag{7.91}$$

where G is the gravitational constant, r is the distance between the two masses, and $\mathbf{e}_r$ is a unit vector along r (pointing outward from M, i.e., in the direction in which r increases) (see Fig. 7.14).

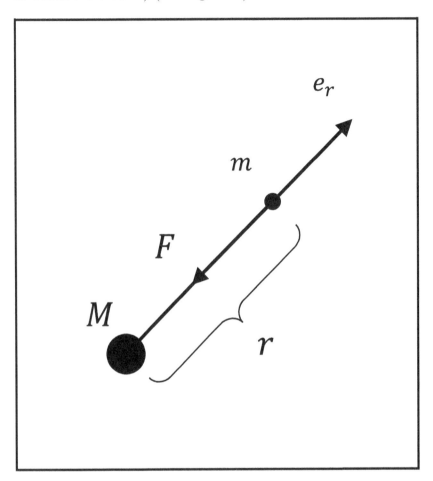

Figure 7.14 Gravitational force exerted by mass M on mass m

Since Equation (7.91) can be rewritten as

$$\mathbf{F} = \left(-\frac{GM}{r^2}\mathbf{e}_r \right) m, \tag{7.92}$$

we introduce the gravitational field generated by the mass M as

$$\mathbf{F} = -\frac{GM}{r^2}\mathbf{e}_r. \tag{7.93}$$

Thus, the gravitational field is the force that acts on a test particle of unit mass at a point due to the presence of the mass M.

The gravitational field admits a potential; i.e., there exists a scalar function Φ so that

$$\mathbf{F} = -\nabla\Phi. \tag{7.94}$$

The minus sign in this equation conforms to the convention that the force exerted by the field on a point particle is always in the direction in which the potential (energy) is decreasing most rapidly.

For the gravitational field of a point mass M, the potential function Φ is

$$\Phi(r) = -\frac{GM}{r}. \tag{7.95}$$

Remark 7.5.2 *Following the standard engineering convention, (see e.g. Standard Mathematical Tables published by CRC), spherical coordinates (see Fig. 7.15) are defined as*

$$x = r\sin\phi\cos\theta \tag{7.96}$$

$$y = r\sin\phi\sin\theta \tag{7.97}$$

$$z = r\cos\phi \tag{7.98}$$

and the expression of the gradient operator is

$$\nabla f = \frac{\partial f}{\partial r}\mathbf{e}_r + \frac{1}{r}\frac{\partial f}{\partial\phi}\mathbf{e}_\phi + \frac{1}{r\sin\phi}\frac{\partial f}{\partial\theta}\mathbf{e}_\theta. \tag{7.99}$$

In our case, Φ is a function of r only.

The superposition principle states that the gravitational field at a point due to two point masses M_1 and M_2 is equal to the (vector) sum of their gravitational fields, which can be written as

$$\mathbf{F}_{total} = \mathbf{F}_1 + \mathbf{F}_2 = -\frac{GM_1}{r_1^2}\mathbf{e}_{r_1} - \frac{GM_2}{r_2^2}\mathbf{e}_{r_2} \tag{7.100}$$

where r_1, r_2 are the distances from M_1 and M_2 to the point under consideration (see Fig. 7.16).

As a corollary, we observe that if Φ_1 and Φ_2 are the potential functions for the gravitational field of the masses M_1 and M_2, respectively, then

$$\mathbf{F}_{total} = -\nabla\Phi_1 - \nabla\Phi_2 = -\nabla(\Phi_1 + \Phi_2) \tag{7.101}$$

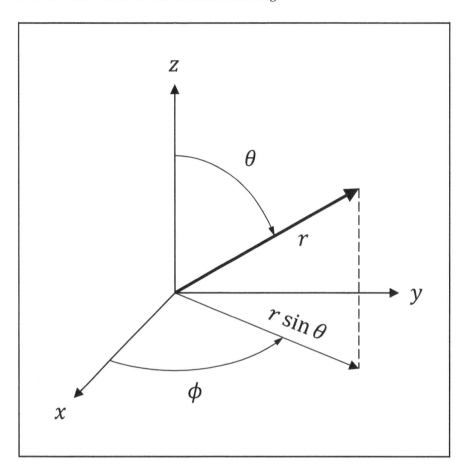

Figure 7.15 Spherical coordinate system

where

$$\Phi_i = -\frac{GM_i}{r_i} \quad i = 1, 2. \tag{7.102}$$

Thus, the potential function of the total gravitational field is given by the (scalar) sum of the individual potential functions.

Similarly, if we are given a finite number of point masses, $M_i, i = 1, ..., n$ with gravitational fields $\mathbf{F}_i$ and potential functions Φ_i, then

$$\mathbf{F}_{total} = \sum_{i=1}^{n} \mathbf{F}_i = -\sum_{i=1}^{n} \nabla\Phi_i = -\nabla\left(\sum_{i=1}^{n} \Phi_i\right). \tag{7.103}$$

Remark 7.5.3 *If the total potential function Φ of a gravitational field is known, then the gravitational field itself is given as $\mathbf{F} = \nabla\Phi$. This is one of the reasons for the introduction of the potential function, since Φ is simply*

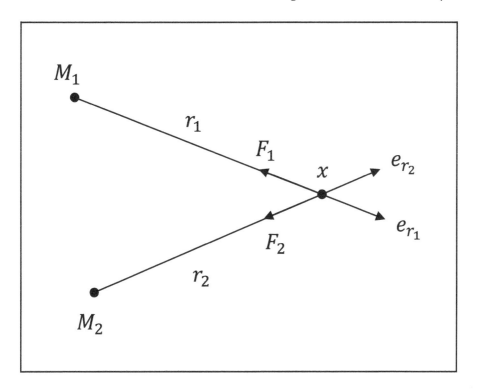

Figure 7.16 Gravitational force due to M_1, M_2 at $\mathbf{x}$

a (scalar) sum of the individual potentials and hence easier to compute than the (vector) sum of the gravitational fields.

Idealizations

1. We assume that the concept of a point mass is valid. As a matter of fact, we note that due to the discrete nature of matter, the notion of a (mathematical) point particle with mass m has no physical meaning.

2. We assume that the field generated by a point particle does not act on itself; otherwise, various contradictions will creep in.

Modeling

To compute the gravitational field due to a continuous mass distribution with mass density $\rho(\mathbf{x})$ in a volume V, we divide the volume into small cells of volume ΔV_i. If we consider each of these cells as a point particle of mass $\rho(\mathbf{x}'_i)\Delta V_i$ (where $\mathbf{x}'_i \epsilon \Delta V_i$), then the gravitational field due to it at a point

$\mathbf{x} = (x, y, z)$ is

$$\Delta \mathbf{F}_i = \frac{-G\rho(\mathbf{x}_i')\Delta V_i}{r_i^2}\mathbf{e}_{r_i} \tag{7.104}$$

where

$$r_i = |\mathbf{x} - \mathbf{x}_i'| = \sqrt{(x - x_i')^2 + (y - y_i')^2 + (z - z_i')^2} \tag{7.105}$$

(see Fig. 7.17).

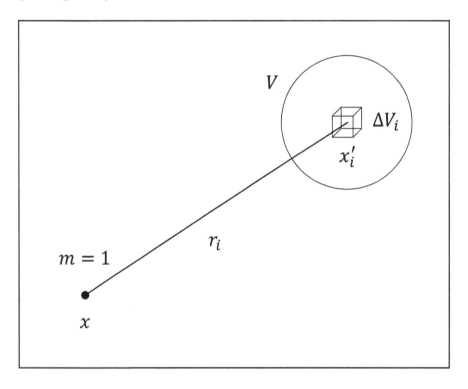

Figure 7.17 Gravitational field at x due to a small volume of the body

The gravitational potential associated with this field is

$$\Delta \Phi_i = -\frac{G\rho(\mathbf{x}_i')\Delta V_i}{r_i}. \tag{7.106}$$

Hence, the potential due to the whole mass is given approximately by

$$\Phi = -\sum_i \Delta \Phi_i = -\sum_i \frac{G\rho(\mathbf{x}_i')\Delta V_i}{r_i}. \tag{7.107}$$

By letting $\Delta V_i \to 0$, the sum in Equation (7.107) will be replaced by the volume integral over V

$$
\begin{aligned}
\Phi(x, y, z) &= -\int_V \frac{G\rho(\mathbf{x}')dV}{r} \\
&= -\int_V \frac{G\rho(x', y', z')dx'dy'dz'}{\sqrt{(x - x')^2 + (y - y')^2 + (z - z')^2}}.
\end{aligned}
\tag{7.108}
$$

Notice that the integral is over the volume of the body whose coordinates are denoted by x', y', and z', whereas the coordinates of the point where the potential is being computed are denoted by x, y, and z.

To find the differential equation satisfied by $\Phi(x, y, z)$, we compute

$$
\nabla\Phi = \left(\frac{\partial\Phi}{\partial x}, \frac{\partial\Phi}{\partial y}, \frac{\partial\Phi}{\partial z} \right)
\tag{7.109}
$$

and $\nabla \cdot (\nabla\Phi) = \nabla^2\Phi$ to show that

$$
\nabla^2\Phi = \frac{\partial^2\Phi}{\partial x^2} + \frac{\partial^2\Phi}{\partial y^2} + \frac{\partial^2\Phi}{\partial z^2} = 0.
\tag{7.110}
$$

This equation is called the **potential or Laplace equation** (in three dimensions).

Compounding

In various applications, we must consider other equations that are closely related to the Laplace equation in three dimensions. Three examples of such equations follow:

1. The Laplace equation in n dimensions is

$$
\nabla^2 u = \sum_{i=1}^{n} \frac{\partial^2 u}{\partial x_i^2} = 0.
\tag{7.111}
$$

2. The Poisson equation is given by

$$
\nabla^2 u = f(\mathbf{x}).
\tag{7.112}
$$

It can be shown that the gravitational field inside a body satisfies such an equation with $f(\mathbf{x}) = 4\pi\rho(\mathbf{x})$. (See Exercise 6.)

3. The Helmholtz equation is

$$\nabla^2 u + k^2 u = 0 \tag{7.113}$$

where k is a constant.

Example 7.5.1 *Compute the gravitational potential of a solid sphere of radius a and a constant mass density ρ.*

Solution 7.5.1 *As noted earlier, in all such problems we first compute the gravitational potential Φ and then evaluate the gravitational field F as $\Delta\Phi$.*

In this problem, because the sphere is isotropic, the gravitational potential must be the same for all points whose distance from the center of the sphere is the same. It follows that if we let the center of the sphere coincide with the origin of the coordinate system, then all points with $r = $ constant must have the same potential. It is enough, therefore, to evaluate Φ at $(0, 0, z)$ with $0 < a < z$. To do so, we infer from Equation (7.108) that

$$\Phi(0,0,z) = -G\rho \int_V \frac{dx'dy'dz'}{\sqrt{(x')^2 + (y')^2 + (z - z')^2}}. \tag{7.114}$$

To compute this integral, we introduce spherical coordinates on x', y' and z'. This transformation is given by

$$
\begin{aligned}
x' &= r \sin\phi \cos\theta \\
y' &= r \sin\phi \sin\theta \\
z' &= r \cos\phi
\end{aligned}
\tag{7.115}
$$

and

$$dV = r^2 \sin\phi \, d\phi dr. \tag{7.116}$$

If we write the volume integral for $\Phi(0, 0, z)$ as an iterated integral, we see that

$$
\begin{aligned}
\Phi(0,0,z) &= -G\rho \int_0^a \int_0^\pi \int_0^{2\pi} \frac{r^2 \sin\phi \, d\theta d\phi dr}{\sqrt{r^2 + z^2 - 2rz\cos\phi}} \\
&= -2\pi G\rho \int_0^a \int_0^\pi \frac{r^2 \sin\phi \, d\phi dr}{\sqrt{r^2 + z^2 - 2rz\cos\phi}}.
\end{aligned}
\tag{7.117}
$$

To compute the integral over ϕ, we introduce the substitution

$$w = r^2 + z^2 - 2rz\cos\phi. \tag{7.118}$$

Remembering that z is a constant in this computation, we obtain

$$
\begin{aligned}
\Phi(0,0,z) &= -\frac{\pi G\rho}{z} \int_0^a r\,dr \int_{(r-z)^2}^{(r+z)^2} w^{-1/2}dw \\
&= -\frac{2\pi G\rho}{z} \int_0^a r\,dr \cdot w^{1/2} \Big|_{(r-z)^2}^{(r+z)^2}.
\end{aligned} \tag{7.119}
$$

But

$$w^{1/2}\Big|_{(r-z)^2}^{(r+z)^2} = (r+z) - |r-z| = 2r \tag{7.120}$$

since $r \le a \le z$, and it follows that

$$\Phi(0,0,z) = -\frac{4\pi a^3 \rho G}{3z} = -\frac{GM}{z} \tag{7.121}$$

where M is the total mass of the body. For a general point whose distance from the origin is R, we obtain (due to symmetry)

$$\Phi(R) = -\frac{GM}{R}. \tag{7.122}$$

Equation (7.122) implies that the potential of a solid sphere of mass M is equivalent to that of a point particle with the same mass situated at its center.

7.5.1 Kirchoff Transformation

In many practical situations, one is called to consider a nonlinear version of Laplace equation in the form

$$\nabla \cdot (D(\phi)\nabla\phi) = 0. \tag{7.123}$$

We can linearize this equation by introducing

$$\psi = \int_a^\phi D(t)dt \tag{7.124}$$

where a is an arbitrary constant. (This is called Kirchoff transformation.) We then have

$$\frac{d\psi}{d\phi} = D(\phi). \tag{7.125}$$

By the chain rule we obtain

$$\nabla\psi = \frac{d\psi}{d\phi}\nabla\phi = D(\phi)\nabla\phi. \tag{7.126}$$

Hence, from Equation (7.123) we infer that

$$\nabla^2\psi = \nabla \cdot (\nabla\psi) = 0; \tag{7.127}$$

i.e., Equation (7.123) has been reduced to Laplace equation. We can recover ϕ from ψ using Equation (7.125).

Exercises

1. Compute the gravitational field of a spherical shell with constant density ρ, inner radius R_1, and outer radius R_2.

   Hint. Consider two cases: one for a point outside the shell and another for a point inside the cavity.

2. Derive an expression for the gravitational potential of a thin metal wire bent to form a circle of radius a if its mass density per unit length is ρ.

3. Derive a model equation for the gravitational field generated by a flat circular ring $a < r < b$ whose mass density per unit area ρ is constant. Consider points in the ring plane only.

4. Compute the expression for the two-dimensional Laplace equation in polar coordinates.

5. Compute the expression for the Laplace equation in three dimensions in cylindrical and spherical coordinates.

6. Derive the Poisson equation as follows:

   (a) Show that the gravitational field for a point of mass m satisfies

   $$\int_S \mathbf{F} \cdot d\mathbf{S} = 4\pi m$$

   where $\mathbf{S}$ is an (arbitrary) sphere around m.

(b) Then deduce that the gravitational field of a continuous mass distribution satisfies

$$\int_S \mathbf{F} \cdot d\mathbf{S} = 4\pi \int_V \rho dV.$$

(c) Use the fact that $\mathbf{F} = \nabla \phi$ and the divergence theorem to obtain the Poisson equation.

7. Show that $\ln(x^2 + y^2)$ and $1/\sqrt{x^2 + y^2 + z^2}$ are solutions of the Laplace equation in two and three dimensions, respectively.

7.6   THE CONTINUITY EQUATION

Objective

Derive a model equation for the traffic flow on a highway without exits and with one entrance and one lane (see Fig. 7.18).

Discussion

One possible approach to modeling the traffic flow is to describe each car as a finite element on the highway and then write a *discrete* model, which describes the motion of each such car. However, if there are many cars on the highway, this approach is not practical, and it is better to construct a *continuous* model, which treats these cars as "smeared out" quantities. We construct such a continuous model in this section.

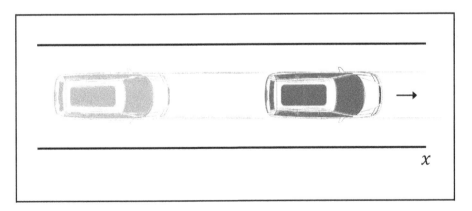

Figure 7.18   Cars on a one lane highway

Approximation and Idealization

1. We assume that the highway is infinite.

2. We define the car density $\rho(x, t)$ as

$$\rho(x, t) \cong \frac{\text{Number of cars on the interval } [x, x + \Delta x] \text{ at time } t}{\Delta x}$$

(7.128)

where Δx must be large compared to a car length. [Otherwise, $\rho(x, t) = 1$ if there is a car at x in time t, or $\rho(x, t) = 0$ if there is none.] In fact, the same approximation is made whenever we define the mass density of a "continuous" body made of discrete atoms and molecules. Hence, the equations we derive in this section apply also to fluid flow in a pipe.

3. We assume that there are no accidents on the highway (or that their number is negligible). Hence, we can formulate the principle of "car conservation" (which is equivalent to that of mass conservation) as follows:

   The rate at which the number of cars on the segment [a,b] is changing equals the rate at which they enter less the rate at which they are leaving.

4. We define the concept of car flux $q(x, t)$ in the same way that we defined this concept in Section 2. However, in this context it is natural to define the flux per lane rather than per unit area. Equivalently, this can be considered as letting the unit length be equal to the width of the lane. Moreover, note that

$$q(x, t) = \rho(x, t)u(x, t)$$

(7.129)

where $u(x, t)$ is the car's speed at x at time t, and the dimension of q is

$$q = \frac{\text{Cars}}{\text{Time}}.$$

(7.130)

Modeling

To derive a model equation for the traffic flow, consider a finite section of the road between a and b. The number n of cars in this segment at time t is

$$n(t, a, b) = \int_a^b \rho(x, t)dx.$$

(7.131)

Hence, the rate of change in this quantity is

$$\frac{dn}{dt} = \int_a^b \frac{\partial \rho(x, t)}{\partial t}dx.$$

(7.132)

This rate of change must equal the flux of cars entering at a less the flux of cars leaving at b. (Remember that the flux was defined per lane!) Therefore,

$$\frac{dn}{dt} = q(a,t) - q(b,t) = -\int_a^b \frac{\partial q}{\partial x}(x,t)dx. \tag{7.133}$$

Thus, we infer from Equations (7.132) and (7.133) that

$$\int_a^b \left(\frac{\partial \rho}{\partial t} + \frac{\partial q}{\partial x} \right) dx = 0. \tag{7.134}$$

And since a and b, are arbitrary, it follows that

$$\frac{\partial \rho}{\partial t} + \frac{\partial q}{\partial x} = 0. \tag{7.135}$$

Using equation (7.129) to substitute for $q(x,t)$, we finally obtain

$$\frac{\partial \rho}{\partial t} + \frac{\partial (\rho u)}{\partial x} = 0. \tag{7.136}$$

This equation is the **continuity equation in one dimension**. Notice, however, that this equation contains two unknown quantities, ρ and u. Therefore, to solve it we must either be able to express $u = u(\rho)$ or find an additional equation that relates these two quantities.

Compounding

To derive the version of the continuity equation in three dimensions, we consider a fluid flow with mass density $\rho(\mathbf{x}, t)$. Let V be a volume contained in the flow. The mass of the fluid in V at time t is given by

$$m(t, V) = \int_V \rho(\mathbf{x}, t)d\mathbf{x}. \tag{7.137}$$

Hence, the rate of change of mass in V is

$$\frac{dm}{dt} = \int_V \frac{\partial \rho}{\partial t}(\mathbf{x}, t)d\mathbf{x}. \tag{7.138}$$

Now let S denote the boundary of V and $\mathbf{n}(\mathbf{x})$ the unit *outward* normal to S at $\mathbf{x}$. The total mass flow rate of the fluid across S in the outward direction is

$$\int_S \mathbf{q} \cdot \mathbf{n}dS = \int_S (\rho\mathbf{u} \cdot \mathbf{n})dS. \tag{7.139}$$

The mass conservation principle implies, however, that the rate of change of mass in V must equal the rate at which the mass is crossing S in the *inward* direction. Therefore,

$$\int_V \frac{\partial \rho}{\partial t}(\mathbf{x}, t) dV = - \int_S \rho \mathbf{u} \cdot \mathbf{n} \, dS. \tag{7.140}$$

To convert the right-hand side of Equation (7.140) into a volume integral, we now invoke the divergence theorem, which states that for any smooth vector field $\mathbf{F}$ in V

$$\int_S \mathbf{F} \cdot \mathbf{n} \, dS = \int_V \operatorname{div} \mathbf{F} \, dV. \tag{7.141}$$

This yields

$$\int_V \left[\frac{\partial \rho}{\partial t} + \operatorname{div}(\rho \mathbf{u}) \right] dV = 0. \tag{7.142}$$

Since V is arbitrary, we infer that the integrand in Equation (7.142) must be zero, or

$$\frac{\partial \rho}{\partial t} + \operatorname{div}(\rho \mathbf{u}) = 0. \tag{7.143}$$

Equation (7.143), which is a first-order partial differential equation, is called the **continuity equation in three dimensions.**

Exercises

1. From your experience, guess the general form of the relationship between u and ρ in a one-lane highway.

2. Derive a model equation for an infinite one-lane highway where cars are entering and leaving the highway at constant rates α and β per mile, respectively. Generalize to the case where $\alpha = \alpha(x, t)$ and $\beta = \beta(x, t)$.

3. Compound the model in Exercise 2 to include accidents at a rate $\alpha(x, t)$ per mile.

4. Consider fluid flow in a long cylindrical pipe with constant cross section A whose axis is along the x-axis. Let $\rho(x, t)$ be the density of the fluid and $q(x, t)$ be its flux. If the walls of the pipe are made of porous material that allows the fluid to leak out at a rate L per unit length, show that

$$\frac{\partial \rho}{\partial t} + \frac{\partial q}{\partial x} + \frac{L}{A} = 0.$$

5. Derive model equations for the car densities $\rho_1(x,t)$ and $\rho_2(x,t)$ in a two-lane infinite highway with no entries or exits where cars are moving from lane 1 to lane 2 at a rate of $a(\rho_2)$ per mile and at a rate of $b(\rho_1)$ per mile from lane 2 to lane 1.

6. Compound the model of Exercise 5 to include entries and exits.

7.7 ELECTROMAGNETISM

7.7.1 Maxwell Equations

Maxwell equations form the core of all electromagnetic theory and applications. These equations (in the rationalized unit system) are:

$$\nabla \cdot \mathbf{D} = \rho \tag{7.144}$$

$$\nabla \cdot \mathbf{B} = 0 \tag{7.145}$$

$$\nabla \times \mathbf{E} = -\frac{\partial \mathbf{B}}{\partial t} \tag{7.146}$$

$$\nabla \times \mathbf{H} = \mathbf{J} + \frac{\partial \mathbf{D}}{\partial t} \tag{7.147}$$

Here

$\mathbf{E}$ - electric field
$\mathbf{B}$ - magnetic field
$\mathbf{D}$ - electric displacement
$\mathbf{H}$ - magnetic intensity
$\mathbf{J}$ - current density
ρ - charge density.

To close these equations we need relations, $\mathbf{D} = \mathbf{D}(\mathbf{E}), \mathbf{H} = \mathbf{H}(\mathbf{B})$, and $\mathbf{J} = \mathbf{J}(\mathbf{E})$. In vacuum or in homogeneous isotropic medium, we have the linear relations

$$\mathbf{D} = \epsilon \mathbf{E}, \quad \mathbf{H} = \frac{\mathbf{B}}{\mu} \tag{7.148}$$

where ϵ and μ are called the permittivity and permeability of the medium. Furthermore, in a homogeneous conducting medium

$$\mathbf{J} = \sigma \mathbf{E} \tag{7.149}$$

where σ is the medium conductivity. In the following, we assume the relations, Equations (10.93) and (10.94) implicitly.

By taking the divergence of Equation (10.92) and using Equation (10.89), we obtain the (electric charge) continuity equation

$$\frac{\partial \rho}{\partial t} + \nabla \cdot \mathbf{J} = 0. \tag{7.150}$$

Remark 7.7.1 *If a point charge ρ is moving with velocity* $\mathbf{v}$, *then the induced current density is* $\mathbf{J} = \rho \mathbf{v}$.

Remark 7.7.2 *The total force exerted by an electromagnetic field on charges and currents in a volume V is given by the Lorentz force equation*

$$\mathbf{F} = \int_V (\rho \mathbf{E} + \mathbf{J} \times \mathbf{B}) dV. \tag{7.151}$$

Remark 7.7.3 *To treat the special electromagnetic fields discussed in the rest of this section, we need the following facts:*

a.

$$\nabla \times (\nabla \times \mathbf{A}) = \nabla(\nabla \cdot \mathbf{A}) - \nabla^2 \mathbf{A} \tag{7.152}$$

b. If $\nabla \times \mathbf{A} = 0$, then $\mathbf{A}$ is conservative and therefore there exists a scalar function ϕ so that

$$\mathbf{A} = \nabla \phi. \tag{7.153}$$

c. If $\nabla \cdot \mathbf{A} = 0$, then $\mathbf{A}$ is solenoidal and, therefore, there exists a vector potential $\mathbf{B}$ so that

$$\mathbf{A} = \nabla \times \mathbf{B}. \tag{7.154}$$

Furthermore, $\mathbf{B}$ can be chosen so that $\nabla \cdot \mathbf{B} = 0$

7.7.2 Electrostatic Fields

When ρ and J are time independent, it follows from Equations (10.89)-(10.92) that the electric and magnetic fields are also time independent. Furthermore, Equations (10.89) and (10.91) decouple from Equations (10.90) and (10.92) and the equations for the electric and magnetic fields can be solved separately.

For the electric field we have

$$\nabla \times \mathbf{E} = 0 \tag{7.155}$$

$$\nabla \cdot \mathbf{E} = \frac{\rho}{\epsilon}. \tag{7.156}$$

From Equation (9.149), it follows that $\mathbf{E}$ is conservative and, hence, there exists a potential function ϕ s.t.

$$\mathbf{E} = -\nabla \phi. \tag{7.157}$$

From Equation (9.150), we obtain

$$\nabla^2 \phi = \frac{-\rho}{\epsilon}. \tag{7.158}$$

This is a Poisson equation whose solution can be obtained in integral form as (see Sec 3)

$$\phi(\mathbf{r}) = \frac{1}{4\pi\epsilon} \int_V \frac{\rho(r')}{|\mathbf{r} - \mathbf{r'}|} d\mathbf{r'}. \tag{7.159}$$

The electric field can be obtained by taking the gradient of this equation

$$\mathbf{E}(\mathbf{r}) = \frac{1}{4\pi\epsilon} \int_V \frac{\rho(r')(\mathbf{r} - \mathbf{r'})}{|\mathbf{r} - \mathbf{r'}|^3} d\mathbf{r'}. \tag{7.160}$$

7.7.3 Multipole Expansion

When $|\mathbf{r} - \mathbf{r'}| >> 1$, we can obtain an approximation to the integral in Equation (9.154) using the expansion

$$\frac{1}{|\mathbf{r} - \mathbf{r'}|^3} = \frac{1}{[r^2 - 2\mathbf{r} \cdot \mathbf{r'} + (r')^2]^{3/2}} = \frac{1}{r^3}\left[1 + \frac{3\mathbf{r} \cdot \mathbf{r'}}{r^2} + \cdots\right]. \tag{7.161}$$

Substituting this expansion in Equation (9.154), we obtain

$$\mathbf{E} = \frac{Q\mathbf{r}}{4\pi\epsilon r^3} + \frac{\mathbf{q}}{4\pi\epsilon r^5} + \cdots \tag{7.162}$$

where

$$Q = \int_V \rho(r') dr' \tag{7.163}$$

$$q_i = \sum_j (3x_i x_j - r^2 \delta_{ij}) p_j, \quad \mathbf{r} = (x_1, x_2, x_3) \tag{7.164}$$

$$p_j = \int_V x'_j \rho(r') dr' \tag{7.165}$$

$\mathbf{p}$ is called the dipole moment of the charge distribution. Observe that when $Q = 0$, the dipole term is the leading term in the expansion Equation (9.156).

7.7.4 Magnetostatic

The equations for the magnetic field in this case are

$$\nabla \cdot \mathbf{B} = 0 \tag{7.166}$$

$$\nabla \times \mathbf{B} = \mu \mathbf{J}. \tag{7.167}$$

From Equation (7.166), we infer that $\mathbf{B}$ is a solenoidal vector field. Hence, there exists a vector potential $\mathbf{A}$ so that

$$\mathbf{B} = \nabla \times \mathbf{A}. \tag{7.168}$$

Hence, from Equation (7.167)

$$\nabla \times (\nabla \times \mathbf{A}) = \mu \mathbf{J}. \tag{7.169}$$

If we choose $\mathbf{A}$ so that $\nabla \cdot \mathbf{A} = 0$, we obtain

$$\nabla^2 \mathbf{A} = -\mu \mathbf{J}, \tag{7.170}$$

i.e., $\mathbf{A}$ satisfies a vector Poisson equation. Hence,

$$\mathbf{A} = \frac{\mu}{4\pi} \int_V \frac{\mathbf{J}(\mathbf{r}')d\mathbf{r}'}{|\,\mathbf{r} - \mathbf{r}'\,|}. \tag{7.171}$$

By taking the curl of this equation, we obtain Biot-Savart law

$$\mathbf{B} = \frac{\mu}{4\pi} \int_V \frac{\mathbf{J}(\mathbf{r}') \times (\mathbf{r} - \mathbf{r}')}{|\,\mathbf{r} - \mathbf{r}'\,|^3}\, d\mathbf{r}'. \tag{7.172}$$

Since the total magnetic charge is always zero, the multipole expansion for $\mathbf{B}$ yields

$$\mathbf{B} = \frac{\mu \mathbf{M}}{4\pi r^5} \tag{7.173}$$

where

$$M_i = \sum_j (3r_i r_j - r^2 \delta_{ij}) m_j \tag{7.174}$$

$$\mathbf{m} = \frac{1}{2} \int_V \mathbf{r}' \times \mathbf{J}(\mathbf{r}')d\mathbf{r}'. \tag{7.175}$$

7.7.5 Electromagnetic Waves

In vacuum (and with no currents)

$$\nabla \cdot \mathbf{E} = 0, \quad \nabla \cdot \mathbf{B} = 0 \tag{7.176}$$

$$\nabla \times \mathbf{E} = -\frac{\partial \mathbf{B}}{\partial t}, \quad \nabla \times \mathbf{B} = \epsilon \mu \frac{\partial \mathbf{E}}{\partial t}. \tag{7.177}$$

By taking the curl of Equation (7.177) and using Equations (10.96) and (7.176), we obtain

$$\frac{\partial^2 \mathbf{E}}{\partial t^2} = c^2 \nabla^2 \mathbf{E}, \quad \frac{\partial^2 \mathbf{B}}{\partial t^2} = c^2 \nabla^2 \mathbf{B} \tag{7.178}$$

where $c^2 = \frac{1}{\epsilon \mu}$. Thus, both $\mathbf{E}$ and $\mathbf{B}$ satisfy the wave equation in three dimensions.

7.7.6 Electromagnetic Energy and Momentum

Consider a point charge ρ moving with velocity $\mathbf{v}$ in an $E - M$ field. The force acting on the charge is (from Equation (10.97))

$$\mathbf{F} = \rho \mathbf{E} + \rho \mathbf{v} \times \mathbf{B}. \tag{7.179}$$

The rate of work done by this force is

$$W = \mathbf{F} \cdot \mathbf{v} = \rho \mathbf{v} \cdot \mathbf{E} = \mathbf{J} \cdot \mathbf{E}. \tag{7.180}$$

We infer then that the rate of work done by an $E - M$ field on a continuous distribution of currents is

$$W = \int \mathbf{J} \cdot \mathbf{E} dV. \tag{7.181}$$

This power (which is converted into thermal or mechanical energy) must be accompanied by an equal rate of decrease in the $E - M$ energy in the volume V. To obtain an explicit expression for the conservation of energy, we eliminate $\mathbf{J}$ in Equation (7.180) using Equation (10.92)

$$\int \mathbf{J} \cdot \mathbf{E} dV = \int \mathbf{E} \cdot \left(\nabla \times \mathbf{H} - \frac{\partial \mathbf{D}}{\partial t} \right) dV. \tag{7.182}$$

Using the vector identity

$$\nabla \cdot (\mathbf{E} \times \mathbf{H}) = \mathbf{H} \cdot (\nabla \times \mathbf{E}) - \mathbf{E} \cdot (\nabla \times \mathbf{H}) \tag{7.183}$$

and Equation (10.91), we have

$$\mathbf{E} \cdot (\nabla \times \mathbf{H}) = -\mathbf{H} \cdot \frac{\partial \mathbf{B}}{\partial t} - \nabla \cdot (\mathbf{E} \times \mathbf{H}). \tag{7.184}$$

Hence,

$$\int_V \mathbf{J} \cdot \mathbf{E} dV = -\int_V \left[\mathbf{H} \cdot \frac{\partial \mathbf{B}}{\partial t} + \mathbf{E} \cdot \frac{\partial \mathbf{D}}{\partial t} + \nabla \cdot (\mathbf{E} \times \mathbf{H}) \right] dV$$
$$= -\int_V \left[\frac{1}{2} \frac{\partial}{\partial t} (\mathbf{H} \cdot \mathbf{B} + \mathbf{E} \cdot \mathbf{D}) + \nabla \cdot (\mathbf{E} \times \mathbf{H}) \right] dV \tag{7.185}$$

where we have used the relations (10.93). Using the divergence theorem, we can rewrite Equation (7.185) as

$$\int_V \mathbf{J} \cdot \mathbf{E} dV = -\int \frac{1}{2} \frac{\partial}{\partial t} (\mathbf{H} \cdot \mathbf{B} + \mathbf{E} \cdot \mathbf{D}) dV - \int_S (\mathbf{E} \times \mathbf{H}) \cdot d\mathbf{S}. \tag{7.186}$$

Since the volume in Equation (7.185) is arbitrary, we can re-express this equation in a differential form

$$\frac{\partial U}{\partial t} + \nabla \cdot \mathbf{P} = -\mathbf{J} \cdot \mathbf{E} \tag{7.187}$$

where

$$U = \frac{1}{2}(\mathbf{E} \cdot \mathbf{D} + \mathbf{B} \cdot \mathbf{H}), \quad \mathbf{P} = \mathbf{E} \times \mathbf{H}. \tag{7.188}$$

U represents the $E - M$ energy density and $\mathbf{P}$, the Poynting vector, is the energy flux density. Observe that only the divergence of $\mathbf{P}$ appears in Equation (7.185). It follows then that we can change the definition of the Poynting vector by adding the curl of an arbitrary vector field without changing the physical contents of the theory.

To treat the conservation of linear momentum, we use Equations (10.89) and (10.92) to eliminate ρ and $\mathbf{J}$ from Equation (10.97)

$$\rho \mathbf{E} + \mathbf{J} \times \mathbf{B} = \mathbf{E}(\nabla \cdot \mathbf{D}) + \left(\nabla \times \mathbf{H} - \frac{\partial \mathbf{D}}{\partial t} \right) \times \mathbf{B}$$
$$= \mathbf{E}(\nabla \cdot \mathbf{D}) + (\nabla \times \mathbf{H}) \times \mathbf{B} - \left[\frac{\partial}{\partial t}(\mathbf{D} \times \mathbf{B}) - \mathbf{D} \times \frac{\partial \mathbf{B}}{\partial t} \right]$$
$$= \mathbf{E}(\nabla \cdot \mathbf{D}) - \mathbf{B} \times (\nabla \times \mathbf{H}) - \mathbf{D} \times (\nabla \times \mathbf{E})$$
$$- \frac{\partial}{\partial t}(\mathbf{D} \times \mathbf{B}). \tag{7.189}$$

By adding $\mathbf{H}(\nabla \cdot \mathbf{B}) = 0$ to the right hand side of this expression, we obtain:

$$\int_V (\rho \mathbf{E} + \mathbf{J} \times \mathbf{B}) dV \quad + \quad \frac{d}{dt} \int_V (\mathbf{D} \times \mathbf{B}) = \int_V [\mathbf{E}(\nabla \cdot \mathbf{D}) - \mathbf{D} \times (\nabla \times \mathbf{E})$$
$$+ \quad \mathbf{H}(\nabla \cdot \mathbf{B}) - \mathbf{B} \times (\nabla \times \mathbf{H})] dV$$
$$= \int_V \nabla_\beta T_{\alpha,\beta} dV . \tag{7.190}$$

Using the divergence theorem, we can rewrite this as

$$\int_V (\rho \mathbf{E} + \mathbf{J} \times \mathbf{B}) dV + \frac{d}{dt} \int_V (\mathbf{D} \times \mathbf{B}) = \int_S \sum_\beta T_{\alpha\beta} n_\beta dS \tag{7.191}$$

where $\mathbf{n}$ is the normal to S and

$$T_{\alpha\beta} = E_\alpha D_\beta + H_\alpha B_\beta - \frac{1}{2}(\mathbf{E} \cdot \mathbf{D} + \mathbf{B} \cdot \mathbf{H}) \delta_{\alpha\beta} \tag{7.192}$$

is Maxwell stress tensor.

The first and second terms on the left hand side of Equation (7.191) represent (respectively) the rate of change in the linear moments of the particles and the $E - M$ field in the volume V. The right hand side of this equation is interpreted as the force per unit area transmitted across S and acting on the fields and particles in V.

7.7.7 Electromagnetic Potential

Maxwell equations (10.89)-(10.92) with the relations Equations (10.93) and (10.94) represent eight equations for the six quantities $\mathbf{E}$ and $\mathbf{B}$. We infer then that a more compact formulation may exist in terms of only four unknowns. To see how this can be done, we first observe that Equation (10.90) can be satisfied identically if

$$\mathbf{B} = \nabla \times \mathbf{A} . \tag{7.193}$$

Substituting this in Equation (10.91), we obtain

$$\nabla \times \left(\mathbf{E} + \frac{\partial \mathbf{A}}{\partial t} \right) = 0, \tag{7.194}$$

i.e., $\mathbf{E} + \frac{\partial \mathbf{A}}{\partial t}$ is a conservative field. It follows then that there exists a scalar field ϕ so that

$$\mathbf{E} = -\nabla \phi - \frac{\partial \mathbf{A}}{\partial t} . \tag{7.195}$$

We observe that the relations, Equations (7.193) and (7.195) remain unchanged under the "gauge" transformations

$$\mathbf{A} \rightarrow \mathbf{A}' = \mathbf{A} + \nabla \wedge \qquad (7.196)$$

$$\phi \rightarrow \phi' = \phi - \frac{\partial \wedge}{\partial t}; \qquad (7.197)$$

i.e., the physical contents of the theory (which is represented by $\mathbf{E}$ and $\mathbf{B}$) remains unchanged under these transformations.

Two particular choices of $\wedge$ are of special interest.

1. Lorentz gauge

    Under this gauge, $\wedge$ is chosen so that

    $$\nabla^2 \wedge = \frac{\partial^2 \wedge}{\partial t^2} . \qquad (7.198)$$

    As a result, the quantity

    $$I = \nabla \cdot \mathbf{A} + \frac{\partial \phi}{\partial t} = \nabla \cdot \mathbf{A}' + \frac{\partial \phi'}{\partial t} \qquad (7.199)$$

    remains unchanged. Choosing the value of I to be zero and substituting Equations (7.193) and (7.195) in Equations (10.89) and (10.92) leads to

    $$\left(\nabla^2 - \frac{\partial^2}{\partial t^2} \right) \mathbf{A} = -\mathbf{J} \qquad (7.200)$$

    $$\left(\nabla^2 - \frac{\partial^2}{\partial t^2} \right) \phi = -\rho/\epsilon. \qquad (7.201)$$

2. Coulomb gauge

    Here we impose on $\wedge$ the condition

    $$\nabla^2 \wedge = 0 . \qquad (7.202)$$

    As a result,

    $$\nabla \cdot \mathbf{A} = \nabla \cdot \mathbf{A}' \qquad (7.203)$$

    and we can choose therefore $\nabla \cdot \mathbf{A} = 0$. With this choice, we substitute Equations (7.193) and (7.195) in Equations (10.89) and (10.92) to obtain

    $$\left(\nabla^2 - \frac{\partial^2}{\partial t^2} \right) \mathbf{A} = -\mathbf{J} + \frac{\partial}{\partial t}(\nabla \phi) \qquad (7.204)$$

    $$\nabla^2 \phi = -\rho. \qquad (7.205)$$

Remark 7.7.4 *Sometimes the Coulomb gauge is referred to as the "Radiation gauge."*

Exercises

1. In free space (i.e., $\rho = 0, \mathbf{J} = \mathbf{0}$), let the electric and magnetic field be given by

$$\mathbf{E} = \mathbf{F}(z - t), \quad \mathbf{F} = (f_1, f_2, f_3)$$

$$\mathbf{B} = \mathbf{G}(z - t), \quad \mathbf{G} = (g_1, g_2, g_3)$$

   (plane waves). Show that to satisfy Maxwell equations we must require that

$$f_3 = g_3 = 0, \quad f_1 = g_2, \quad f_2 = -g_1.$$

2. Show that for

$$f = \tilde{f}(t)e^{-i\mathbf{k}\cdot\mathbf{r}}, \quad \mathbf{F} = \tilde{\mathbf{F}}(t)e^{-i\mathbf{k}\cdot\mathbf{r}}$$

$$\nabla f = -i\mathbf{k}f, \quad \nabla \cdot \mathbf{F} = -i\mathbf{k} \cdot \mathbf{F}, \quad \nabla \times \mathbf{F} = -i\mathbf{k} \times \mathbf{F}.$$

3. Use Ex. 2 to show that if $\mathbf{E} = \mathbf{E}_0 e^{i(\omega t - \mathbf{k}\cdot\mathbf{r})}$, then $\mathbf{B} = \mathbf{B}_0 e^{i(\omega t - \mathbf{k}\cdot\mathbf{r})}$, then $\mathbf{B} = (\mathbf{k} \times \mathbf{E})/\omega$, and $\mathbf{k} \cdot \mathbf{E} = 0$; i.e., the electric and magnetic fields are orthogonal to each other and $\mathbf{k}$ (the propagation direction) is orthogonal to both.

4. Consider EM field in a homogeneous conducting medium where $\mathbf{J} = \sigma\mathbf{E}$, and suppose $\rho = 0$ (change density). Show that

$$\frac{\partial^2 \mathbf{E}}{\partial t^2} + \frac{\sigma}{\epsilon}\frac{\partial \mathbf{E}}{\partial t} = c^2\nabla^2\mathbf{E}, \quad c^2 = \frac{1}{\epsilon\mu}.$$

CHAPTER 8

Solutions of Partial Differential Equations

CONTENTS

8.1 METHOD OF SEPARATION OF VARIABLES

When it comes to solving boundary value problems involving partial differential equations, a number of approaches are available. The method of separation of variables is very convenient because it draws on many well-known mathematical concepts and frequently works well. In this section, we introduce this method and illustrate its application by considering various systems that were modeled in the last chapter by partial differential equations. In particular, we concentrate on the heat, wave, and potential equations that are important in many scientific and engineering applications.

The general objective of this method is to reduce the solution of a given partial differential equation into the solution of a number of ordinary differential equations. Very often these ordinary differential equations are well known, and their solutions are easily found. The whole process follows a logical step-by-step development that terminates in the evaluation of the Fourier coefficients of a Fourier-type series. We suggest that you carefully catalog each step and its position in the stairway to a successful conclusion.

8.1.1 Method of Separation of Variables By Example

Although the method of separation of variables contains many steps, it does follow a set pattern as we move toward the solution. Probably the best way to understand the method is through observing a number of examples. We start with a straightforward heat flow problem.

Example 8.1.1 *For an introduction to this process, we will seek the solution of the following boundary value problem:*

$$\frac{\partial u}{\partial t} = k\frac{\partial^2 u}{\partial x^2} \quad 0 < x < L, \ 0 < t$$

$$u(0, t) = 0 \quad 0 < t$$

$$u(L, t) = 0 \quad 0 < t$$

$$u(x, 0) = x \quad 0 < x < L.$$

This is a one-dimensional heat flow problem in a rod of length L. We are to find the temperature $u(x, t)$ if the temperature at the right- and left-hand ends is always zero and the initial temperature in the rod is x.

Solution 8.1.1 : *Step 1. We seek "elementary" solutions of the partial differential equation in the special form $u(x,t) = X(x)T(t)$. What we are suggesting is that the variable x occurs only in the function X, whereas T is a function of t only. This device does not always work, but it does solve many engineering problems and is usually a good method to use as a first approach.*

Substituting $u(x,t) = X(x)T(t)$ into the differential equation leads to

$$X(x)T'(t) = kX''(x)T(t) \tag{8.1}$$

Notice that we are able to use the "prime" notation for the derivatives because each factor depends only on one variable; that is,

$$T'(t) = \frac{dT(t)}{dt}$$

$$X''(x) = \frac{d^2X}{dx^2}.$$

Step 2. We next see if it is possible to separate the variables. Is it possible to get all the functions dependent on x's on one side of the equation and all those dependent on t's on the other side? Since we are not looking for trivial solutions where either X or T is identically zero, we can divide both sides of Equation (12.30) by $X(x)T(t)$ provided $X(x)T(t) \neq 0$. Thus,

$$\frac{1}{k}\frac{XT'}{XT} = \frac{X''T}{XT}$$

or

$$\frac{1}{k}\frac{T'(t)}{T(t)} = \frac{X''(x)}{X(x)}. \tag{8.2}$$

Now pick a fixed value $t = t_0$. Then

$$\frac{1}{k}\frac{T'(t_0)}{T(t_0)} = constant = \frac{X''(x)}{X(x)}.$$

But this is true for any x on the right-hand side of the equation and, therefore (since the left hand side is a constant),

$$\frac{X''(x)}{X(x)} = constant.$$

In the same way we can show that

$$\frac{T'(t)}{T(t)} = constant$$

by fixing $x = x_0$. Observe that this constant is the same for both sides of Equation (12.31). In this case (for convenience) let the constant be equal to $-\lambda$. Equation (12.31) becomes

$$\frac{1}{k}\frac{T'(t)}{T(t)} = \frac{X''(x)}{X(x)} = -\lambda. \tag{8.3}$$

Step 3. From Equation (12.32) we see that

$$X''(x) + \lambda X(x) = 0 \tag{8.4}$$

$$T'(t) + \lambda k T(t) = 0 \tag{8.5}$$

which are two ordinary differential equations that are easily solved.

Comment. *If $X(x)$ or $T(t)$ is zero at $x = x_0$ and $t = t_0$, respectively, the result in Step 3 is still valid. Suppose $X(x_0) = 0$. Returning to Equation (12.30) (before we divide by XT) and substituting $X(x_0) = 0$ into Equation (ref4.1), we have*

$$0 = kX''(x_0)T(t).$$

Since this equation must hold for all $t > 0$, and $T(t)$ is not identically zero, there must be some $t = t_0$ so that $T(t_0) \neq 0$. Since $k > 0$, it follows that $X''(x_0) = 0$, and we can state that

$$X''(x_0) + \lambda X(x_0) = 0$$

is satisfied. The argument for showing that

$$T'(t_0) + \lambda k T(t_0) = 0$$

for $T(t_0) = 0$ is similar. □

*Step 4. Before solving Equations (8.4)-(8.5), let us find the boundary conditions that go with these ordinary differential equations. Now $u(0,t) = X(0)T(t) = 0$. If $T(t) = 0$ for all t, then $u(x,t) = X(x)T(t) = X(x) \cdot 0 = 0$. But if $u(x,t) \equiv 0$, it is known as the **trivial solution**. Since we are looking for nontrivial solutions, we must avoid setting $T(t) = 0$. Therefore,*

$$X(0) = 0$$

is one boundary condition.

In a similar fashion we can show that

$$X(L) = 0$$

is another boundary condition. In this problem there are no initial conditions for the first-order equation in t.

Warning. *Do not attempt this type of argument on a nonhomogeneous boundary condition. For example, the initial condition $u(x, 0) = X(x)T(0) = x$ cannot be solved uniquely for $T(0)$. Only when the product of two quantities equals zero can we conclude that one or both of the factors is zero.*

Step 5. Now we have the following ordinary differential equations plus boundary conditions

$$X'' + \lambda X = 0 \ \ X(0) = 0 \ \ X(L) = 0$$

$$T' + \lambda kT = 0.$$

*The first of these two equations with the two boundary conditions is known as an **eigenvalue** problem and is solved first. Unfortunately, λ is unknown to us at the moment, and therefore we must look at three problems because the character of the solution changes with each case. We shall look at the eigenvalue problem for $\lambda > 0$, $\lambda = 0$, and $\lambda < 0$.*

Case 1. $\lambda > 0$. Let $\alpha^2 = \lambda > 0$ (again for convenience). Then the general solution to $X''(x) + \alpha^2 X(x) = 0$ is

$$X(x) = A \cos \alpha x + B \sin \alpha x.$$

Now since $X(0) = 0$ we have

$$X(0) = 0 = A \cdot 1 + B \cdot 0$$

which tells us that

$$A = 0$$

and

$$X(x) = B \sin \alpha x.$$

Using the second boundary condition $X(L) = 0$, we have

$$X(L) = 0 = B \sin \alpha L.$$

Now if $B = 0$, then since $A = 0$, $X(x) \equiv 0$, which implies $u(x,t) = X(x)T(t) \equiv 0$, which is the trivial solution we are trying to avoid. There-fore, $\sin \alpha L = 0$. But recall from trigonometry that $n\pi$ $(n = 0, \pm 1, \pm 2, \ldots)$ are angles that make the sine zero, and therefore we write

$$\alpha L = n\pi$$

so that

$$\alpha = \frac{n\pi}{L}$$

and

$$\lambda = \lambda_n = \alpha_n^2 = \frac{n^2\pi^2}{L^2} \quad \text{for } n = \pm 1, \pm 2, \ldots.$$

Note. The value $n = 0$ is dropped since this implies $\alpha = 0$, and we are assuming $\alpha > 0$. This case is covered next.

*The values of λ in Case 1 are called the **eigenvalues** of the boundary value problem.*

Now the solutions of the ordinary differential equation and its boundary conditions corresponding to these eigenvalues λ_n are

$$X_n = \sin \frac{n\pi x}{L} \quad n = \pm 1, \pm 2, \ldots.$$

*These solutions are called **eigenfunctions**.*

Case 2. $\lambda = 0$. The general solution of $X'' = 0$ is

$$X(x) = A + Bx.$$

Since

$$X(0) = 0 = A$$

and

$$X(L) = BL = 0$$

implies $B = 0$, we see that the only solution to this problem is $X(x) \equiv 0$, which is the trivial solution. $\lambda = 0$ is not an eigenvalue.

Note. *Eigenfunctions corresponding to an eigenvalue λ must be **nontrivial**.*

Case 3. $\lambda < 0$. Let $\lambda = -\alpha^2 < 0$. The differential equation $X'' + \alpha^2 X = 0$ has the general solution

$$X(x) = C_1 e^{\alpha x} + C_2 e^{-\alpha x}.$$

Although this solution may be used, it is more convenient (in this case) to write the solution in terms of hyperbolic functions instead. Since

$$\sinh \alpha x = \frac{e^{\alpha x} - e^{-\alpha x}}{2}$$

and

$$\cosh \alpha x = \frac{e^{\alpha x} - e^{-\alpha x}}{2},$$

which are linear independent functions, we can rewrite the general solution as

$$X(x) = A \cosh \alpha x + B \sinh \alpha x.$$

Now, since $X(0) = 0$ we have

$$X(0) = 0 = A \cdot 1 + B \cdot 0$$

which shows that

$$A = 0$$

and $X(x) = B \sinh \alpha x$.

Using $X(L) = B \sinh \alpha L = 0$, we find either $B = 0$, which leads to the trivial solution or $\sinh \alpha L = 0$. But $\sinh 0 = 0$ is the only zero of the hyperbolic sine function. Therefore, $\alpha L = 0$. And since $L \neq 0$, it follows that $\alpha = 0$. But this is a contradiction since we are assuming $-\alpha^2 < 0$.

There are no nontrivial solutions under this case, which tells us that there are no negative eigenvalues.

This long discourse completes the solution of the eigenvalue problem posed in Step 5.

Step 6. We can now shift to the differential equation

$$T' + \lambda k T = T' + \frac{n^2 \pi^2}{L^2} k T = 0. \tag{8.6}$$

Notice that the eigenvalues are now known to us from Step 5, and it is possible to solve Equation (12.33) up to an arbitrary constant. The general solution to the first-order equation is

$$T_n(t) = C_n exp(-\frac{k n^2 \pi^2}{L^2} t) \quad n = \pm 1, \pm 2, \dots.$$

In solving boundary value problems by this approach, we find it convenient to set all C_ns$= 1$ because we will be multiplying each solution by an arbitrary constant later on.

Step 7. To continue with the solution, we multiply X_n by T_n because in Step 1 we assumed a solution of the partial differential equation of the form $u(x, t) = X(x)T(t)$. Therefore, we can write

$$u_n(x, t) = exp(-\frac{kn^2\pi^2}{L^2}t)\sin\frac{n\pi x}{L} \quad n = \pm 1, \pm 2, \dots.$$

For $n = \pm 1, \pm 2, \dots, u_n(x, t)$ satisfies the partial differential equation and the two boundary conditions.

We still have to match the initial condition $u(x, 0) = x$. To do this, we form an infinite linear combination of all solutions $u_n(x, t)$. This is made possible by the fact that the heat equation is linear and the superposition principle applies. It follows then that the general solution to the partial differential equation and boundary conditions can be written as

$$u(x, t) = \sum_{n=1}^{\infty} b_n u_n(x, t) = \sum_{n=1}^{\infty} b_n exp(-\frac{kn^2\pi^2}{L^2}t)\sin\frac{n\pi x}{L}. \tag{8.7}$$

Note. When dealing with sines and cosines, one may question whether it is necessary to consider n for negative integers. However, since $\sin(-n\pi x/L) = -\sin(n\pi x/L)$ and $\cos(-n\pi x/L) = \cos(n\pi x/L)$, each solution derived from a negative n can be combined with the corresponding solution derived from a positive n. From here on we will assume this is done and let $n = 1, 2, \dots$.

Step 8. Finally, using the condition $u(x, 0) = x$, $0 < x < a$, we must satisfy

$$u(x, 0) = x = \sum_{n=1}^{\infty} b_n \sin\frac{n\pi x}{L}$$

since the time exponential is 1. But this is just a Fourier sine series when x on $(0, L)$ is extended into $(-L, 0)$ as an odd function, and it follows that

$$b_n = \frac{2}{L}\int_0^L x\sin\frac{n\pi x}{L}dx \tag{8.8}$$

$$= \frac{2L^2}{Ln^2\pi^2}\left\{\sin\frac{n\pi x}{L} - \frac{n\pi x}{L}\cos\frac{n\pi x}{L}\right\}|_0^L$$

$$= \frac{2L}{n\pi}\{-cosn\pi\} = \frac{-2L}{n\pi}(-1)^n$$

or

$$b_n = (-1)^{n+1}\frac{2L}{n\pi}. \tag{8.9}$$

We have now determined all the constants b_n. Substituting their value from Equation (12.35) in Equation (12.34), we see that the final solution is

$$u(x,t) = \frac{2L}{\pi}\sum_{n=1}^{\infty}\frac{(-1)^{n+1}}{n}exp\left(-\frac{kn^2\pi^2}{L^2}t\right)\sin\frac{n\pi x}{L}. \tag{8.10}$$

It is possible to garner some information about the solution $u(x,t)$ without too much effort. True, if we have access to a computer, we could numerically evaluate a large number of terms of the series to construct a table of values or even plot a graph of an approximation to the solution. Of course, how well this can be done depends on the convergence properties of the series which represents the solution.

*We observe that the solution we found for $u(x,t)$ is a **transient** solution; that is, $u(x,t) \to 0$ as $t \to +\infty$. This follows easily from the fact that each term in the series is of the form $e^{-\alpha t}$ where $\alpha = kn^2\pi^2/L^2$. Carrying this idea a bit further, we notice that n enters into the exponent as n^2, which indicates that $e^{-\alpha t}$ decreases very rapidly as n increases. Therefore, most of the information about the solution is carried in the first few terms of the series.*

For example, if we let $k = 1$, $L = \pi$, and $t = 1$, and compare the $e^{-\alpha t}$ terms for $n = 1, 2, 3$, we shall find that the magnitude of consecutive terms in the series is decreasing exponentially.

We now look at an example in which $\lambda = 0$ is an eigenvalue.

Example 8.1.2 : *Consider a high-frequency transmission line whose length is 40 meters. Suppose the rate of change of voltage with respect to x at both ends is zero. If the initial voltage is $f(x)$ and the rate of change of voltage with respect to t is zero, find the solution as a function of x and t. This problem can be written formally as*

$$\frac{\partial e}{\partial x}(0,t) = 0 \ \ 0 < t$$

$$\frac{\partial e}{\partial x}(40,t) = 0 \ \ 0 < t$$

$$e(x, 0) = f(x) \ 0 < x < 40$$

$$\frac{\partial e}{\partial t}(x, 0) = 0 \ 0 < x < 40.$$

Solution 8.1.2 *The differential equation is*

$$\frac{\partial^2 e}{\partial x^2} = LC \frac{\partial^2 e}{\partial t^2} \ 0 < x < 40, \ 0 < t.$$

Steps 1 and 2. Let $e(x, t) = X(x)T(t)$. Substituting into the differential equation, we find

$$\frac{X''}{X} = LC \frac{T''}{T} = -\lambda.$$

Step 3. This result leads to the two ordinary differential equations

$$X'' + \lambda X = 0$$

$$T'' + \frac{\lambda}{LC} T = 0.$$

Step 4. The three homogeneous boundary and initial conditions yield the following conditions for the ordinary differential equations:

$$X'(0) = 0, \ X'(40) = 0, \ T'(0) = 0.$$

Steps 5 and 6. When we solve the differential equation in x, we again consider three cases: $\lambda > 0$, $\lambda = 0$, and $\lambda < 0$.

Case 1. $\lambda > 0$. Let $\alpha^2 = \lambda > 0$. The general solution of $x'' + \alpha^2 X = 0$ is then

$$X(x) = A \cos \alpha x + B \sin \alpha x$$

and

$$X'(x) = -\alpha A \sin \alpha x + \alpha B \cos \alpha x.$$

Applying the boundary conditions, we find

$$X'(0) = -\alpha A \cdot 0 + \alpha B \cdot 1 = 0$$

for which

$$\alpha B = 0.$$

However, since $\alpha > 0$, $B = 0$, and $X'(x) = -\alpha A \sin \alpha x$, then $X'(40) = -\alpha A \sin 40\alpha = 0$. Neither α nor A can equal zero, so we can write

$$\sin 40\alpha = 0$$

from which it follows that

$$40\alpha = n\pi \quad n = 1, 2, \ldots$$

$$\alpha = \frac{n\pi}{40}.$$

The positive eigenvalues are

$$\lambda_n = \alpha^2 = \left[\frac{n\pi}{40}\right]^2$$

and the eigenfunctions are

$$X_n(x) = \cos\frac{n\pi x}{40}.$$

Solving the other differential equation, we have

$$T_n(t) = D_n \cos\frac{n\pi t}{40\sqrt{LC}} + E_n \sin\frac{n\pi t}{40\sqrt{LC}}$$

and

$$T_n'(t) = -D_n\frac{n\pi}{40\sqrt{LC}}\sin n\pi t 40\sqrt{LC} + E_n\frac{n\pi}{40\sqrt{LC}}\cos\frac{n\pi t}{40\sqrt{LC}}.$$

Using the condition

$$T_n'(0) = E_n\frac{\pi n}{40\sqrt{LC}} = 0$$

we observe that

$$E_n = 0.$$

Therefore,

$$T_n(t) = \cos\frac{n\pi t}{40\sqrt{LC}} \quad n = 1, 2,$$

Combining X_n and T_n we find

$$e_n(x, t) = \cos\frac{n\pi x}{40}\cos\frac{n\pi t}{40\sqrt{LC}} \quad n = 1, 2, \ldots.$$

Case 2. $\lambda = 0$. In this case, $X'' = 0$ and $X(x) = A + Bx$. It follows that

$$X'(x) = B$$

and

$$X'(0) = B = 0.$$

But now $X'(x) = 0$, for all x and in particular for $x = 40$; that is, $X'(40) = 0$. Therefore, both boundary conditions are satisfied: $\lambda = 0$ is an eigenvalue and $X(x) = A$ (a constant) is an eigenfunction. We find it convenient to let $A = 1$ and use $X(x) = 1$ as our eigenfunction. To identify the eigenvalue and eigenfunction associated with $\lambda = 0$, we write $\lambda_0 = 0$ and $X_0 = 1$. Now solving $T'' = 0$ we have $T(t) = Dt + E$ and $T'(t) = D$. Using the initial condition $T'(0) = 0$, we find $D = 0$ and $T_0 = 1$. Therefore, $e_0(x, t) = X_0 T_0 = 1$.

Case 3. $\lambda < 0$. In a way similar to Case 3 of Example 1 we can show that there are no negative eigenvalues.

Step 7. To finish the problem we form an infinite linear combination of all the solutions e_0 and e_n, which yields

$$e(x, t) = \frac{a_0}{2} e_0(x, t) + \sum_{n=1}^{\infty} a_n e_n(x, t) = \frac{a_0}{2} + \sum_{n=1}^{\infty} a_n \cos \frac{n \pi x}{40} \cos \frac{n \pi t}{40 \sqrt{LC}}. \quad (8.11)$$

Step 8. We use our final condition, which tells us that

$$e(x, 0) = f(x) = \frac{a_0}{2} + \sum_{n=1}^{\infty} a_n \cos \frac{n \pi x}{40} \quad \text{on } (0, 40).$$

Since this is just a Fourier cosine series, we know that

$$a_0 = \frac{1}{20} \int_0^{40} f(x) dx$$

and

$$a_n = \frac{1}{20} \int_0^{40} f(x) \cos \frac{n \pi x}{40} dx.$$

Substituting these values back into Equation (12.36), we have the solution to our problem:

$$e(x, t) = \frac{1}{40} \int_0^{40} f(x) dx + \quad (8.12)$$

$$\frac{1}{20} \sum_{n=1}^{\infty} \left[\int_0^{40} f(s) \cos \frac{n \pi s}{40} ds \right] \cos \frac{n \pi x}{40} \cos n \pi t 40 \sqrt{LC}.$$

We recognize immediately that the solution $e(x, t)$ is an even function in both x and t. The integral used for evaluating a_0 is the average value for $f(x)$

over the half Fourier interval whose length is 40. Half the value of this integral, that is, $a_0/2$, is equal to the DC component of the series answer.

The frequency f_n of the components of the voltage $e(x,t)$ with respect to time is

$$f_n = \frac{n\pi}{2\pi(40\sqrt{LC})} = \frac{n}{80\sqrt{LC}}$$

Therefore, the fundamental frequency is $1/80\sqrt{LC}$.

The previous two examples cover two of the three major classes of second-order partial differential equations: the parabolic type and the hyperbolic type. We now investigate the solution of the elliptic type, which introduces us to some new techniques used in solving boundary value problems.

Example 8.1.3 *Recall that the temperature in a plate insulated above and below must satisfy the differential equation*

$$\frac{\partial u}{\partial t} = k\left[\frac{\partial^2 u}{\partial x^2} + \frac{\partial^2 u}{\partial y^2}\right].$$

If the temperature is independent of t (i.e., steady state), this differential equation becomes

$$\frac{\partial^2 u}{\partial x^2} + \frac{\partial^2 u}{\partial y^2} = 0$$

which is Laplace's equation in two variables.

Now suppose we have a rectangular plate whose dimensions are a by b and whose boundary conditions are given by

$$u(x,0) = f_1(x) \ \ 0 < x < a$$

$$u(a,y) = 0 \ \ 0 < y < b$$

$$u(x,b) = f_2(x) \ \ 0 < x < a$$

$$u(0,y) = 0 \ \ 0 < y < b.$$

Find the temperature u as a function of x and y (see Figure 8.1).

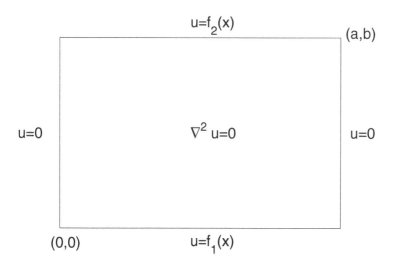

Figure 8.1 Steady state heat conduction in a plate

Solution 8.1.3 *In order to carry out specific operations, let us examine the special case where* $f_1(x) = 0$ *and* $f_2(x) = x$. *Steps 1,2, and 3. Assume* $u(x, y) = X(x)Y(y)$. *Computing the necessary derivatives and substituting into the differential equation, we have*

$$X''Y + XY'' = 0$$

or

$$\frac{X''}{X} = -\frac{Y''}{Y} = -\lambda$$

which yields the two ordinary differential equations

$$X''(x) + \lambda X(x) = 0$$

$$Y''(y) - \lambda Y(y) = 0.$$

Steps 4 and 5. From our assumption and the given boundary conditions, we arrive at the following boundary conditions associated with the ordinary differential equations:

$$X(0) = 0, \quad X(a) = 0$$

and

$$Y(0) = 0.$$

Case 1. $\lambda > 0$. *Let* $\lambda = \alpha^2 > 0$. *Then we can show as before that the eigenvalues are*

$$\lambda_n = \alpha_n^2 = \frac{n^2\pi^2}{a^2} \quad n = 1, 2, \ldots$$

and the eigenfunctions are

$$X_n(x) = \sin\frac{n\pi x}{a}.$$

Cases 2 and 3. $\lambda = 0$ *and* $\lambda < 0$. *It is easily shown that there are no non-positive eigenvalues.*

Step 6. One new idea introduced in this problem occurs when we wish to solve the second differential equation,

$$Y'' - \frac{n^2\pi^2}{a^2}Y = 0.$$

Normally a beginning student would solve this equation using exponentials as follows

$$Y_n(y) = C_n exp\frac{n\pi y}{a} + D_n exp\left(\frac{-n\pi y}{a}\right) \tag{8.13}$$

where $exp\ y = e^y$. *Because neither exponential in Equation (12.37) vanishes, it is necessary to solve a system of two equations to find D in terms of C. Therefore, it is much more convenient to use hyperbolic functions, and we can write*

$$Y_n(y) = C_n \cosh\frac{n\pi y}{a} + D_n \sinh\frac{n\pi y}{a}.$$

Using the boundary condition $Y(0) = 0$, *we see that*

$$Y_n(0) = 0 = C_n \cdot 1 + D_n \cdot 0$$

or

$$C_n = 0$$

and that

$$Y_n(y) = \sinh\frac{n\pi y}{a}$$

is a solution to our ordinary differential equation in y and its one boundary condition. For convenience we have again set the constant $D = 1$.

Step 7. Combining the two families of solutions $X_n(x)$ and $Y_n(y)$, we construct solutions to the partial differential equation and the three homogeneous boundary conditions that take the form

$$u_n(x, y) = X_n(x)Y_n(y) = \sin \frac{n\pi x}{a} \sinh \frac{n\pi y}{a}.$$

Step 8. We still have to meet the final inhomogeneous boundary condition $u(x, b) = x$. In order to do this we use the superposition principle and form the linear combination

$$u(x, y) = \sum_{n=1}^{\infty} b_n \sin \frac{n\pi x}{a} \sinh \frac{n\pi y}{a}, \tag{8.14}$$

expecting that we can find values for b_n such that the inhomogeneous condition is met. Letting $y = b$ in Equation (9.69), we see that

$$u(x, b) = x = \sum_{n=1}^{\infty} b_n \sinh \frac{n\pi b}{a} \sin \frac{n\pi x}{a}. \tag{8.15}$$

If we let

$$B_n = b_n \sinh \frac{n\pi b}{a}$$

Equation (9.70) becomes

$$x = \sum_{n=1}^{\infty} B_n \sin \frac{n\pi x}{a}$$

which is a Fourier sine series. Therefore,

$$B_n = b_n \sinh \frac{n\pi b}{a} = \frac{2}{a} \int_0^a x \sin \frac{n\pi x}{a} dx \tag{8.16}$$

$$= \frac{2}{a} \frac{a^2}{n^2\pi^2} \left[\sin \frac{n\pi x}{a} - \frac{n\pi x}{a} \cos \frac{n\pi x}{a} \right] \Big|_0^a$$

$$= \frac{2a}{n^2\pi^2} [-n\pi \cos n\pi] = (-1)^{n+1} \frac{2a}{n\pi}.$$

The constants b_n become

$$b_n = (-1)^{n+1} \frac{2a}{n\pi \sinh \frac{n\pi b}{a}}$$

and the final answer can be written as

$$u(x, y) = \frac{2a}{\pi} \sum_{n=1}^{\infty} (-1)^{n+1} \frac{\sin(n\pi x/a) \sin h(n\pi y/a)}{n \sin h(n\pi b/a)}.$$

Remark.

When solving the ordinary differential equation

$$y''(x) - \alpha^2 y(x) = 0$$

it is useful to remember that there are three practical ways in which to express the solution. If one of the boundary conditions is of the form

$$y(0) = 0 \ \ or \ \ y'(0) = 0$$

then the general solution

$$y(x) = A \cosh \alpha x + B \sinh \alpha x \tag{8.17}$$

is most convenient to use because the value of A or B is quickly determined. On the other hand, if

$$y(L) = 0 \ \ or \ \ y'(L) = 0$$

is given as a boundary condition, one of the following forms of the solution is useful:

$$y(x) = E \sinh \alpha (x - d) \ \ for \ y(L) = 0$$
$$y(x) = E \cosh \alpha (x - d) \ \ for \ y'(L) = 0$$

where E and d are arbitrary constants. These forms of the solutions are not as well known as the form in Equation (9.71), but they can easily be shown to be equivalent for the particular boundary condition given using the well-known hyperbolic identities

$$\sinh(\alpha - \beta) = \sinh \alpha \cosh \beta - \cosh \alpha \sinh \beta$$

and

$$\cosh(\alpha - \beta) = \cosh \alpha \cosh \beta - \sinh \alpha \sinh \beta.$$

Finally, the classic solution

$$y(x) = A e^{\alpha x} + B e^{-\alpha x}$$

finds its greatest use in solving boundary value problems over semi-infinite intervals and the study of Fourier integrals.

Exercises

1. The temperature $u(x,t)$ in a laterally insulated rod of length L satisfies the following boundary value problem:

$$\frac{\partial^2 u}{\partial x^2} = \frac{1}{k}\frac{\partial u}{\partial t} \quad 0 < x < L, \ 0 < t \tag{8.18}$$

$$u(0,t) = 0 \quad 0 < t$$

$$u(L,t) = 0 \quad 0 < t$$

$$u(x,0) = 100 \quad 0 < x < L.$$

Use the technique of separation of variables to find $u(x,t)$.

2. Use the technique of separation of variables to solve for the temperature $u(x,t)$ if

$$\frac{\partial^2 u}{\partial u^2} = \frac{1}{k}\frac{\partial u}{\partial t} 0 < x < 10, \ 0 < t \tag{8.19}$$

$$\frac{\partial u}{\partial x}(0,t) = 0 \quad 0 < t$$

$$\frac{\partial u}{\partial x}(10,t) = 0 \quad 0 < t$$

$$u(x,0) = 1 - x \quad 0 < x < 10.$$

3. The voltage $e(x,t)$ along a submarine cable 3000 kilometers long satisfies the boundary value problem

$$e_{xx} = RCe_t \quad 0 < x < 3000, \ 0 < t \tag{8.20}$$

$$e(0,t) = 0 \quad 0 < t$$

$$e(3000,t) = 0 \quad 0 < t$$

$$e(x,0) = \sin\frac{x}{100} \quad 0 < x < 3000.$$

Use the technique of separation of variables to find $e(x,t)$.

4. The current $i(x,t)$ along a submarine cable of length L satisfies

$$i_{xx} = RCi_t \quad 0 < x < L, \ 0 < t \tag{8.21}$$

$$\frac{\partial i}{\partial x}(0,t) = 0 \quad 0 < t$$

$$\frac{\partial i}{\partial x}(L,t) = 0 \quad 0 < t$$

$$i(x,0) = 2 \quad 0 < x < L.$$

Use the technique of separation of variables to find the current $i(x,t)$ in the cable.

5. Find the temperature $u(x,t)$ by the technique of separation of variables in a laterally insulated rod of length 20 meters that has a heat source given by $\delta u(x,t)$ joules/m^3. The other conditions are

$$u(0,t) = 0 \qquad\qquad (8.22)$$
$$u(20,t) = 0 \qquad\qquad (8.23)$$
$$u(x,0) = x(20 - x). \qquad\qquad (8.24)$$

6. The voltage $e(x,t)$ satisfies the differential equation $e_{xx} = RCe_t + Ae_x$. Using separation of variables find $e(x,t)$ if

$$\frac{\partial e}{\partial x}(0,t) = 0 \ \ 0 < t \qquad\qquad (8.25)$$
$$\frac{\partial e}{\partial x}(L,t) = 0 \ \ 0 < t$$
$$e(x,0) = 50 \ \ 0 < x < L.$$

7. (a) A vibrating string is fastened to air bearings situated on vertical rods at $x = 0$ and $x = 2$. Find the displacement $y(x,t)$ if the conditions are

$$y_x(0,t) = 0 \ \ 0 < t \qquad\qquad (8.26)$$
$$y_x(2,t) = 0 \ \ 0 < t$$
$$y(x,0) = x \ \ 0 < x < 2$$
$$\frac{\partial y}{\partial t}(x,0) = 0 \ \ 0 < x < 2$$

(b) Sketch $y(x,t)$ over $0 < x < 2$ for $t = 0$, $t = 1$, and $t = 2$.

Hint. Use only the first two terms of the series solution.

8. The pressure $p(x,t)$ in an organ pipe satisfies the differential equation

$$p_{xx} = \frac{1}{c^2} p_{tt}.$$

If the pipe is L meters long and open at both ends, find the pressure $p(x,t)$ if $p(x,t) = 0$ and $(\partial p/\partial t)(x,0) = 40$.

9. (a) Given the telegraph equation for finding voltage $e(x,t)$ on a transmission line of length a along with the boundary and initial conditions, we can write

$$e_{xx} = LCe_{tt} + (RC + GL)e_t + RG_ei, \qquad (8.27)$$

$$0 < x < a, \ 0 < t$$

$$\frac{\partial e}{\partial x}(0,t) = 0, \ \ 0 < t$$

$$\frac{\partial e}{\partial x}(a,t) = 0,$$

$$e(x,0) = 100, \ \ 0 < x < a$$

$$\frac{\partial e}{\partial t}(x,0) = 0 \ \ 0 < x < a.$$

Solve for $e(x,t)$ using separation of variables.

(b) What is the frequency of the third harmonic?

10. The length of a guitar string is 65 centimeters. If the string is plucked 15 centimeters from the bridge (i.e., at the end of the wire) by raising it 3 millimeters, find the displacement $u(x,t)$ using separation of variables.

11. The length of a piano string is 1 meter. When a pupil strikes a key, the following velocity is imparted to the string:

$$\frac{\partial u}{\partial t} = velocity = \begin{cases} 0 & 0 < x < 49 \ cm \\ 1 & 49 < x < 51 \ cm \\ 0 & 51 < x < 100 \ cm. \end{cases} \qquad (8.28)$$

Find the displacement $u(x,t)$ using separation of variables. Do not evaluate the Fourier coefficients of the final solution.

8.1.2   Non Cartesian Coordinate Systems

Examples 1, 3, and 4 were chosen especially so that you would see the method of separation of variables applied to the three classes of second-order differential equations: parabolic, hyperbolic, and elliptic. All were set in terms of rectangular coordinates. In Example 5 we will examine a boundary value problem that is more easily solved in polar coordinates.

Example 8.1.4 : *Laplace Equation on a Disk*

Let us consider a problem similar to Example 4 except that the shape of the plate is circular rather than rectangular. We wish to find the steady-state temperature $u(r, \theta)$ throughout a circular plate of radius c that is insulated laterally. The temperature on the circumference is $100°C$ over one semicircle and $0°$ over the other (see Figure 8.2).

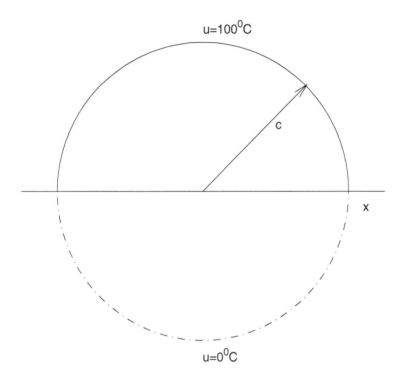

Figure 8.2 Laplace equation on a disk of radius c

Expressed formally, $u(r, \theta)$ must satisfy

$$r^2\frac{\partial^2 u}{\partial r^2} + r\frac{\partial u}{\partial r} + \frac{\partial^2 u}{\partial \theta^2} = 0, \quad -\pi < \theta < \pi$$

$$u(c, \theta) = \begin{cases} 100 & 0 < \theta < \pi \\ 0 & -\pi < \theta < 0. \end{cases} \qquad (8.29)$$

Solution 8.1.4 *To begin with we look for "elementary solutions" to this problem in the form*

$$u(r, \theta) = R(r)\Theta(\theta).$$

Substituting in the differential equation and using the method of separation of variables, we are led to the following two ordinary differential equations

$$\Theta'' + \lambda\Theta = 0 \tag{8.30}$$

$$r^2 R'' + rR' - \lambda R = 0. \tag{8.31}$$

When we attempt to solve Equation (9.72) (which is similar to the ones derived previously in Examples 1, 3, and 4), we notice we have no boundary conditions since there are no exposed radial edges. There are, however, "implicit boundary conditions" due the periodic nature of the θ coordinate. We require therefore that

$$u(r, -\pi) = u(r, \pi) \tag{8.32}$$
$$\frac{\partial u}{\partial \theta}(r, -\pi) = \frac{\partial u}{\partial \theta}(r, \pi) \quad 0 < r < c.$$

Such boundary conditions are called **periodic boundary conditions**.

As before, we must consider the three cases $\lambda > 0$, $\lambda = 0$, and $\lambda < 0$. If we let $\lambda = \alpha^2 > 0$, the differential equation in Equation (9.72) becomes

$$\Theta'' + \alpha^2\Theta = 0$$

where the solution is

$$\Theta(\theta) = A\cos\alpha\theta + B\sin\alpha\theta \tag{8.33}$$

and

$$\Theta'(\theta) = -\alpha A\sin\alpha\theta + \alpha B\cos\alpha\theta.$$

From the periodic boundary conditions in Equation (9.74) we see that

$$\Theta(-\pi) = \Theta(\pi), \quad \Theta'(-\pi) = \Theta'(\pi). \tag{8.34}$$

Substituting the solutions in the Equations (8.33) into the Equations (8.34), we have

$$A\cos\alpha(-\pi) + B\sin\alpha(-\pi) = A\cos\alpha\pi + B\sin\alpha\pi \tag{8.35}$$
$$-\alpha A\sin\alpha(-\pi) + \alpha B\cos\alpha(-\pi) = -\alpha A\sin\pi + \alpha B\cos\alpha\pi$$

which yields

$$B \sin \alpha \pi = 0 \ \ and \ \ A \sin \alpha \pi = 0. \qquad (8.36)$$

Since A and B cannot both be zero, it follows that

$$\sin \alpha \pi = 0$$

or

$$\alpha = n \ \ n = 1, 2, \ldots$$

or

$$\lambda_n = n^2$$

and

$$\Theta_n = A_n \cos n\theta + B_n \sin n\theta.$$

If $\lambda = 0$, then $\Theta'' = 0$ has the solution

$$\Theta = A\theta + B$$

Substituting in the boundary conditions in Equation (8.34), we find

$$A(-\pi) + B = A\pi + B$$

or $A = -A$, which implies $A = 0$. Therefore, $\lambda = 0$ is an eigenvalue, and the corresponding eigenfunction is an arbitrary constant that we will choose as 1.

It can be shown that there are no negative eigenvalues.

We now move to the solution of the other ordinary differential equation

$$r^2 R'' + r R' - n^2 R = 0. \qquad (8.37)$$

We recognize that this is a Cauchy-Euler differential. To obtain a solution to this differential equation, we attempt a solution of the form $R(r) = r^\alpha$, from which it follows that $R'(r) = \alpha r^{\alpha-1}$ and $R''(r) = \alpha(\alpha - 1)r^{\alpha-2}$. Substituting these values in Equation (8.37), we find that

$$r^\alpha [\alpha(\alpha - 1) + \alpha - n^2] = r^\alpha [\alpha^2 - n^2] = 0.$$

This equation must hold for all r, $0 < r < c$; therefore,

$$\alpha^2 - n^2 = 0 \ \ or \ \ \alpha = \pm n.$$

Since r^n and r^{-n} are linearly independent solutions to Equation (8.37), the general solution of the Cauchy-Euler equation is

$$R(r) = Cr^n + Dr^{-n} \quad n = 1, 2, 3, \ldots.$$

Once again there is no explicit boundary condition, but we observe that since $r = 0$ is the center of our circular plate, the

$$\lim_{r \to 0^+} r^{-n} = +\infty.$$

Since we are only considering bounded solutions, we must set $D = 0$ and

$$R_n(r) = r^n, \quad n = 1, 2, \ldots.$$

For $\lambda = 0$, the Cauchy-Euler equation becomes

$$r^2 R'' + r R' = 0 \tag{8.38}$$

whose solution is

$$R(r) = C + D lnr.$$

Since we are considering only bounded solutions, D must equal zero since

$$\lim_{r \to 0} lnr = -\infty$$

and therefore we take $R_0 = 1$ as the solution of Equation (8.38).

The solution to the partial differential equation is

$$u(r, \theta) = \frac{A_0}{2} + \sum_{n=1}^{\infty} r^n (A_n \cos n\theta + B_n \sin n\theta).$$

Using the inhomogeneous condition we can write

$$u(c, \theta) = \frac{A_0}{2} + \sum_{n=1}^{\infty} c^n (A_n \cos n\theta + B_n \sin n\theta)$$

where

$$A_0 = \frac{1}{\pi} \int_0^{\pi} 100 d\theta = 100 \tag{8.39}$$

$$c^n A_n = \frac{1}{\pi} \int_0^{\pi} 100 \cos n\theta d\theta = \frac{100}{n\pi} \sin n\theta \Big|_0^{\pi} = 0$$

and

$$c^n B_n = \frac{1}{\pi} \int_0^\pi 100 \sin n\theta d\theta = \frac{-100}{n\pi} \cos n\theta \Big|_0^\pi \qquad (8.40)$$

$$= \frac{100}{n\pi}[1 - (-1)^n]$$

$$= \begin{cases} 200/n\pi & n \text{ odd} \\ 0 & n \text{ even} \end{cases}$$

The solution to our problem is

$$u(r, \theta) = 50 + \frac{200}{\pi} \sum_{n=1,3,5}^\infty \left(\frac{r}{c}\right)^n \frac{\sin n\theta}{n}.$$

If we set $\theta = 0$ or π, we see immediately that the temperature $u(r, \theta)$ along the diameter $\theta = 0$ or π is 50 degrees. At points of symmetry with respect to this diameter, the temperature equals 50 plus or minus the value of the series. Notice that only odd harmonics of the fundamental frequency $\sin \theta$ exist.

Example 8.1.5 : *Flow Around A Cylinder*

*The steady-state flow of a fluid in a direction transverse to a long metal cylinder can be approximated at low velocities by Laplace's equation in two dimensions. We will see that the boundary conditions are of the Neumann type. If φ is the **velocity potential**, then*

$$\nabla^2 \varphi = r^2 \frac{\partial^2 \varphi}{\partial r^2} + r \frac{\partial \varphi}{\partial r} + \frac{\partial^2 \varphi}{\partial \theta^2} = 0$$

where the velocity $\mathbf{v} = grad\ \varphi = \nabla\varphi$. For our purposes it is natural to locate the cylinder at the origin. Its equation is then $x^2 + y^2 = c^2$. Far from the cylinder (i.e., as $r^2 = x^2 + y^2 \to +\infty$), the velocity $\mathbf{v}$ of the fluid is constant and parallel to the x-axis and can be written as $\mathbf{v}(x, y) = a\mathbf{i}$ as $r^2 \to +\infty$ (see Figure 8.3).

Solution 8.1.5 *To solve this problem we introduce polar coordinates and note that in these coordinates the velocity is given by*

$$\mathbf{v} = \mathbf{v}\varphi = \varphi_r \mathbf{e}_r + \frac{1}{r}\varphi_\theta \mathbf{e}_\theta \qquad (8.41)$$

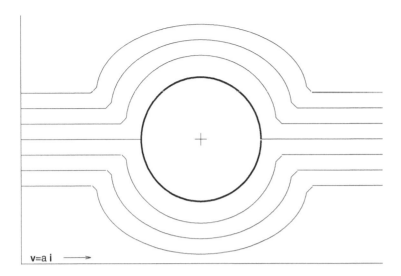

Figure 8.3 Flow around a cylinder

where $\mathbf{e}_r$ is the unit vector outward from the origin and $\mathbf{e}_\theta$, a unit vector, is perpendicular to $\mathbf{e}_r$ in a counterclockwise direction.

Looking at Figure 8.4 we observe that the boundary conditions at $r = +\infty$

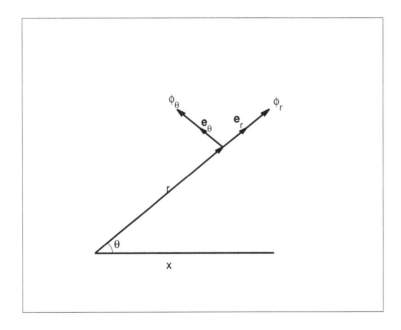

Figure 8.4 Radial coordinate system and the components of ϕ

are

$$\varphi_r = a\cos\theta, \quad \frac{1}{r}\varphi_\theta = -a\sin\theta. \tag{8.42}$$

Since the fluid cannot penetrate the cylinder, we must have $\varphi_r(c,\theta) = 0$. Finally, since the velocity at θ and $\theta + 2\pi$ is the same, we must have from Equation (8.41)

$$\varphi_r(r, \theta + 2\pi) = \varphi_r(r, \theta), \tag{8.43}$$
$$\varphi_\theta(r, \theta + 2\pi) = \varphi_\theta(r, \theta).$$

The method used in this problem is similar to that used in Example 5. We assume $\varphi(r, \theta) = R(r)\Theta(\theta)$, which (after substituting into Laplace's equation in polar form) yields the two ordinary differential equations and boundary conditions

$$\Theta'' + \lambda\Theta = 0, \quad \Theta(\theta + 2\pi) = \Theta(\theta), \quad \Theta'(\theta + 2\pi) = \Theta'(\theta) \tag{8.44}$$

$$r^2 R'' + R' - \lambda R = 0, \quad R'(c) = 0. \tag{8.45}$$

If $\lambda = \alpha^2 > 0$, then the general solution of $\Theta'' + \alpha^2\Theta = 0$ is

$$\Theta(\theta) = A\cos\alpha\theta + \beta\sin\alpha\theta$$

and

$$\Theta'(\theta) = -\alpha A\sin\alpha\theta + \alpha B\cos\alpha\theta.$$

Substituting these two equations into the periodic conditions in Equation (8.42), we find (after some effort) that

$$\alpha = n, \quad n = 1, 2, 3, \dots$$

or

$$\lambda_n = n^2$$

and

$$\Theta_n(\theta) = A_n\cos n\theta + B_n\sin n\theta.$$

With these eigenvalues the differential equation in Equation (8.45) becomes

$$r^2 R'' + r R' - n^2 R = 0 \quad n = 1, 2, \dots$$

whose solution is

$$R_n(r) = C_n r^n + D_n r^{-n}.$$

Using the boundary conditions in Equation (8.45), we have

$$R'(c) = 0 = n C_n c^{n-1} - n D_n c^{-n-1}$$

from which it follows that

$$D_n = c^{2n} C_n.$$

For $\lambda = 0$, $\Theta'' = 0$, and its solution is

$$\Theta_0 = A_0 + B_0 \theta.$$

The solution to the differential equation in Equation (8.45) is

$$R_0 = C_0 \ln r + D_0.$$

Now since $\ln r \to +\infty$ as $r \to +\infty$, we must set $C_0 = 0$ and we can let

$$R_0 = D_0 = 1.$$

Since there are no negative eigenvalues, we expect the solution to take the form

$$\varphi(r, \theta) = A_0 + B_0 \theta + \sum_{n=1}^{\infty} [r^n + c^{2n} r^{-n}][A_n \cos n\theta + B_n \sin n\theta]. \qquad (8.46)$$

We still have two conditions [Equation (8.42] that must be met. Differentiating Equation (8.46) with respect to r and θ, we have

$$\varphi_r = \sum_{n=1}^{\infty} [n r^{n-1} - n c^{2n} r^{-n-1}][A_n \cos n\theta + B_n \sin n\theta] \qquad (8.47)$$

and

$$\varphi_\theta = B_0 + \sum_{n=1}^{\infty} [r^n + c^{2n} r^{-n}] n [-A_n \sin n\theta + B_n \cos n\theta]. \qquad (8.48)$$

Now the only term in Equation (8.47) that is bounded occurs when $n = 1$; therefore, A_n, $B_n = 0$ for $n = 2, 3, \ldots$ and

$$\lim_{r \to \infty} \varphi_r(r, \theta) = a \cos \theta = A_1 \cos \theta + B_1 \sin \theta$$

or

$$A_1 = a$$

$$B_1 = 0.$$

Under these conditions Equation (8.48) becomes

$$\frac{1}{r}\varphi_\theta = \frac{B_0}{r} + \frac{1}{r}[r + c^2 r^{-1}][-A_1 \sin\theta] \tag{8.49}$$

$$= \frac{B_0}{r} + \left[1 + \frac{c^2}{r^2}\right][-a\sin\theta].$$

As $r \to +\infty$,

$$\frac{1}{r}\varphi_\theta = -a\sin\theta$$

and the second condition in Equation (8.42) is satisfied.

Combining all these conditions in Equation (8.46), we write the solution as

$$\varphi(r,\theta) = A_0 + B_0\theta + a\left[r + \frac{c^2}{r}\right]\cos\theta$$

where A_0 and B_0 are arbitrary constants.

As we investigate this solution, we must recall that $\varphi(r,\theta)$ is the velocity potential, not the velocity. The velocity field is given by

$$\mathbf{v} = \varphi_r \mathbf{e}_r + \frac{1}{r}\varphi_\theta \mathbf{e}_\theta = a\left[1 - \frac{c^2}{r^2}\right]\cos\theta\mathbf{e}_r - a\left[1 + \frac{c^2}{r^2}\right]\sin\theta\mathbf{e}_\theta.$$

When we measure the velocity far from the axis of the cylinder,

$$\mathbf{v} \approx a\cos\theta\mathbf{e}_r - a\sin\theta\mathbf{e}_\theta = a\mathbf{i}.$$

In other words, the effect on the velocity due to the cylinder becomes less noticeable as we move away from the axis. On the other hand, when we are near the cylindrical obstruction, that is $r \approx c$,

$$\mathbf{v} \approx -2a\sin\theta\mathbf{e}_\theta$$

which shows that the flow follows the shape of the cylinder when $r \approx c$ and approaches zero as we get nearer to the x-axis.

Example 8.1.6 : *Vibrations in a Sphere*

In all our examples up to this point it has been relatively easy to evaluate the eigenvalues and eigenfunctions. We will now examine a problem in which this is not the case. We wish to study the vibrations or pressure of air within a sphere of radius c which might be caused by an exploding firecracker, for example.

The pressure equation in space is given by

$$(\rho^2 p_\rho)_\rho + \frac{1}{\sin^2 \theta} p_{\varphi\varphi} + \frac{1}{\sin \theta}(p_\theta \sin \theta)_\theta = \frac{\rho^2}{a^2} p_{tt}. \tag{8.50}$$

Solution 8.1.6 *If we assume the pressure is independent of θ and φ (i.e., in the radial direction only), Equation (8.50) becomes*

$$\frac{1}{\rho^2}(\rho^2 p_\rho)_\rho = \frac{1}{a^2} p_{tt}. \tag{8.51}$$

Since the pressure under consideration is within the sphere, the directional derivative of p in the direction ρ must be zero on the boundary, or

$$p_\rho(c,t) = 0 \ \ t > 0.$$

Choosing a trial solution of the form $p(\rho, t) = R(\rho)T(t)$, we arrive at the two ordinary differential equations using the method of separation of variables on Equation (8.51):

$$\rho^2 R'' + 2\rho R' + \lambda \rho^2 R = 0 \ \ R'(c) = 0 \tag{8.52}$$

$$T'' + a^2 \lambda T = 0. \tag{8.53}$$

The general solution of Equation (8.52) for $\lambda = \alpha^2 > 0$ is

$$R(\rho) = A\frac{\cos \alpha\rho}{\rho} + B\frac{\sin \alpha\rho}{\rho}.$$

This solution can be found by replacing $R(\rho)$ by $S(\rho)/\rho$, which transforms the differential equation into one with constant coefficients (see Exercise 16). Since $p(\rho, t)$ must be bounded in the sphere, it is necessary that $A = 0$. trigonometric equation is difficult to solve in order to find the eigenvalues. We can easily

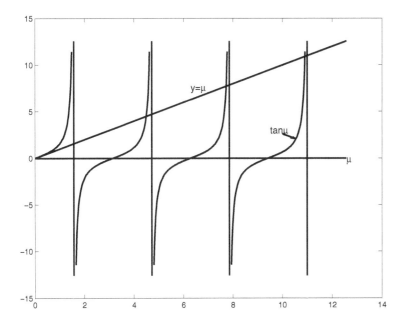

Figure 8.5 Eigenvalues μ_n are given by the intersections of $y = \mu$ with $tan\mu$

see that there are an infinite number of eigenvalues by solving the equation graphically, as shown in Figure 8.5.

The first nonzero root is 4.49, the second 7.73, and so on. Notice in particular that as $\mu = \alpha c \to \infty$, the intersection points approach $(2n + 1)\pi/2$, $n = 1, 2, \ldots$. Table 1 lists the first eight nonzero solutions done numerically. Using these values for μ we see that the eigenvalues λ satisfy

Table 8.1

Table 1	Solutions	of	$\tan \mu$	$= \mu$			
1	2	3	4	5	6	7	8
4.49	7.73	10.90	14.07	17.22	20.37	23.52	26.66

$$\lambda_n = \alpha_n^2 = \frac{\mu_n^2}{c^2} \ \ n = 1, 2, \ldots$$

and the eigenfunctions are

$$R_n(\rho) = \frac{\sin(\mu_n \rho/c)}{\rho}.$$

To see if $\lambda = 0$ is an eigenvalue, we solve the differential equation

$$R'' + \frac{2}{\rho}R' = 0$$

whose general solution is

$$R(\rho) = \frac{A}{\rho} + B.$$

Once again since $R(\rho)$ must be bounded in the sphere, $A = 0$ and $R(\rho) = B$. Since $R'(\rho) = 0$ for any ρ, it certainly follows that $R'(c) = 0$. Therefore, $\lambda = 0$ is an eigenvalue and its corresponding eigenfunctions $R_0(\rho) = 1$. This ends our search for eigenvalues, for you can show there are no negative eigenvalues.

We now proceed to the solution of Equation (8.53). For $\lambda = \mu_n^2/c^2$, $n = 1, 2, \ldots$, its solution is

$$T_n(t) = C_n \cos \frac{a\mu_n t}{c} + D_n \sin \frac{a\mu_n t}{c}$$

and for $\lambda = 0$,

$$T_0(t) = C_0 + D_0 t.$$

Combining the solutions R and T, we see that the answer to our boundary value problem takes the form

$$p(\rho, t) = C_0 + D_0 t + \sum_{n=1}^{\infty} \frac{\sin(\mu_n \rho/c)}{\rho} \qquad (8.54)$$

$$\times \left(C_n \cos \frac{a\mu_n t}{c} + D_n \sin \frac{a\mu_n t}{c} \right).$$

In order to solve for constants C_n and D_n, $n = 0, 1, 2, \ldots$, we need two initial conditions:

$$p(\rho, 0) = f(\rho) \ \text{ and } \ \frac{\partial p(\rho, 0)}{\partial t} = g(\rho) \ \ 0 < \rho < c.$$

When we attempt to solve for the coefficients, the series is not the Fourier series. But the family of functions does possess an orthogonal property, and we can solve for the coefficients as follows:

$$C_0 = \frac{3}{c^3} \int_0^c \rho^2 f(\rho) d\rho, \ \ D_0 = \frac{3}{c^3} \int_0^c \rho^2 g(\rho) d\rho \qquad (8.55)$$

$$C_n = \frac{2(1 + \mu_n^2)}{c\mu_n^2} \int_0^c \rho f(\rho) \sin \frac{\mu_n \rho}{c} d\rho$$

$$D_n = \frac{2(1 + \mu_n^2)}{a\mu_n^3} \int_0^c \rho g(\rho) \sin \frac{\mu_n \rho}{c} d\rho.$$

Exercises

1. You wish to find the temperature $u(\rho, \theta)$ in a laterally insulated pie-shaped region of radius c. The temperature satisfies the differential equation

$$\rho^2 u_{\rho\rho} + \rho u_\rho + u_{\theta\theta} = 0$$

and the boundary conditions

$$u(\rho, 0) = 0 \quad 0 < \rho < c$$
$$u\left(\rho, \frac{\pi}{6}\right) = 0 \quad 0 < \rho < c$$
$$u(c, \theta) = \theta \quad 0 < \theta < \frac{\pi}{6}.$$

Use the technique of separation of variables to find the temperature $u(\rho, \theta)$.

2. Find the steady-state temperature $u(\rho, \theta)$ in a circular plate insulated laterally of radius 10 if the temperature on the circumference is $3 - \theta$.

3. Find the electrostatic potential $\Phi(\rho, \theta)$ on the half disk plate with radius $25cm$ with the boundary conditions on the potential Φ as shown in Figure 8.6.

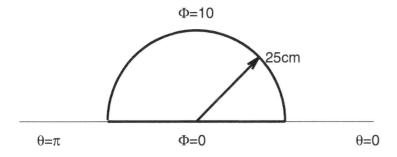

Figure 8.6 Laplace equation on the half disk plate

4. Show that Equation (8.52) can be transformed into one with constant coefficients by using the substitution $R(\rho) = S(\rho)/\rho$.

5. Evaluate and graph the velocity $\mathbf{v}$ of the fluid around the cylinder discussed in the previous section for $a = 1$, $r = 2c$, $\theta = 0$, $\pi/4$, $\pi/2$ radians.

6. Apply the method of separation of variables to Laplace's equation in Example 5 to find the two ordinary differential equations in Equations (9.72) and (9.73).

7. Show that there are no negative eigenvalues to be found in the boundary value problem in Equation (8.44).

8. Prove that the solutions of Equation (8.50) satisfy

$$\int_0^c \rho \sin \frac{\mu_n \rho}{c} d\rho = 0$$

where $\tan \mu_n = \mu_n$.

9. Show that

(a) $\int_0^c \sin \frac{\mu_n \rho}{c} \sin \frac{\mu_m \rho}{c} d\rho = 0 \ \ m \neq n$

(b) $\int_0^c \sin^2 \frac{\mu_n \rho}{c} d\rho = \frac{c}{2} \left(\frac{\mu m^2}{1+\mu m^2} \right) \ \ n = 1, 2, \ldots$

where $\tan \mu_n = \mu_n$.

10. Using the series in Equation (8.54) and the fact that $p(\rho, 0) = f(\rho)$, justify C_0 in Equation (8.55).

   Hint. Multiply both sides of Equation (8.53) by $\rho^2 d\rho$ and integrate from 0 to c.

11. Using the conditions in Exercise 22, justify C_n in Equation (8.55).

   Hint. Multiply both sides of Equation (8.54) by $\rho \sin(\mu_m \rho / c) d\rho$ and integrate from 0 to c.

12. The coefficients D_n can be found easily from the equation for C_n in Equation (8.55). Using the condition $\partial p(\rho, 0)/\partial t = g(\rho)$, prove D_n in Equation (8.55).

13. Find $p(\rho, t)$ if $c = 1$ and

$$p_\rho(1, t) = 0 \ \ t > 0$$

$$p(\rho, 0) = 0 \ \ 0 < \rho < 1$$

$$p_t(\rho, 0) = \begin{cases} \rho & 0 < \rho < \frac{1}{2} \\ 0 & \frac{1}{2} < \rho < 1. \end{cases}$$

14. Write out the boundary value problem for finding the pressure $p(\rho, t)$ in a spherical region if the pressure is zero at 100 meters from the center and the initial pressure is zero, and the rate of change of pressure with respect to t is $h(p)$.

15. Consider two concentric spheres of radius a and b, $a < b$, respectively. If the pressure on the inner sphere is 1 and the rate of change of pressure with respect to ρ on the outer sphere is 0, and the initial pressure is zero and the rate of change of pressure with respect to t is ρ, write the partial differential equation and boundary and initial conditions satisfied.

8.1.3 Boundary Value Problems with General Initial Conditions

We will now examine the method of solution of boundary value problems where there are several initial conditions and two or more are inhomogeneous. There are two well-known methods for solving such a problem.

The first approach consists of solving two boundary value problems, as we have done previously. Suppose we wish to solve a vibrating string problem where both ends are fixed and initially the string is not only displaced but a velocity is imparted to the string (see Figure 8.7).

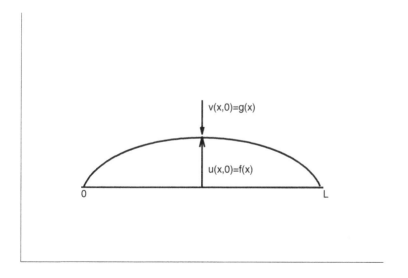

Figure 8.7 Vibrating string, u is the displacement $v = \frac{\partial u}{\partial t}$

Formally, this problem is written as

$$\frac{\partial^2 u}{\partial t^2} = c^2 \frac{\partial^2 u}{\partial x^2}, \quad 0 < x < L, \; 0 < t \tag{8.56}$$
$$u(0,t) = 0, \; 0 < t, \quad u(L,t) = 0, \; 0 < t,$$
$$u(x,0) = f(x), \; 0 < x < L, \quad \frac{\partial u}{\partial t}(x,0) = g(x), \; 0 < x < L$$

We now assume that the solution to this problem can be written in the form $u(x,t) = v(x,t) + w(x,t)$ where v and w satisfy the following boundary value problems, respectively:

$$\frac{\partial^2 v}{\partial t^2} = c^2 \frac{\partial^2 v}{\partial x^2} \quad \frac{\partial^2 w}{\partial t^2} = c^2 \frac{\partial^2 w}{\partial x^2}$$

$$v(0,t) = 0 \quad w(0,t) = 0 \quad 0 < t$$

$$v(L,t) = 0 \quad w(L,t) = 0 \quad 0 < t$$

$$v(x,0) = f(x) \quad w(x,0) = 0 \quad 0 < x < L$$

$$\frac{\partial v}{\partial t}(x,0) = 0 \quad \frac{\partial w}{\partial t}(x,0) = g(x) \quad 0 < x < L.$$

It follows from the method of superposition that if v and w satisfy their boundary value problem, u satisfies the problem in Equation (8.56).

The second way of solving might be called a straightforward attack. We solve Equation (8.56) directly as we have done in earlier examples (see Example 3). The eigenvalue problem will be the same as before, but when we come to solve the other ordinary differential equation, we will have no initial conditions. When we multiply X_n by T_n and then form the infinite linear combination to find $u(x,t)$, we will find it necessary to solve for two sets of arbitrary constants. Using one initial condition at a time, we are led to solve Fourier series problems from which we evaluate our two sets of constants.

Example 8.1.7 : *Consider a vibrating string fastened to air bearings that move along two parallel rods 4 meters apart. Find the displacement $u(x,t)$ if the initial displacement is 1 meter and the initial velocity is x meter, per second.*

Solution 8.1.7 *The formal statement of this problem is given as*

$$\frac{\partial^2 u}{\partial x^2} = c^2 \frac{\partial^2 u}{\partial t^2} \quad 0 < x < 4, \ 0 < t$$

$$\frac{\partial u}{\partial x}(0, t) = 0 \quad 0 < t$$

$$\frac{\partial u}{\partial x}(4, t) = 0 \quad 0 < t$$

$$u(x, 0) = 1 \quad 0 < x < 4$$

$$\frac{\partial u}{\partial t}(x, 0) = x \quad 0 < x < 4.$$

Using the method of separation of variables with $u(x, t) = X(x)T(t)$, we see easily that

$$X'' + \lambda X = 0, \quad X'(0) = 0, \quad X'(4) = 0 \tag{8.57}$$

$$T'' + \frac{\lambda}{c^2} T = 0. \tag{8.58}$$

Solving the eigenvalue problem for Equation (8.57), we find the eigenvalues are

$$\lambda_n = \frac{n^2 \pi^2}{16}$$

and the eigenfunctions are

$$X_n = \cos \frac{n \pi x}{4} \quad n = 0, 1, 2, \ldots.$$

Our next step is to solve the differential equation in Equation (8.58). Since λ is known, the solution of this equation is

$$T_0(t) = A_0 + B_0 t$$

$$T_n(t) = A_n \cos \frac{n \pi t}{4c} + B_n \sin \frac{n \pi t}{4c} \quad n = 1, 2, \ldots.$$

Unlike previous examples, no initial conditions are attached to the differential equation in Equation (8.58) and therefore we cannot evaluate the $A_n s$ or $B_n s$ at this stage. We continue by assuming a solution of the form

$$u(x, t) = \frac{1}{2}(A_0 + B_0 t) + \sum_{n=1}^{\infty} \left[A_n \cos \frac{n \pi t}{4c} + B_n \sin \frac{n \pi t}{4c} \right] \cos \frac{n \pi x}{4} \tag{8.59}$$

from which it follows that

$$\frac{\partial u}{\partial t}(x,t) = \frac{B_0}{2} + \sum_{n=1}^{\infty} \left(\frac{n\pi}{4c}\right)\left(-A_n \sin\frac{n\pi t}{4c} + B_n \cos\frac{n\pi t}{4c}\right)\cos\frac{n\pi x}{4}.$$

Using the initial condition $u(x,0) = 1$, we have

$$1 = \frac{A_0}{2} + \sum_{n=1}^{\infty} A_n \cos\frac{n\pi x}{4}$$

which is a Fourier cosine series whose coefficients are $A_0 = 2$ and $A_n = 0$. In the same way, knowing that $(\partial u/\partial t)(x,0) = x$, we can write

$$x = \frac{B_0}{2} + \sum_{n=1}^{\infty} \frac{n\pi}{4c} B_n \cos\frac{n\pi x}{4}$$

on the interval $(0,4)$. This is a Fourier cosine series representing an even function by extending x into $(-4,0)$ as $-x$. Our calculations yield

$$B_0 = 4 \quad and$$

$$B_n = \begin{cases} \frac{-64c}{n^3\pi^3} & n \text{ odd} \\ 0 & n \text{ even.} \end{cases}$$

Substituting these coefficients into Equation (8.59), the solution to our problem is

$$u(x,t) = 1 + 2t - \frac{64c}{\pi^3}\sum_{1,3,5}^{\infty}\frac{1}{n^3}\sin\frac{n\pi t}{4c}\cos\frac{n\pi x}{4}. \tag{8.60}$$

We see from the solution to Equation (8.60) that the center of the string, $x = 2$, has the displacement $u(2,t) = 1 + 2t$. This fact tells us that once the string is put in motion, the center moves away from $u(2,0) = 1$ at a constant velocity of 2 meters per second. When $t = 4cq$, where $q = 0, 1, 2, \ldots$, the string becomes parallel to the x-axis.

The factor $\sin(n\pi t/4c)$ in Equation (8.60) allows us to determine the frequency of a vibrating string. The period λ_n of the individual terms in the series (8.60) is given by

$$\lambda_n = period\,of\,oscillations = \frac{2\pi}{n\pi}4a = \frac{8c}{n}$$

from which we can find the frequency f_n which is related to the period by

$$f_n = \frac{1}{\lambda_n} = \frac{n}{8c}.$$

Now the smallest n allowed in the series (8.60), in this case $n = 1$, determines the **fundamental frequency**, *that is*

$$f_1 = \frac{1}{8c}.$$

Notice that all higher frequencies are integer multiples of the fundamental frequency and are called **harmonics**. *The value $n = 2$ is the second harmonic, $n = 3$ the third harmonic, and so on.*

8.1.4 Boundary Value Problems with Inhomogeneous Equations

In some boundary value problems the differential equation and the boundary conditions may be inhomogeneous. In general, there is no straightforward way to solve these problems. However, if the inhomogeneous part of the differential equation is a function of x and the boundary conditions are constants, there is a step-by-step way to solve such a problem. The method is best shown by example.

Example 8.1.8 *Suppose we wish to find the temperature $u(x,t)$ in a laterally insulated rod of length π whose initial temperature is $f(x)$. For $t > 0$, the left end of the rod is fixed at $500°$ and the right end is fixed at a temperature of $100°$. Furthermore, for $t > 0$, an electric current is made to pass through the rod heating it, which introduces the $\sin x$ term in the partial differential equation.*

Solution 8.1.8 *Formally stated, this boundary value problem looks like*

$$\frac{\partial u}{\partial t} = k\frac{\partial^2 u}{\partial x^2} + \sin x$$

$$u(0, t) = 500 \ \ 0 < t$$

$$u(\pi, t) = 100 \ \ 0 < t$$

$$u(x, 0) = f(x) \ \ 0 < x < \pi.$$

We assume the solution can be broken into two parts; that is,

$$u(x,t) = v(x,t) + h(x). \tag{8.61}$$

Our plan of attack is to choose constants of iteration so that eventually the $v(x,t)$ term will be a solution of a **homogeneous boundary value problem**.

Substituting the right-hand side of Equation (8.61) into the differential equation, we have

$$\frac{\partial v}{\partial t} = k\frac{\partial^2 v}{\partial x^2} + kh''(x) + \sin x.$$

Therefore, to follow our plan, $kh''(x) = -\sin x$. Solving this simple differential equation, we find

$$h(x) = C_1 x + C_2 + \frac{\sin x}{k}. \tag{8.62}$$

Next we look at the boundary conditions, which can be written as

$$v(0,t) + h(0) = 500$$

$$v(\pi,t) + h(\pi) = 100.$$

In order that $v(0,t)$ and $v(\pi,t)$ equal zero,

$$h(0) = 500 \ \text{ and } \ h(\pi) = 100.$$

We use these conditions to find the specific values of C_1 and C_2 in Equation (8.62). Therefore,

$$h(0) = 500 = C_1(0) + C_2 + \frac{\sin 0}{k}$$

which implies that $C_2 = 500$. Then,

$$h(\pi) = 100 = C_1\pi + 500 + \frac{\sin \pi}{k}$$

or

$$C_1 = \frac{-400}{\pi}.$$

Substituting these constants in the solution [Equation (8.62)], it follows that

$$h(x) = \frac{-400}{\pi}x + 500 + \frac{\sin x}{k}.$$

We have completely determined $h(x)$. Our next task is to find $v(x,t)$, which now satisfies the boundary value problem

$$\frac{\partial v}{\partial t} = k\frac{\partial^2 v}{\partial x^2}$$

$$v(0,t) = 0 \ \ 0 < t$$

$$v(\pi,t) = 0 \ \ 0 < t$$

$$v(x,0) = u(x,0) - h(x) = f(x) - \frac{\sin x}{k} + \frac{400}{\pi}x - 500 \ \ 0 < x < \pi.$$

If $f(x) = (\sin x/k) + 500$, then $v(x,0) = (400/\pi)x$. Using the method of separation of variables and setting $v(x,t) = X(x)T(t)$, we are led to the two differential equations and boundary conditions

$$X''(x) + \lambda X(x) = 0 \ \ X(0) = X(\pi) = 0$$

$$T''(t) + \lambda k T(t) = 0.$$

The eigenvalues are $\lambda = n^2$, $n = 1, 2, \ldots$, and the eigenfunctions are $X_n(x) = \sin nx$. Solving the other equation, find $T_n(t) = e^{-kn^2 t}$.

Combining this information, we expect our solution to be of the form

$$v(x,t) = \sum_{n=1}^{\infty} b_n e^{-kn^2 t} \sin nx.$$

And since

$$v(x,0) = \frac{400}{\pi}x = \sum_{n=1}^{\infty} b_n \sin nx$$

we see that

$$b_n = \frac{2}{\pi} \int_0^{\pi} \frac{400x}{\pi} \sin nx \tag{8.63}$$

$$= (-1)^{n+1}\frac{800}{n\pi}.$$

The solution to the homogeneous boundary value problem is

$$v(x,t) = \frac{800}{\pi} \sum_{n=1}^{\infty} \infty \frac{(-1)^n}{n} e^{-kn^2 t} \sin nx.$$

The solution to the inhomogeneous problem using the principle of superposition is

$$u(x,t) = \frac{800}{\pi} \sum_{n=1}^{\infty} \frac{(-1)^n}{n} e^{-kn^2 t} \sin nx - \frac{400}{\pi}x + 500 + \frac{\sin x}{k}.$$

This problem is one that commonly occurs in heat flow problems. We observe that the answer for $u(x,t)$ consists of two parts: the infinite series, which depends on x and t, and the remaining part $h(x)$, which depends only on x. Because of the exponential term in $v(x,t)$ we see that as $t \to +\infty$ the value of the series approaches zero. This part of the solution is called the **transient** solution because it passes away quickly. The other part of the solution $h(x)$ does not vary with time and is called the **steady-state** solution.

Once again we notice that because the factor n^2 appears in the exponential term in the series, the terms decrease quite rapidly in size for even small values of t. Therefore, we can often get a good approximation of the transient term by taking only one or two terms.

Exercises

1. Given the wave equation $y_{tt} = c^2 y_{xx}$ and the conditions

$$y(0,t) = 0 \ \ 0 < t$$

$$y(L,t) = 0 \ \ 0 < t$$

$$y(x,0) = 10 \ \ 0 < x < L$$

$$y_t(x,0) = -5 \ \ 0 < x < L$$

use the technique of separation of variables to find $y(x,t)$.

2. Find the voltage $e(x,t)$ on a high-frequency line of length 20 centimeters if both ends of the line are shorted. The initial voltage $e(x,0)$ is $20(1-x)$; $e_t(x,0)$ is $20x$.

3. Find the longitudinal displacement of a rod of length L whose ends are free if the initial displacement is $x(L - x)$ and $u_t(x,0)$ is 2.

4. Given the wave equation and the conditions

$$u_{tt} = c^2 u_{xx} + e^{-x}$$

$$u(0,t) = 100 \ \ 0 < t$$

$$u(L,t) = 50 \ \ 0 < t$$

$$u(x,0) = 0 \quad 0 < x < L$$

$$u_t(x,0) = x \quad 0 < x < L$$

and letting $u(x,t) = v(x,t) + \Phi(x)$, solve for Φ completely and write (but do not solve) the boundary value problem for $v(x,t)$.

5. The voltage $e(x,t)$ satisfies the differential equation $e_{xx} = RCe_t + Ax$. Using separation of variables, find e if

$$\frac{\partial e}{\partial x}(0,t) = 0 \quad 0 < t$$

$$\frac{\partial e}{\partial x}(L,t) = 0 \quad 0 < t$$

$$e(x,0) = 50 \quad 0 < x < L.$$

6. The current in a submarine cable of length 1000 meters is given by

$$i_{xx} = RCi_t.$$

If the conditions are

$$i(0,t) = 2a \quad 0 < t$$

$$i(1000,t) = 0.1a \quad 0 < t$$

$$i(x,0) = 0a \quad 0 < x < 1000$$

find the current $i(x,t)$.

7. Solve the heat equation with decomposition

$$u_{xx} - ku_t + A = 0$$

with conditions

$$u(0,t) = 0 \quad 0 < t$$

$$u(L,t) = 0 \quad 0 < t$$

$$u(x,0) = 0 \quad 0 < x < L$$

8. The equation for a vibrating string with external force is given by $y_{tt} = c^2 y_{xx} + F/\delta$, where $F = \delta x$ and where the left-hand end $x = 0$ is fixed and the slope of the right-hand end $x = L$ is zero. The initial displacement is $f(x)$ while initial velocity is zero. Let $y(x,t) = z(x,t) + \Phi(x)$. Solve for $\Phi(x)$ completely and write (but do not solve) the boundary value problem for $z(x,t)$.

9. Poisson's equation is given by $u_{xx} + u_{yy} = y$. If the conditions are

$$u(0,y) = \frac{5}{6} + \frac{y^3}{6} \quad 0 < y < 1$$

$$u(1,y) = \frac{y^3}{6} - \frac{1}{6} \quad 0 < y < 1$$

$$u(x,0) = \frac{-1}{6} \quad 0 < x < 1$$

$$u(x,1) = 0 \quad 0 < x < 1$$

solve for $u(x,y)$.

Hint. Let $u(x,y) = w(x,y) + g(y)$.

10. The equation of a vibrating string with an external force is given by $u_{tt} = c^2 u_{xx} + \sin x$. The boundary and initial conditions are $u(0,t) = 0$, $u_x(L,t) = -1$, $u(x,0) = 0$, and $u_t(x,0) = f(x)$. Let $u(x,t) = w(x,t) + \Phi(x)$. Find $\Phi(x)$ so that $w(x,t)$ satisfies a homogeneous boundary value problem. Write out the differential equation and conditions satisfied by $w(x,t)$, but do not solve the equation for w.

8.2 GREEN'S FUNCTIONS

When boundary value problems are solved using separation of variables, the solution is represented by an infinite series. However, a series solution is neither the only possible approach to the solution of boundary value problems nor always the best. In this appendix we introduce a method that represents the solution in terms of an integral. This method is usually referred to as "Green's function method."

Example 8.2.1 *Consider Poisson equation*

$$\nabla^2 u(\mathbf{x}) = f(\mathbf{x}) \tag{8.64}$$

on a domain D of R^2 having a piecewise smooth boundary with the boundary conditions

$$u(\mathbf{x})|_{\partial D} = g(\mathbf{x}) \tag{8.65}$$

where ∂D is the boundary of D. We want to find a function $G(\mathbf{x}, \boldsymbol{\eta})$, $\boldsymbol{\eta} \in R^2$, which is called the **Green's function** *for the problem and depends only on the differential operator and the domain D so that the solution of Equations (8.64) and (8.65) can be represented as*

$$u(\mathbf{x}) = \int_D G(\mathbf{x}, \boldsymbol{\eta}) f(\boldsymbol{\eta}) d\boldsymbol{\eta} + \int_{\partial D} g(\boldsymbol{\eta}) \frac{\partial G}{\partial \boldsymbol{\eta}} ds \tag{8.66}$$

where

$$\frac{\partial G}{\partial \boldsymbol{\eta}} = grad G \cdot \mathbf{n} = \left[\frac{\partial G}{\partial \eta_1}, \frac{\partial G}{\partial \eta_2} \right] \cdot \mathbf{n}$$

and $\mathbf{n}$ is the outward normal to the boundary ∂D.

To see some of the advantages for using such an integral representation of the solution, we note that Equation (8.66) holds for all possible $f(\mathbf{x})$ and $g(\mathbf{x})$, thus providing us with the solution to Equations (8.64) and (8.65) even when the method of separation of variables is not applicable. Furthermore, the dependence of the solution on this function is explicit and in closed form. Therefore, it enables us to investigate the dependence of the solution on these elements of the problem directly.

From an applied point of view, Green's functions have a natural physical interpretation as the influence of a unit source located at $\boldsymbol{\eta}$ on the point $\mathbf{x}$. Accordingly, we can refer to $G(\mathbf{x}, \boldsymbol{\eta})$ as the "influence function." We emphasize this interpretation of the Green's functions throughout this presentation and use it to infer some of their properties.

The major undertaking in this solution methodology is the computation of the Green's function, which appears as a kernel in the integral representation of the solution.

Green's Function for the Laplace Operator

Definition: The solution of

$$\nabla^2 G(\mathbf{x}, \boldsymbol{\eta}) = \delta(\mathbf{x} - \boldsymbol{\eta}) \tag{8.67}$$

on a domain D in R^n satisfying the boundary condition

$$G(\mathbf{x}, \boldsymbol{\eta})|_{\partial D} = 0 \tag{8.68}$$

is called the Green's function for the Laplace operator in D. Here $\delta(\mathbf{x})$ denotes the Dirac delta function.

Theorem 1 G is symmetric in $\mathbf{x}$ and $\boldsymbol{\eta}$; that is

$$G(\mathbf{x}, \boldsymbol{\eta}) = G(\boldsymbol{\eta}, \mathbf{x}).$$

Proof. To prove this theorem we use Green's lemma, which states that under proper restrictions on ϕ, ψ, and D, the following formula holds:

$$\int_D (\psi \nabla^2 \phi - \phi \nabla^2 \psi) dA = \int_{\partial D} \left(\psi \frac{\partial \phi}{\partial \mathbf{n}} - \phi \frac{\partial \psi}{\partial \mathbf{n}} \right) ds. \tag{8.69}$$

Substituting $\psi = G(\mathbf{x}, \boldsymbol{\eta})$ and $\phi = G(\mathbf{x}, \boldsymbol{\eta}^*)$, where $\boldsymbol{\eta}, \boldsymbol{\eta}^*$ are arbitrary but fixed points in D, in Equation (8.69) we obtain

$$\int_D [G(\mathbf{x}, \boldsymbol{\eta}) \nabla^2 G(\mathbf{x}, \boldsymbol{\eta}^*) - G(\mathbf{x}, \boldsymbol{\eta}^*) \nabla^2 G(\mathbf{x}, \boldsymbol{\eta})] dA =$$
$$\int_{\partial D} [G(\mathbf{x}, \boldsymbol{\eta}) \frac{\partial G}{\partial \mathbf{n}}(\mathbf{x}, \boldsymbol{\eta}^*) - G(\mathbf{x}, \boldsymbol{\eta}^*) \frac{\partial G}{\partial \mathbf{n}}(\mathbf{x}, \boldsymbol{\eta})] ds. \tag{8.70}$$

Using Equations (8.67),(8.68) this yields

$$\int_D [G(\mathbf{x}, \boldsymbol{\eta}) \delta(\mathbf{x} - \boldsymbol{\eta}^*) - G(\mathbf{x}, \boldsymbol{\eta}^*) \delta(\mathbf{x} - \boldsymbol{\eta})] dA = 0.$$

Hence

$$G(\mathbf{x}, \boldsymbol{\eta}) = G(\boldsymbol{\eta}, \mathbf{x}).$$

Since $\boldsymbol{\eta}, \boldsymbol{\eta}^*$ are arbitrary points in D this proves the theorem. **Theorem 2.**

$\frac{\partial G}{\partial \mathbf{n}}(\mathbf{x}, \boldsymbol{\eta})$ has a discontinuity at $\mathbf{x} = \boldsymbol{\eta}$. Moreover,

$$\lim_{\epsilon \to 0} \int_{C_\epsilon} \frac{\partial G}{\partial \mathbf{n}} ds = 1 \tag{8.71}$$

where C_ϵ is the circle of radius ϵ around $\mathbf{x} = (x_1, x_2)$; that is

$$(x_1 - \eta_1)^2 + (x_2 - \eta_2)^2 = \epsilon^2.$$

Proof. Let D_ϵ be the domain bounded by the circle C_ϵ. From (8.67) we infer that

$$\int_{D_\epsilon} \nabla^2 G dA = \int_{D_\epsilon} \delta(\mathbf{x} - \boldsymbol{\eta}) = 1.$$

Hence, using the divergence theorem we infer that

$$1 = \int_{D_\epsilon} \nabla^2 G dA = \int_{C_\epsilon} \nabla G \cdot \mathbf{n} ds = \int_{C_\epsilon} \frac{\partial G}{\partial \mathbf{n}} ds.$$

This shows that $\partial G / \partial \mathbf{n}$ has a discontinuity at $\mathbf{x} = \boldsymbol{\eta}$ since otherwise (i.e. if $\partial G / \partial \mathbf{n}$ was continuous) the integral in Equation (8.71) would have to be equal to 0.

Theorem 3. The solution of the Dirichlet problem in Equations (8.64), (8.65) is given by Equation (8.66) where G is the Green's function for the Laplace operator.

Proof. We prove that if $u(\mathbf{x})$ satisfies Equations (8.64), (8.65) then it must satisfy Equation (8.66). To prove this statement we use Green's lemma with

$$\psi(\boldsymbol{\eta}) = G(\boldsymbol{\eta}, \mathbf{x}), \quad \phi(\boldsymbol{\eta}) = u(\boldsymbol{\eta}).$$

We obtain

$$\int_D [G(\mathbf{x}, \boldsymbol{\eta}) \nabla^2 u(\boldsymbol{\eta}) - u(\boldsymbol{\eta}) \nabla^2 G(\boldsymbol{\eta}, \mathbf{x})] d\boldsymbol{\eta} =$$
$$\int_{\partial D} [G(\boldsymbol{\eta}, \mathbf{x}) \frac{\partial u(\boldsymbol{\eta})}{\partial \mathbf{n}} - u(\boldsymbol{\eta}) \frac{\partial G}{\partial \mathbf{n}}(\mathbf{x}, \boldsymbol{\eta})] ds. \qquad (8.72)$$

Using Equations (8.64),(8.65),(8.67),(8.68) this reduces to

$$\int_D [G(\boldsymbol{\eta}, \mathbf{x}) f(\boldsymbol{\eta}) - u(\boldsymbol{\eta}) \delta(\boldsymbol{\eta} - \mathbf{x})] d\boldsymbol{\eta} = - \int_{\partial D} g(\boldsymbol{\eta}) \frac{\partial G}{\partial \mathbf{n}} ds. \qquad (8.73)$$

The desired result follows now from the symmetry of $G(\boldsymbol{\eta}, \mathbf{x})$ (Theorem 1) and the definition of the Dirac δ function.

At this juncture it is natural to inquire into the intuitive (or physical) meaning of G and Equation (8.66). To answer this question we note that the solution of Poisson equation (8.64) can be interpreted as the gravitational potential due to a source distribution $f(\mathbf{x})$. It follows then that Green's function which satisfies Equation (8.67) represents the gravitational potential at $\mathbf{x}$ due to a unit source at $\boldsymbol{\eta}$. Equation (8.66) can be interpreted simply as a restatement of the superposition principle: that is the total gravitational potential due to a volume distribution $\mathbf{x}$ and surface distribution g is equal to the sum (represented by the integral) of the pointwise contributions of these sources.

Computation of Green's Function

Different techniques exist for the computation of Green's function for a given differential operator on a domain D. In this section we give two examples for the computation of this function for the Laplace operator.

Example 8.2.2 *To begin with we compute Green's function for the Laplace operator when $D = R^2$. In this case the domain has no boundaries and Green's function has to satisfy only Equation (8.67). This function is referred to as the "infinite space Green's Function".*

Solution 8.2.1 *To compute G in this case we have to solve*

$$\nabla^2 G = \delta(\mathbf{x} - \boldsymbol{\eta}). \tag{8.74}$$

Since the right-hand side of Equation (8.74) depends only on $\mathbf{x} - \boldsymbol{\eta}$ we attempt to find G in the form

$$G(x, y, \eta_1, \eta_2) = G(r)$$

where $r^2 = (x - \eta_1)^2 + (y - \eta_2)^2$. However, since $\delta(\mathbf{x} - \boldsymbol{\eta})$ is zero for $\mathbf{x} \neq \boldsymbol{\eta}$ it follows that G must satisfy

$$\nabla^2 G = 0, \quad for \ \ \mathbf{x} \neq \boldsymbol{\eta}.$$

Using the expression of the Laplace operator in polar coordinates this equation reduces to

$$\frac{1}{r}\frac{\partial}{\partial r}\left(r\frac{\partial G}{\partial r}\right) = 0, \quad r \neq 0 \tag{8.75}$$

whose solution is

$$G = C_1 + C_2 \ln r$$

where C_1, C_2 are constants. To determine C_2 we use Equation (8.71). We obtain

$$1 = \lim_{\epsilon \to 0} \int_{C_\epsilon} \frac{\partial G}{\partial \mathbf{n}} ds = \lim_{\epsilon \to 0} \int_0^{2\pi} \frac{C_2}{r} r d\theta = 2\pi C_2.$$

Hence

$$G = C_1 + \frac{1}{2\pi} \ln r.$$

Although C_1 remains arbitrary, we set it to zero for convenience.

Example 8.2.3 *Compute the Green's function for the Laplace operator on the unit disk in R^2.*

Solution 8.2.2 *The usual technique to compute Green's function on a domain with boundary is to rewrite it as a sum*

$$G = G_1 + v$$

where G_1 is the infinite space Green's function. For the Laplace operator we have

$$\nabla^2 G = \nabla^2 G_1 + \nabla^2 v = \delta(\mathbf{x} - \boldsymbol{\eta}).$$

However G_1 satisfies Equation (8.74); hence, v must satisfy

$$\nabla^2 v = 0.$$

Moreover, from (8.68) we infer that on the boundary

$$v|_{\partial D} = -G_1|_{\partial D}.$$

It follows then that v is a regular solution of Laplace equation whose values on ∂D are equal to those of $-G_1$ on this boundary.

To solve the boundary value problem

$$\frac{\partial^2 v}{\partial x^2} + \frac{\partial^2 v}{\partial y^2} = 0 \tag{8.76}$$

$$v \big|_{x^2+y^2} = -\frac{1}{2\pi} \ln r \big|_{x^2+y^2}$$

where $r^2 = (x - \eta_1)^2 + (y - \eta_2)^2$ (note that $\boldsymbol{\eta}$ is considered as a fixed but *arbitrary point on the disk). We now introduce polar coordinates for x, y and η_1, η_2:*

$$x = \rho \cos \theta, \quad y = \rho \sin \theta$$

$$\eta_1 = \sigma \cos \phi, \quad \eta_2 = \sigma \sin \phi.$$

Solving Equation (8.76) by the method of separation of variables we have

$$v = \frac{a_0}{2} + \sum_{n=1}^{\infty} \rho^n (a_n \cos n\theta + b_n \sin n\theta). \tag{8.77}$$

To satisfy the boundary condition we must have (using the cosine rule)

$$v \mid_{\rho=1} = -\frac{1}{4\pi} \ln[1 + \sigma^2 - 2\sigma \cos(\theta - \phi)] = \frac{1}{2\pi} \sum_{n=1}^{\infty} \frac{\sigma^n \cos(\theta - \phi)}{n}. \tag{8.78}$$

From Equations(8.77), (8.78) it follows that:

$$a_0 = 0, \quad a_n = \frac{\sigma^n \cos n\phi}{2\pi n}, \quad b_n = \frac{\sigma^n \sin n\phi}{2\pi n}. \tag{8.79}$$

Combining Equation (8.77) and Equation (8.79) we have

$$v(\rho, \theta, \sigma, \phi) = \frac{1}{2\pi} \sum_{n=1}^{\infty} \frac{(\rho\sigma)^n}{n} \cos n(\theta - \phi) \tag{8.80}$$

$$= -\frac{1}{4\pi} \ln[1 + (\rho\sigma)^2 - 2\rho\sigma \cos(\theta - \phi)].$$

Hence

$$G = \frac{1}{4\pi} \ln[\rho^2 + \sigma^2 - 2\rho\sigma \cos(\theta - \phi)] - \frac{1}{4\pi} \ln[1 + (\rho\sigma)^2 - 2\rho\sigma \cos(\theta - \phi)]. \tag{8.81}$$

Corollary. The solution of Laplace equation

$$\nabla^2 u(\rho, \theta) = 0 \tag{8.82}$$

on the unit disk subject to the boundary condition

$$u \mid_{\rho=1} = g(\theta) \tag{8.83}$$

is

$$u(\rho, \theta) = \int_0^{2\pi} P(\rho, \theta - \phi) g(\phi) d\phi \tag{8.84}$$

where P is the Poisson kernel

$$P(\rho, \theta - \phi) = \frac{1 - \rho^2}{1 + \rho^2 - 2\rho\cos(\theta - \phi)}.$$

Proof. From Equation (8.66) we know that the solution of Equations (8.82), (8.83) can be expressed as

$$u(\rho, \theta) = \int_{\partial D} g(\boldsymbol{\eta}) \frac{\partial G}{\partial \mathbf{n}} ds. \tag{8.85}$$

In this case, however,

$$\frac{\partial G}{\partial \mathbf{n}}\Big|_{\partial D} = \left(\frac{\partial G}{\partial \sigma}\right)\Big|_{\sigma = 1} = \frac{1 - \rho^2}{1 + \rho^2 - 2\rho\cos(\theta - \phi)}$$

which yields the desired result. Note that in Equation (8.85) ρ, θ are fixed and η_1, η_2 are variables. Therefore we must compute $\frac{\partial G}{\partial \mathbf{n}}$ with respect to these variables.

Remark: The formula that has been used in Equation (8.78) follows from the identity

$$1 + \sum_{n=1}^{\infty} r^n \cos(nx) = \frac{1 - r\cos x}{1 + r^2 - 2\cos x}, \quad 0 \le r \le 1. \tag{8.86}$$

To prove this identity we use the fact that

$$\cos\theta = \frac{e^{i\theta} + e^{-i\theta}}{2}.$$

This enables us to rewrite the left hand side of (8.86) as

$$\frac{1}{2}\left[1 + \sum_{n=1}^{\infty}(re^{ix})^n + 1 + \sum_{n=1}^{\infty}(re^{-ix})^n\right] = \frac{1}{2}\left[\frac{1}{1 - re^{ix}} + \frac{1}{1 - re^{-ix}}\right].$$

The desired identity follows by simplifying the right hand side of this equation.

To apply this identity in Equation (8.78) we differentiate the log function in this equation with respect to σ, use the identity above, and then integrate the result.

8.3 LAPLACE TRANSFORM

In some instances the Laplace transform method provides another technique for the solution of boundary value problems by converting the partial differential equation into an ordinary differential equation.

We start with a review of the Laplace transform and its basic properties.

8.3.1 Basic Properties of the Laplace Transform

.

The Laplace transform $\mathcal{L}$ of a function $f(t)$ is defined as

$$\mathcal{L}(f)(s) = \int_0^\infty e^{-st} f(t)dt.$$

Here we assumed implicitly that the improper integral in this definition converges. Obviously the Laplace transform is a linear operator that is:

$$\mathcal{L}(af + bg) = a\mathcal{L}(f) + b\mathcal{L}(g)$$

where a, b are constants.

Example 8.3.1 *Compute the Laplace transform of $f(x) = \delta(x - a)$.*

Solution 8.3.1 *By definition,*

$$\mathcal{L}(f)(s) = \int_0^\infty e^{-st} \delta(t - a)dt = e^{-sa}.$$

Example 8.3.2 *Compute the Laplace transform of $f(t) = t^k$ for $k \geq -1$.*

Solution 8.3.2

$$\mathcal{L}(t^k)(s) = \int_0^\infty t^k e^{-st} dt = \frac{1}{s^{k+1}} \int_0^\infty r^k e^{-r} dr = \frac{\Gamma(k+1)}{s^{k+1}},$$

where we made the substitution $r = st$.

Example 8.3.3 *Compute the Laplace transform of the translated Heaviside function*

$$H(t - a) = \begin{cases} 1, & t \geq a \\ 0, & t < a. \end{cases} \tag{8.87}$$

Solution 8.3.3

$$\mathcal{L}[(H(t - a)])(s) = \int_0^\infty H(t - a)e^{-st} dt = \int_a^\infty e^{-st} dt = \frac{e^{-as}}{s}.$$

The most important property of the Laplace transform is the relation between $\mathcal{L}(f)$ and the Laplace transform of the derivatives of f, i.e. $\mathcal{L}(f')$, $\mathcal{L}(f'')$, and so on.

Theorem 1:

$$\mathcal{L}(f') = s\mathcal{L}(f) - f(0), \tag{8.88}$$

$$\mathcal{L}(f'') = s^2\mathcal{L}(f) - sf(0) - f'(0), \tag{8.89}$$

and so on.

This theorem can be proved by repeated use of the formula for integration by parts, e.g.

$$\mathcal{L}(f') = \int_0^\infty e^{-st}f'(t)dt = f(t)e^{-st}\big|_0^\infty + s\int_0^\infty e^{-st}f(t)dt = s\mathcal{L}(f) - f(0).$$

The final step in the application of the Laplace transform to ordinary differential equations requires the inversion of the transform. This is usually done with the aid of a table of Laplace transforms and some "factor theorems." We quote some of these:

Theorem 2: If $\mathcal{L}(f)(s) = g(s)$, then

$$\mathcal{L}[e^{at}f(t)](s) = g(s - a) \tag{8.90}$$

$$\mathcal{L}[H(t - a)f(t - a)](s) = e^{-as}g(s) \tag{8.91}$$

$$\mathcal{L}[tf(t)](s) = -\frac{dg}{ds}(s). \tag{8.92}$$

Example 8.3.4 *Compute*

$$\mathcal{L}^{-1}\left[\frac{s + a}{(s + a)^2 + k^2}\right].$$

Solution 8.3.4 *From a table we have,*

$$\mathcal{L}(\cos kt) = \frac{s}{s^2 + k^2} = g(s).$$

Therefore,

$$\frac{s + a}{(s + a)^2 + k^2} = g(s + a).$$

Hence from Theorem 2 we infer that

$$\mathcal{L}^{-1}\left[\frac{s + a}{(s + a)^2 + k^2}\right] = e^{-at}\cos kt.$$

Another important property of the Laplace transform is related to the convolution of two functions.

Definition: The convolution of two functions f, g is defined as

$$f * g = \int_0^t f(\tau)g(t - \tau)d\tau.$$

Theorem 3: Let $F(s)$, $G(s)$ be the Laplace transforms of f and g respectively then

$$\mathcal{L}(f * g) = F(s)G(s).$$

Thus the transform of the convolution equals the multiplication of the two transforms.

Example 8.3.5 *Use the Laplace transform to solve the following initial value problem:*

$$y'' + 4y' + 3y = 0, \quad y(0) = 2, \quad y'(0) = -4. \tag{8.93}$$

Solution 8.3.5 *Applying the Laplace transform to Equation (8.93) and using Equations (8.88), (8.89), we obtain for $g(s) = \mathcal{L}(y)(s)$.*

$$g(s) = \frac{2s + 4}{s^2 + 4s + 3} = \frac{1}{s + 1} + \frac{1}{s + 3}.$$

From a table of Laplace transforms we have, however,

$$\mathcal{L}(e^{at})(s) = \frac{1}{s - a}.$$

Hence, applying the inverse Laplace transform to $g(s)$ we have

$$y(t) = \mathcal{L}^{-1}[g(s)] = \mathcal{L}^{-1}\left(\frac{1}{s + 1}\right) + \mathcal{L}^{-1}\left(\frac{1}{s + 3}\right) = e^{-t} + e^{-3t}.$$

Note that this is the solution of the differential equation and the initial conditions.

8.3.2 Applications to the Heat Equation

In this section we present the solution of two boundary value problems related to the heat equation in one space dimension using the Laplace transform

Example 8.3.6 *We consider here the heat conduction in a rod which cooled down by its surroundings due to the fact that it is not insulated laterally ("a one dimensional radiator"). According to Newton's law of cooling the equation which governs the heat conduction in this rod is*

$$\rho c \frac{\partial u}{\partial t} = \kappa \frac{\partial^2 u}{\partial x^2} - a(u - T_0) \tag{8.94}$$

where ρ, c, κ, a, T_0 are all constants (T_0 is the ambient temperature). Making the transformation $\bar{u} = u - T_0$, dividing by ρc and then dropping the bar on $\bar{u}$, Equation (8.94) becomes

$$\frac{\partial u}{\partial t} = k \frac{\partial^2 u}{\partial x^2} - bu. \tag{8.95}$$

We now want to solve this equation for a semi-infinite rod, $0 \le x < \infty$, subject to the following initial and boundary conditions:

$$u(x, 0) = 0, \quad u(0, t) = C = constant. \tag{8.96}$$

Furthermore we assume (on physical grounds and the modeling assumptions that lead to the heat equation)that $u(x,t)$ remains bounded as $x \to \infty$ for all t.

Solution 8.3.6 *Let $g(x, s)$ be the Laplace transform of $u(x, t)$ with respect to t:*

$$g(x, s) = \mathcal{L}[u(x, t)] = \int_0^\infty e^{-st} u(x, t) dt.$$

It follows then that:

$$\mathcal{L}\left[\frac{\partial^2 u}{\partial x^2}\right] = \frac{\partial^2}{\partial x^2} \mathcal{L}(u) = \frac{\partial^2 g}{\partial x^2}.$$

From Equation (8.88) and the initial condition in Equation (8.96) we obtain

$$\mathcal{L}\left[\frac{\partial u}{\partial t}\right] = sg(x, s) - u(x, 0) = sg(x, s). \tag{8.97}$$

Hence the application of Laplace transform to Equation (8.95) yields

$$kg_{xx}(x, s) - (s + b)g(x, s) = 0. \tag{8.98}$$

Moreover the boundary conditions on $g(x, s)$ are

$$g(0, s) = \frac{C}{s}, \quad \lim_{x \to \infty} g(x, s) = 0.$$

Equation (8.98) is an ordinary differential equation with constant coefficients in x (s is considered to be a constant) whose solution subject to the boundary conditions above is

$$g(x, s) = \frac{C}{s} e^{-\alpha x},$$

where

$$\alpha = \left(\frac{s + b}{k}\right)^{1/2}.$$

Thus

$$\frac{\partial u}{\partial t} = \mathcal{L}^{-1}[s g(x, s)] = \mathcal{L}^{-1}(C e^{-\alpha x}).$$

From a table of Laplace transforms and Equation (8.90) we infer that

$$\mathcal{L}^{-1}(e^{-\alpha x}) = \frac{x}{2\sqrt{\pi k t^3}} exp\left[-bt - \frac{x^2}{4kt}\right].$$

The required solution of our boundary value problem is

$$u(x, t) = \frac{Cx}{2\sqrt{\pi k}} \int_0^t \frac{exp\left(-b\tau - \frac{x^2}{4k\tau}\right)}{\tau^{3/2}} d\tau.$$

Example 8.3.7 *The heat equation for a thin homogeneous and insulated rod in which heat is being generated at the rate of $r(t)$ per unit volume is*

$$\frac{\partial u}{\partial t} = k \frac{\partial^2 u}{\partial x^2} + r(t). \tag{8.99}$$

Solution 8.3.7 *We now solve this equation for a semi-infinite rod $x \geq 0$ subject to the following boundary conditions*

$$u(0, t) = 0, \quad u(x, 0) = 0$$

and assuming that $u(x, t)$ is bounded for all x and t. (These boundary conditions imply that heat will flow out at $x = 0$.)

Applying the Laplace transform to Equation (8.99) and using the notations of the previous example we obtain

$$k g_{xx}(x, s) - s g(x, s) = R(s) \tag{8.100}$$

where $R(s) = \mathcal{L}[r(t)]$. The boundary conditions on g are

$$g(0, s) = 0, \quad \lim_{x \to \infty} g(x, s) = 0.$$

The solution of Equation (8.100) subject to this boundary conditions is

$$g(x, s) = \frac{R(s)}{s}(1 - e^{-x\sqrt{s/k}}).$$

Hence

$$\frac{\partial u}{\partial t} = \mathcal{L}^{-1}[sg(x, s)] = \mathcal{L}^{-1}[R(s)] - \mathcal{L}^{-1}\left\{R(s)exp\left[-x\sqrt{\frac{s}{k}}\right]\right\}.$$

Using the convolution theorem we obtain

$$\frac{\partial u}{\partial t} = r(t) - r(t) * \left[\frac{x}{2\sqrt{\pi k t^3}}exp\left(-\frac{x^2}{4kt}\right)\right].$$

Hence

$$u(x, t) = \int_0^t \left\{r(\tau) - r(\tau) * \left[\frac{x}{2\sqrt{\pi k \tau^3}}exp\left(-\frac{x^2}{4k\tau}\right)\right]\right\} d\tau.$$

8.4 NUMERICAL SOLUTIONS OF PDES

8.4.1 Finite Difference Schemes

In this section we demonstrate the application of the finite difference approximation that was introduced in a previous chapter to the solution of various partial differential equations.

8.4.2 Numerical Solutions for the Poisson Equation

To solve numerically the Dirichlet problem for the Poisson or Laplace equations on a rectangular $D \subset R^2$, that is

$$\nabla^2 u = f(\mathbf{x}) \quad on \ D \tag{8.101}$$

and

$$u|_{\partial D} = g(\mathbf{x})$$

we introduce on D an equispaced grid with step size h. Using the finite difference formula in Equation (8.101) for $\partial^2 u/\partial x^2$ and its equivalent for $\partial^2 y/\partial y^2$, we approximate the differential equation at each interior grid point (x_i, y_j) by

$$\frac{u_{i+1,j} + u_{i-1,j} - 2u_{ij}}{h^2} + \frac{u_{i,j+1} + u_{i,j-1} - 2u_{ij}}{h^2} = f(x_i, y_j), \qquad (8.102)$$

that is,

$$u_{i+1,j} + u_{i-1,j} - 4u_{ij} + u_{i,j+1} + u_{i,j-1} = h^2 f(x_i, y_j). \qquad (8.103)$$

Thus, for each interior point we obtain a linear equation in the unknowns u_{ij}. Furthermore, since the values of u on the boundary are known, the total number of equations equals the number of the unknowns; that is, the boundary value problem has been reduced to a system of linear equations.

Example 8.4.1 *Derive a system of linear equations for the solution of*

$$\nabla^2 u = 0$$

on

$$D = \{(x, y) \ 0 \le x \le 1, 0 \le y \le 1\}$$

with the boundary conditions

$$u(x, 0) = 1, \quad u(1, y) = 1, \quad u(0, y) = u(x, 1) = 0$$

and $h = \frac{1}{3}$ (see Figure 8.8).

Solution 8.4.1 *From the boundary conditions we have*

$$u_{1,2} = u_{1,3} = u_{2,4} = u_{3,4} = 1$$

and

$$u_{2,1} = u_{3,1} = u_{4,2} = u_{4,3} = 0.$$

Hence, to solve this problem we have to compute only $u_{22}, u_{23}, u_{32},$ and u_{33}.

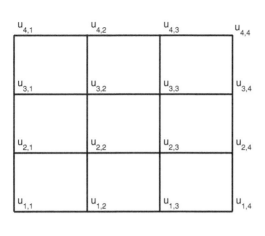

Figure 8.8 Computational Grid for example 8.4.1

From Equation (8.103) we obtain the following equations for the four inner grid points:

$$u_{2,3} + u_{2,1} - 4u_{2,2} + u_{3,2} + u_{1,2} = 0 \qquad (8.104)$$

$$u_{2,2} + u_{2,4} - 4u_{2,3} + u_{3,3} + u_{1,3} = 0$$

$$u_{3,1} + u_{3,3} - 4u_{3,2} + u_{4,2} + u_{2,2} = 0$$

$$u_{3,2} + u_{3,4} - 4u_{3,3} + u_{4,3} + u_{2,3} = 0.$$

Using the boundary conditions this yields

$$u_{2,3} - 4u_{2,2} + u_{3,2} = 1 \qquad (8.105)$$

$$u_{2,2} - 4u_{2,3} + u_{3,3} = -2$$

$$u_{3,3} - 4u_{3,2} + u_{2,2} = 0$$

$$u_{3,2} - 4u_{3,3} + u_{2,3} = -1$$

which is the required system of equations.

8.4.2.1 Other Boundary Conditions

When the boundary value problem is of Neumann or mixed type, the number of equations derived using Equation (8.103) at the inner grid points will not be equal to the number of the unknowns since the values of u on the boundary are not known. To overcome this difficulty we add additional grid points outside the region whenever the boundary conditions are given in terms of the derivatives and then use a finite difference formula to equalize the number of unknown to the number of equations. The details of this "trick" are demonstrated in the following example.

Example 8.4.2 *Solve*

$$\nabla^2 u = 0$$

on

$$D = \{(x, y) \ 0 \le x \le 1, 0 \le y \le 1\}$$

with the boundary conditions

$$\frac{\partial u}{\partial y}(x, 0) = 1, \quad u(1, y) = 1, \quad u(0, y) = u(x, 1) = 0$$

and $h = \frac{1}{3}$.

Solution 8.4.2 : *Since the boundary conditions along $y = 0$ are given in terms of $\partial u / \partial y$, we add the grid points $u_{0,2}$ and $u_{0,3}$ and use Equation (8.103) at the four inner grid points and the two grid points along $x = 0$ [i.e., $(\frac{1}{3}, 0)$ and $(\frac{3}{2}, 0)$]. We obtain six equations in eight unknowns. These equations consist of the four given by the system in Equation (8.104) and*

$$u_{1,1} + u_{1,3} - 4u_{1,2} + u_{2,2} + u_{0,2} = 0 \qquad (8.106)$$
$$u_{1,2} + u_{1,4} - 4u_{1,3} + u_{2,3} + u_{0,3} = 0$$

(See Figure 8.9).

   But

$$u_{3,1} = u_{2,1} = u_{1,1} = u_{4,2} = u_{4,3} = 0$$

and

$$u_{1,4} = u_{2,4} = u_{3,4} = 1.$$

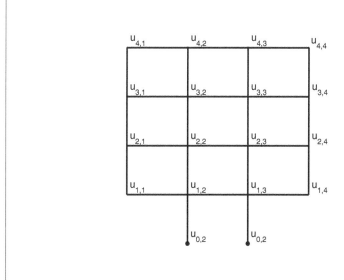

Figure 8.9   Computational Grid for example 8.4.3

Hence,

$$u_{2,3} - 4u_{2,2} + u_{3,2} + u_{1,2} = 0$$
$$u_{2,2} - 4u_{2,3} + u_{3,3} + u_{1,3} = -1$$
$$u_{3,3} - 4u_{3,2} + u_{2,2} = 0$$
$$u_{3,2} - 4u_{3,3} + u_{2,3} = -1$$
$$u_{1,3} - 4u_{1,2} + u_{2,2} + u_{0,2} = 0$$
$$u_{1,2} - 4u_{1,3} + u_{2,3} + u_{0,3} = -1.$$

To solve the system we need two additional equations. These equations are obtained from the boundary condition at $\left(\frac{1}{3}, 0\right)$ and $\left(\frac{2}{3}, 0\right)$ using the approximation

$$\left(\frac{\partial u}{\partial y}\right)_{ij} = \frac{u_{i,j+1} - u_{i,j-1}}{2h}.$$

This yields

$$u_{2,2} - u_{0,2} = \frac{2}{3}, \quad u_{2,3} - u_{0,3} = \frac{2}{3}.$$

The system now consists of eight equations in eight unknowns.

8.4.3 Irregular Regions

When the domain D has an irregular shape, the finite difference forumlas for the derivatives will not be applicable to grid points whose distance from the boundary is less than h. Under these circumstances we must introduce appropriate approximation formulas for the derivatives (see Chapter 3). Thus, using the notation shown in Figure 8.10, we obtain the following finite difference approximations at the grid point **a**:

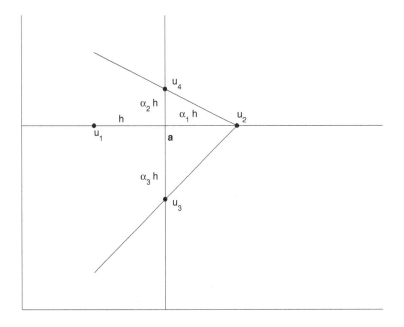

Figure 8.10 Computational Grid near the boundary of an irregular domain

$$\left(\frac{\partial u}{\partial x}\right)_{\mathbf{a}} = \frac{u_2 - u_1}{h(1 + \alpha_1)}, \quad \left(\frac{\partial u}{\partial y}\right)_{\mathbf{a}} = \frac{u_4 - u_3}{h(\alpha_2 + \alpha_3)} \qquad (8.107)$$

$$\left(\frac{\partial^2 u}{\partial x^2}\right)_{\mathbf{a}} = \frac{2}{h^2} \frac{u_2 + \alpha_1 u_1 - (1 + \alpha_1)u_a}{\alpha_1(1 + \alpha_1)}$$

$$\left(\frac{\partial^2 u}{\partial y^2}\right)_{\mathbf{a}} = \frac{2}{h^2} \frac{\alpha_2 u_4 + \alpha_3 u_3 - (\alpha_2 + \alpha_3)u_a}{\alpha_2 \alpha_3(\alpha_2 + \alpha_3)}$$

where $u_{\mathbf{a}} = u(\mathbf{a})$. The finite difference approximation scheme for $\nabla^2 u = f$ at **a** is given by

$$\frac{u_2 + \alpha_1 u_1 - (1 + \alpha_1)u_a}{\alpha_1(1 + \alpha_1)} + \frac{\alpha_2 u_4 + \alpha_3 u_3 - (\alpha_2 + \alpha_3)u_a}{\alpha_2 \alpha_3(\alpha_2 + \alpha_3)} = \frac{h^2}{2} f(\mathbf{a}).$$

Symmetry Considerations

In practical applications of the finite difference method, the number of equations to be solved can be very large; for example, a two-dimensional grid with 50 divisions along the $x-$ and $y-$axes will yield 2500 equations. The same grid in three dimensions will lead to 625,000 equations! This number can be reduced, however, when the region D, the boundary conditions, and the differential equation are invariant under certain asymmetry operations.

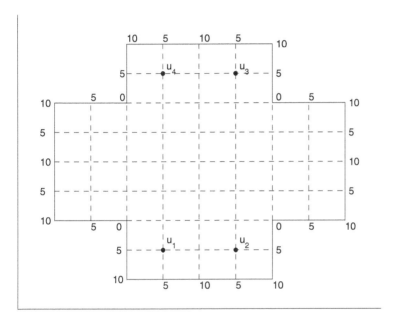

Figure 8.11 Using symmetry to reduce the number of unknowns

Example 8.4.3 *Since the region and the boundary conditions as shown in Figure 8.11 are invariant under reflections with respect to the x and y axes, it follows that the solutions of $\nabla^2 u = 0$, which is also invariant with respect to these operations, will satisfy*

$$u_1 = u_2 = u_3 = u_4$$

and so on.

Exercises

1. Derive a finite difference approximation scheme for the Poisson equation in three dimensions.

2. For a boundary value problem with circular symmetry in R^2, it is natural to use polar rather than Cartesian coordinates. Derive a finite difference approximation to $\nabla^2 u = f$ in these coordinates.

   Hint: use a grid with constant $\triangle\theta$ and $\triangle r$. Note also that

$$\nabla^2 u = \frac{\partial^2 u}{\partial r^2} + \frac{1}{r}\frac{\partial u}{\partial r} + \frac{1}{r^2}\frac{\partial^2 u}{\partial \theta^2}$$

3. Solve

$$\nabla^2 u = r\cos^2\theta \quad u(1,\theta) = 0$$

   on the unit disk using $\triangle\theta = \pi/6$ and $\triangle r = 0.2$. Compare with the exact solution.

4. Solve

$$\nabla^2 u = x^2 - y^2$$

   on

$$D = \{(x,y)\ 0 \le x \le 1, 0 \le y \le 1\}$$

   with the boundary conditions

$$\frac{\partial u}{\partial x}(0,y) = 1, \quad u(x,0) = 1, \quad u(1,y) = u(x,1) = 0.$$

   Use different step sizes and compare the solutions obtained.

5. Complete the solutions of Examples 3 ,4, and 5. Use several different h.

6. Derive a finite difference scheme to solve

$$\nabla^2 u + (x^2 + y^2)\frac{\partial u}{\partial x} = f(\mathbf{x}).$$

7. Solve numerically the differential equations that appear in the previous exercise if

$$f(\mathbf{x}) = x^2 + y^2$$

   and the boundary conditions are the same as in Example 3.

8.4.4 Numerical Solutions for the Heat and Wave Equations

To obtain numerical solutions for the heat equation in one dimension,

$$\frac{\partial^2 u}{\partial x^2} = \frac{1}{k}\frac{\partial u}{\partial t} \tag{8.108}$$

with the boundary conditions

$$u(x,0) = f(x), \quad u(0,t) = g_1(t), \quad u(a,t) = g_2(t)$$

on $0 \le x \le a$ and $0 \le t \le T$. We introduce a grid with step size $\triangle x, \triangle t$ in x and t, respectively, and use finite difference formulas to approximate Equation (8.108). Denoting $u(x_i, t_j)$ by u_{ij} leads to the following finite difference equation at the grid point (ij):

$$\frac{u_{i+1,j} - 2u_{ij} + u_{i-1,j}}{(\triangle x)^2} = \frac{1}{k}\frac{u_{i,j+1} - u_{i,j}}{\triangle t} \tag{8.109}$$

or

$$u_{i,j+1} = \frac{k\triangle t}{(\triangle x)^2}(u_{i+1,j} + u_{i-1,j}) + \left(1 - \frac{2k\triangle t}{(\triangle x)^2}\right)u_{i,j}. \tag{8.110}$$

We observe that in Equation (8.109) the forward difference formula is used to approximate $\partial u/\partial t$ since otherwise (for the central difference formula) the values of u at $t \pm \triangle t$ are needed to evaluate u at $t + \triangle t$, and these data are not available.

The formula in Equation (8.109) expresses u at t_{j+1} in terms of its values at time t_j; hence, since u at time $t = 0$ is given, we can compute u on $[0, T]$ consecutively, that is, at $\triangle t, 2\triangle t$, and so on (there is no need to solve systems of equations). Furthermore, Equation (8.110) can be simplified by choosing $r = k\triangle t/(\triangle x)^2$ to be equal to $\frac{1}{2}$, which leads to

$$u_{i,j+1} = \frac{1}{2}(u_{i+1,j} + u_{i-1,j}). \tag{8.111}$$

The meaning of this formula is that on such a grid the value of $u(x_i, t_{j+1})$ is the average of $u(x_{i-1}, t_j)$ and $u(x_{i+1}, t_j)$.

The algorithm defined by Equation (8.110) or Equation (8.111) is called the **explicit method** for the numerical solution of the heat equation. Its drawback is that in order to ensure its stability, that is, to prevent the accumulated numerical error from becoming too large and therefore rendering the numerical solution meaningless, we must choose $r \leq \frac{1}{2}$. This condition places a restriction on $\triangle t$ that must satisfy $\triangle t \leq (\triangle x^2)/2k$ and, therefore, increases the computational effort needed to obtain the required solution.

To overcome this restriction, Crank and Nicholson observed that the forward difference formula

$$\left(\frac{\partial u}{\partial t}\right)_{ij} = \frac{u_{i,j+1} - u_{i,j}}{\triangle t}$$

can be interpreted as a central difference formula at time $t_j + \frac{1}{2}\triangle t$ and, hence, a consistent approximation scheme must evaluate $\partial^2 u/\partial x^2$ at this time. To accomplish this we average the approximations for $\partial^2 u/\partial x^2$ at this time. To accomplish this we average the approximations for $\partial^2 u/\partial x^2$ at t_j and t_{j+1} and, thus, obtain the following finite difference formula for Equation (8.108) at (ij):

$$\frac{u_{i+1,j} - 2u_{i,j} + u_{i-1,j}}{2(\triangle x)^2} + \tag{8.112}$$

$$\frac{u_{i+1,j+1} - 2u_{i,j+1} + u_{i-1,j+1}}{2(\triangle x)^2} = \frac{1}{k}\frac{u_{i,j+1} - u_{i,j}}{\triangle t}$$

that is

$$-ru_{i+1,j+1} + (2+2r)u_{i,j+1} - ru_{i-1,j+1} \tag{8.113}$$
$$= ru_{i+1,j} + (2-2r)u_{i,j} + ru_{i-1,j}.$$

The algorithm represented by Equation (8.113) is called the **Crank-Nicholson algorithm**. It requires the solution of a system of linear equations at each time step, but it is stable for all values of r.

As for the wave equation,

$$\frac{\partial^2 u}{\partial x^2} = \frac{1}{c^2}\frac{\partial^2 u}{\partial t^2} \tag{8.114}$$

with the boundary conditions

$$u(x,0) = f(x), \quad \frac{\partial u}{\partial t}(x,0) = g(x)$$

$$u(0,t) = u(a,t) = 0.$$

We use Equation (8.101) to derive a finite difference approximation for Equation (8.114). We obtain

$$\frac{u_{i+1,j} - 2u_{i,j} + u_{i-1,j}}{(\triangle x)^2} = \frac{1}{c^2}\frac{u_{i,j+1} - 2u_{i,j} + u_{i,j-1}}{(\triangle t)^2}.$$

Hence,

$$u_{i,j+1} = w^2(u_{i+1,j} + u_{i-1,j}) + 2(1 - w^2)u_{i,j} - u_{i,j-1} \tag{8.115}$$

where $w = (c\triangle t)/\triangle x$. Furthermore, if we set $w = 1$, the algorithm formulas will be simplified and we obtain

$$u_{i,j+1} = u_{i+1,j} + u_{i-1,j} - u_{i,j-1}. \tag{8.116}$$

We infer from Equation (8.115) or Equation (8.116) that we can calculate the values of u at t_{j+1} if these values are known at t_j and t_{j-1}. It follows then that in order to begin the computation of u at $\triangle t$, its values at $-\triangle t$ must be known. We can easily overcome this difficulty, however, if we observe that $(\partial u/\partial t)(x, 0) = g(x)$ implies

$$\frac{u_{i,1} - u_{i,-1}}{2\triangle t} = g(x_i). \tag{8.117}$$

Thus, using Equations (8.116) and (8.117) we obtain for $u_{i,1}$,

$$\begin{aligned} u_{i,1} &= \tfrac{1}{2}(u_{i+1,0} + u_{i-1,0}) + g(x_i)\triangle t \\ &= \tfrac{1}{2}[f(x_{i+1}) + f(x_{i-1})] + g(x_i)\triangle t. \end{aligned} \tag{8.118}$$

The algorithm given by Equation (8.115) is stable for $w \leq 1$. However, unexpectedly the numerical results obtained by setting $w = 1$ are better than those for $w < 1$; that is, decreasing the time step will not bring the numerical solution closer to the exact solution.

Exercises

1. Solve the heat equation in one dimension with the following boundary conditions using the explicit and Crank-Nicholson algorithms. Compare the accuracy of the solutions and the computational effort.

$$\begin{aligned} (a) & u(x,0) = x, & u(0,t) = T, & u(a,t) = 0 \\ (b) & u(x,0) = \sin x, & u(0,t) = T, & u(a,t) = 0 \\ (c) & u(x,0) = \cos x, & \tfrac{\partial u}{\partial x}(0,t) = c, & u(a,t) = 0. \end{aligned}$$

2. Solve the wave equation (8.114) in one dimension with $w = 1$ and $w = \tfrac{1}{2}$

using the following initial conditions:

$$(a)u(x,0) = \sin x, \qquad \frac{\partial u}{\partial t}(x,0) = 0$$
$$(b)u(x,0) = 0, \qquad \frac{\partial u}{\partial t} = \sin x$$
$$(c)u(x,0) = \cos 2x, \qquad \frac{\partial u}{\partial t}(x,0) = \sin x.$$

Always assume that $u(0,t) = u(a,t) = 0$. Compare the numerical and exact solutions.

3. Derive a finite difference algorithm to solve

$$\frac{\partial^2 u}{\partial x^2} + b(x)\frac{\partial u}{\partial x} = \frac{1}{k}\frac{\partial u}{\partial t}$$

subject to the appropriate boundary conditions.

4. Solve numerically the differential equation that appears in Exercise 3 if the boundary conditions on u are $u(x,0) = \sin 2x$, $u(0,t) = T$, $u(a,t) = 0$, and $b(x) = x$.

5. Derive a finite difference scheme for

$$\frac{\partial^2 u}{\partial x^2} + b(x)\frac{\partial^2 u}{\partial x \partial t} = \frac{1}{c^2}\frac{\partial^2 u}{\partial t^2}$$

if $u(x,0) = \sin 5x$, $(\partial u/\partial t)(x,0) = 0$, and $u(0,t) = u(a,t) = 0$.

CHAPTER 9

Variational Principles

CONTENTS

9.1 EXTREMA OF FUNCTIONS.

It is well known from elementary calculus that the local extrema of a smooth function $f = f(x)$ in one variable coincides with the points x_i at which $f'(x_i) = 0$. Furthermore, an extremal point x_i is a local maximum or minimum if $f''(x_i) < 0$ or $f''(x_i) > 0$ respectively.

Similar criteria exist naturally for multivariable functions. Thus, for functions in two variables $F = F(x, y)$ we have the following:

Definition: Let $F(x, y)$ be C^2 (= twice differentiable with continuous derivatives) in some region R. The Hessian of F at $(x, y) \epsilon R$ is defined as

$$H(x, y) = \begin{vmatrix} F_{xx}(x, y) & F_{xy}(x, y) \\ F_{yx}(x, y) & F_{yy}(x, y) \end{vmatrix}. \tag{9.1}$$

Remark: This definition can be extended naturally to functions with n variables. Thus if $F = F(\mathbf{x}) = F(x_1, \ldots, x_n)$, then

$$H(\mathbf{x}) = \left| \frac{\partial^2 F(\mathbf{x})}{\partial x_i \partial x_j} \right|. \tag{9.2}$$

Definition: $\mathbf{a} = (a_1, \ldots, a_n)$ is said to be a critical point of $F(\mathbf{x})$ if

$$\frac{\partial F}{\partial x_i}(\mathbf{a}) = 0; \quad i = 1, \ldots, n. \tag{9.3}$$

We quote the following theorem without proof:

Theorem 9.1.1 *Let $F(x, y)$ be C^2 in a domain D and let (a, b) be a critical point of F in the interior of D. If*

1. *$H(a, b) > 0$ and $F_{xx}(a, b) < 0$ then F has a maximum at (a, b).*

2. *$H(a, b) > 0$ and $F_{xx}(a, b) > 0$ then F has a minimum at (a, b).*

3. *$H(a, b) < 0$ then F has neither a maximum nor a minimum at (a, b).*

Example: Let $F(x, y) = x^2 - y^2$. Then $F_x = 2x$, $F_y = -2y$, and hence at $(0, 0)$ $F_x(0, 0) = F_y(0, 0) = 0$ (i.e. $(0, 0)$ is a critical point of F). However, $H(0, 0) = -4 < 0$ and, therefore, $(0, 0)$ is not a maximum or a minimum point of F. To see what it is, we observe that when we approach $(0, 0)$ along the x-axis $(y = 0)$, $F(x, 0) = x^2$ and hence $(0, 0)$ "looks like" a minimum. On the other hand, if we approach $(0, 0)$ along the y-axis then $F(0, y) = -y^2$ and $(0, 0)$ "looks like" a maximum. Thus, $(0, 0)$ is a saddle point.

9.2 CONSTRAINTS AND LAGRANGE MULTIPLIERS

In many practical applications one wants to find the extremum points of $F(\mathbf{x})$ subject to a finite number of constraints on the value of $\mathbf{x}$,

$$g_k(bfx) = c_k, \quad k = 1, \ldots, m \tag{9.4}$$

where c_k are constants.

To solve such problems one uses Lagrange multipliers. The algorithm is described by the following:

Theorem 9.2.1 : *The extremum points of $F(\mathbf{x})$ subject to the constraints Equation (9.4) are the solutions $x_1, \ldots, x_n$, $\lambda_1, \ldots, \lambda_m$ of the $m+n$ equations*

$$\frac{\partial f}{\partial x_i} = 0, \quad i = 1, \ldots, n \tag{9.5}$$

$$g_k = c_k, \quad k = 1, \ldots, m \tag{9.6}$$

where

$$f(x, \lambda) = F - \lambda_1 g_1 - \ldots - \lambda_k g_k. \tag{9.7}$$

Remark: $\lambda_1, \ldots, \lambda_m$ are called the Lagrange multipliers and f the auxiliary function of the problem.

Example: Find the points on the ellipse $2x^2 + 2xy + y^2 = 1$ that are closest to the origin.

Solution: we want to minimize

$$F(x, y) = x^2 + y^2 \tag{9.8}$$

(which is equivalent to minimizing $(x^2 + y^2)^{1/2}$) subject to the constraint

$$g_1(x, y) = 2x^2 + 2xy + y^2 = 1. \tag{9.9}$$

The auxiliary function $f(x, y, \lambda)$ is therefore

$$f(x, y, \lambda) = x^2 + y^2 - \lambda(2x^2 + 2xy + y^2). \tag{9.10}$$

Hence the extremum points have to satisfy the equations

$$2x - 4\lambda x - 2\lambda y = 0 \tag{9.11}$$

$$2y - 2\lambda x - 2\lambda y = 0 \tag{9.12}$$

$$2x^2 + 2xy + y^2 = 1. \tag{9.13}$$

Solving Equation (9.11) for y yields

$$y = -\frac{x(-1 + 2\lambda)}{\lambda}. \tag{9.14}$$

Substituting this expression for y in Equation (12.9) and solving for λ we obtain

$$\lambda = \frac{3 \pm \sqrt{5}}{2}.$$

Using this expression of λ in Equation (9.14) and then substituting for y in Equation (12.10) we obtain the following solutions for (x, y):

$$(x, y) = (\pm 0.528, \pm 0.325), \quad (x, y) = (\mp 0.850, \pm 1.376).$$

The first two pairs represent the points with minimum distance from the origin while the last two pairs have a maximum distance from the origin.

Remark: It should be observed from this example that the method of Lagrange multipliers is "easiest to apply" when F is at most a quadratic polynomial and g_k are linear functions since in this case the solution of the resulting system of equations for the extremum points is straightforward.

Exercises

1. Plot the ellipse $2x^2 + 2xy + y^2 = 1$ and identify the points closest to the origin.

2. Find the point on the ellipse $x^2 - 2xy + 2y^2 = 1$ with maximum distance from the origin.

3. Find the dimensions of the right circular cylinder of fixed total surface area A (including top and bottom) with maximum volume.

4. Find the relative extrema of

$$F(x, y, z) = xz + yz$$

which lie on the intersection of the surfaces

$$x^2 + y^2 = 2, \quad yz = 2$$

(two constraints).

5. Show that

    (a) Among all triangles with the same perimeter the equilateral triangle has the greatest area.

    (b) Among all rectangles with the same perimeter the square encloses the greatest area.

9.3 CALCULUS OF VARIATIONS

The basic problem of the calculus of variations can be stated as follows: Find the function $y(x)$ defined on $[x_1, x_2]$ and satisfying the boundary conditions

$$y(x_1) = y_1, \quad y(x_2) = y_2 \tag{9.15}$$

so that the value of the integral

$$I(y) = \int_{x_1}^{x_2} f(x, y, y')dx \tag{9.16}$$

is at extremum.

    The following examples show how such problems arise in actual applications.

Example 1: Formulate a variational principle to find the arc of minimum length that passes through given end points $(x_1, y_1), (x_2, y_2)$.

Solution: The differential of the distance in two dimensions is

$$(ds)^2 = (dx)^2 + (dy)^2,$$

hence our problem is to find $y(x)$ which minimizes the integral

$$I(y) = \int ds = \int \sqrt{(dx)^2 + (dy)^2} = \int_{x_1}^{x_2} \sqrt{1 + (y')^2} \, dx \tag{9.17}$$

with

$$y(x_1) = y_1, \quad y(x_2) = y_2. \tag{9.18}$$

Example 2: Find the plane curve between (x_1, y_1), (x_2, y_2), $y_1, y_2 > 0$ which generates the smallest surface area when revolved around the x-axis.

Solution: As is well known, the surface area generated by revolving a curve $y = y(x)$ around the x-axis is

$$S(y) = \int_{x_1}^{x_2} 2\pi y \sqrt{1 + (y')^2} \, dx. \tag{9.19}$$

We have to find the function $y(x)$ which minimizes the value of this integral and satisfies the boundary conditions Equation (12.26).

To solve the basic variational problem (or at least recast it in terms of differential equations) we proceed as follows: let $y(x)$ be the desired optimal solution (unknown as yet). Consider the following set of functions:

$$Y(x, \epsilon) = y(x) + \epsilon \eta(x) \tag{9.20}$$

where $\eta(x)$ is an arbitrary but fixed function which satisfies the boundary conditions

$$\eta(x_1) = \eta(x_2) = 0. \tag{9.21}$$

For a function $Y(x, \epsilon)$ in this family the variational integral (Equation (3.2)) takes the form

$$I(\epsilon) = \int_{x_1}^{x_2} f(x, Y, Y')dx. \tag{9.22}$$

Thus the variational integral "degenerates" into a one variable function in ϵ. Moreover, since we know that at $\epsilon = 0$ $I(\epsilon)$ has an extremum (according to our assumption on $y(x)$), it follows that $I'(0) = 0$, i.e.

$$I'(0) = \int_{x_1}^{x_2} \left[\frac{\partial f}{\partial Y} \cdot \frac{dY}{d\epsilon} + \frac{df}{dY'} \cdot \frac{dY'}{d\epsilon} \right] \bigg|_{\epsilon=0} dx = 0. \tag{9.23}$$

But from Equation (12.28)

$$\frac{dY}{d\epsilon} = \eta, \quad \frac{dY'}{d\epsilon} = \eta'$$

and $Y(x, 0) = y(x)$, $Y'(x, 0) = y'(x)$; therefore, Equation (10.35) reduces to

$$\int_{x_1}^{x_2} \left[\frac{\partial f}{\partial y} \eta + \frac{\partial f}{\partial y'} \cdot \eta' \right] dx = 0. \tag{9.24}$$

Integrating the last term in Equation (10.36) by parts we obtain

$$\left[\frac{\partial f}{\partial y'} \cdot \eta \right] \bigg|_{x_1}^{x_2} + \int_{x_1}^{x_2} \left[\frac{\partial f}{\partial y} - \frac{d}{dx} \left[\frac{\partial f}{\partial y'} \right] \right] \eta(x) dx = 0 \tag{9.25}$$

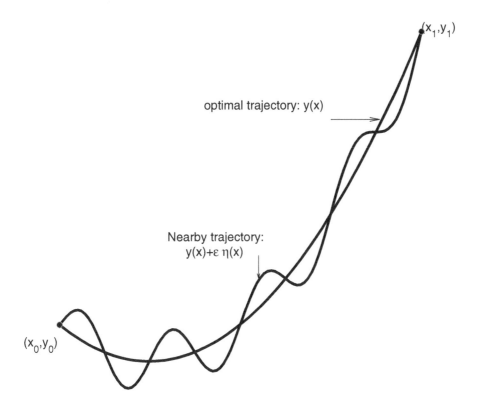

Figure 9.1 Optimal trajectory and another one close by

and hence by Equation (9.21)

$$\int_{x_1}^{x_2} \left[\frac{\partial f}{\partial y} - \frac{d}{dx} \left[\frac{\partial f}{\partial y'} \right] \right] \eta(x) dx = 0. \qquad (9.26)$$

Equation (9.26) must hold true for any choice of $\eta(x)$ which satisfies Equation (10.46); hence we infer that the extremum function $y(x)$ must satisfy the partial differential equation

$$\frac{\partial f}{\partial y} - \frac{d}{dx} \frac{\partial f}{\partial y'} = 0. \qquad (9.27)$$

This equation is called the Euler-Lagrange equation.

We observe that when f has no explicit dependence on x, i.e. $f = f(y, y')$, then Equation (10.39) can be written as

$$\frac{d}{dx} \left[y' \frac{\partial f}{\partial y'} - f \right] = 0. \qquad (9.28)$$

In fact

$$
\frac{d}{dx}\left[y'\frac{\partial f}{\partial y'} - f\right] = y''\frac{\partial f}{\partial y'} + y'\frac{d}{dx}\left[\frac{\partial f}{\partial y'}\right] - y'\frac{\partial f}{\partial y} - y''\frac{\partial f}{\partial y'}
$$

$$
= -y'\left[\frac{\partial f}{\partial y} - \frac{d}{dx}\left[\frac{\partial f}{\partial y'}\right]\right] = 0. \tag{9.29}
$$

Example 1 (cont.): For this problem

$$
f(x, y, y') = \sqrt{1 + (y')^2}. \tag{9.30}
$$

Hence from Equation (10.39) we obtain

$$
-\frac{d}{dx}\left[\frac{y'}{\sqrt{1 + (y')^2}}\right] = 0, \tag{9.31}
$$

i.e.

$$
\frac{y'}{\sqrt{1 + (y')^2}} = \text{const} = c \tag{9.32}
$$

or

$$
y' = \sqrt{\frac{c^2}{1 - c^2}} = A \text{ (const.)}. \tag{9.33}
$$

Hence

$$
y = Ax + B \tag{9.34}
$$

as we expected!

Example 3: Find the plane curve along which a particle of mass m starting from rest will slide without friction from (x_1, y_1) to $(x_2, y_2), y_2 < y_1$, in the shortest time.

Remarks:

1. From a historical perspective this was the original problem that led Euler to "invent" the calculus of variations. The solution curve to this problem is called the "brachistochrone."

2. To simplify the algebra we can assume without loss of generality that $(x_1, y_1) = (0, 0)$.

Solution: Since the speed of the particle along the curve is

$$
v = \frac{ds}{dt}, \tag{9.35}
$$

the total time of descent from 0 to (x_2, y_2) is

$$I(y) = \int_0^{x_2} dt = \int_0^{x_2} \frac{ds}{v} = \int_0^{x_2} \frac{\sqrt{1 + (y')^2}}{v} \, dx. \tag{9.36}$$

However by conservation of energy we have

$$\frac{1}{2}mv^2 - \frac{1}{2}mv_1^2 = mg(y - 0). \tag{9.37}$$

Since $v_1 = 0$ we infer that

$$I(y) = \frac{1}{\sqrt{2g}} \int_0^{x_2} \frac{\sqrt{1 + (y')^2}}{\sqrt{y}}, \tag{9.38}$$

i.e.

$$f(y, y') = \frac{1}{\sqrt{2g}} \frac{\sqrt{1 + (y')^2}}{\sqrt{y}}. \tag{9.39}$$

From Equation (10.40) we infer that

$$\frac{(y')^2}{\sqrt{y}\sqrt{1 + (y')^2}} - \frac{\sqrt{1 + y'^2}}{\sqrt{y}} = \bar{c}. \tag{9.40}$$

Solving this for y' and integrating we obtain

$$x = \int \frac{\sqrt{y}}{\sqrt{2c - y}} \, dy \tag{9.41}$$

where $\bar{c} = \frac{1}{\sqrt{2c}}$. Substituting

$$y = 2c \sin^2 \frac{\theta}{2} \tag{9.42}$$

we finally obtain a parametric representation of the required curve

$$x = c(\theta - \sin\theta) \tag{9.43}$$

$$y = c(1 - \cos\theta), \tag{9.44}$$

which is a cycloid.

Example 4: Find the shortest path between two points on a sphere of radius R.

Remarks: A curve on a surface that provides the shortest path between two points is called a "geodesic."

Solution: A curve on a sphere is given by

$$x = R \sin \phi \cos \theta, \qquad y = R \sin \phi \sin \theta, \quad z = R \cos \phi \tag{9.45}$$

where $\theta = \theta(t)$, $\phi = \phi(t)$. Hence

$$(ds)^2 = (dx)^2 + (dy)^2 + (dz)^2 = R^2 (d\phi)^2 + R^2 \sin^2 \phi (d\theta)^2. \tag{9.46}$$

The variational problem is therefore to minimize

$$I = \int_{t_1}^{t_2} ds = R \int_{\theta_1}^{\theta_2} \sqrt{(\phi')^2 + \sin^2 \phi} \, d\theta \tag{9.47}$$

where $\phi' = \frac{d\phi}{d\theta}$. Hence

$$f(\theta, \theta') = R \sqrt{(\phi')^2 + \sin^2 \phi}. \tag{9.48}$$

Since f does not have explicit dependence on θ, we can use Equation (10.40) to obtain

$$\frac{R(\phi')^2}{\sqrt{(\phi')^2 + \sin^2 \phi}} - R \sqrt{(\phi')^2 + \sin^2 \phi} = c_1. \tag{9.49}$$

Solving for (ϕ') and integrating we obtain

$$\phi = c_1 \int \frac{d\phi}{\sqrt{R^2 \sin^4 \phi - c_1^2 \sin^2 \phi}} = -\arcsin \left\{ \frac{\cot \phi}{b} \right\} - c_2 \tag{9.50}$$

where $c_1 \, c_2$ are integration constants and $b = \sqrt{R/c_1^2 - 1}$. Hence

$$(\sin c_2) R \sin \phi \cos \theta + (\cos c_2) R \sin \phi \sin \theta - \frac{R \cos \phi}{b} = 0 \tag{9.51}$$

or in Cartesian coordinates

$$x \sin c_2 + y \cos c_2 - \frac{z}{b} = 0. \tag{9.52}$$

Thus the shortest arc that connects (θ_1, ϕ_1) and (θ_2, ϕ_2) on the sphere is the intersection between the sphere and a plane that passes through its center. Thus the arc is part of the great circle connecting the points on the sphere.

9.3.1 Natural Boundary Conditions

In some variational problems the boundary values of y at one or both end points may not be specified. Since $y(x_i)$ remains arbitrary, it is impossible to choose η to be zero at such an end point (Equation (9.21)). However, it is still possible to derive the Euler-Lagrange equation for this problem if we note that $\eta(x_i) = 0$ was used only to "discard" $\dfrac{\partial f}{\partial y'}\eta\Big|_{x_1}^{x_2}$ in Equation (9.25). This can be also be done, however, by choosing

$$\frac{\partial f}{\partial y'} = 0 \text{ at } x = x_i, \ \ i = 1, 2 \tag{9.53}$$

(if $y(x_i)$ is not specified). This condition is then called the natural boundary condition of the problem.

9.3.2 Variational Notation

Many books on variational problems introduce the concept of variation $\delta y(x)$ to replace $\epsilon\eta(x)$ that we used above. This implies

$$\delta y'(x) = \epsilon\eta'(x) \tag{9.54}$$

and

$$\frac{d}{dx}(\delta y) = \epsilon\eta' = \delta y'. \tag{9.55}$$

Moreover, given a function $F(x, y, y')$ one defines the variation of F as

$$\begin{aligned} \delta F &= \Delta F = F(x, y + \epsilon\eta, y' + \epsilon\eta') - F(x, y, y') \\ &\cong \frac{\partial F}{\partial y}(\epsilon\eta) + \frac{\partial F}{\partial y'}(\epsilon\eta') = \frac{\partial F}{\partial y}\delta y + \frac{\partial F}{\partial y'}\delta y'. \end{aligned} \tag{9.56}$$

It is easy to verify the following:

$$\delta(F^n) = nF^{n-1}\delta F \tag{9.57}$$

$$\delta(aF_1 + bF_2) = a\delta F_1 + b\delta F_2, \ \ a, b, \ \ \text{constants} \tag{9.58}$$

$$\delta(F_1 \cdot F_2) = F_1\delta F_2 + F_2\delta F_1. \tag{9.59}$$

The variational principle Equation (12.24) can now be rewritten as

$$\delta I = \Delta I = I(y + \epsilon\eta) - I(y) = \int_{x_1}^{x_2} \delta f(x, y, y')dx. \tag{9.60}$$

Exercises

1. Fermat principle for the motion of a light ray in a medium states that "light will travel between two points along the path which minimizes the transmission time." Formulate the corresponding variational principle for the motion of light in a medium in which the speed of light at any point in any direction is a function of the position only.

2. Show the Euler-Lagrange equation for a variational principle which contains second order derivatives of y, viz.

$$I = \int_{x_1}^{x_2} f(x, y, y', y'')$$

   is

$$\frac{\partial f}{\partial y} - \frac{d}{dx}\left[\frac{\partial f}{\partial y'}\right] + \frac{d^2}{dx^2}\left[\frac{\partial f}{\partial y''}\right] = 0.$$

   Hint: Note that the boundary conditions on $y(x)$ are

$$y(x_1) = a_1 \quad y(x_2) = a_2$$

$$y'(x_1) = b_1 \quad y'(x_2) = b_2$$

   where a_i, b_i are constant.

9.4 EXTENSIONS

When the function f in the variational integral is a function of more than one dependent variable, e.g.

$$f = f(x, y, \dot{x}, \dot{y}, t),$$

the extremum of the variational integral

$$I = \int_{t_1}^{t_2} f(x, y, \dot{x}, \dot{y}, t)dt \tag{9.61}$$

is achieved by the functions $x(t), y(t)$ satisfying the following Euler-Lagrange equations.

$$\frac{\partial f}{\partial x} - \frac{d}{dt}\left[\frac{\partial f}{\partial \dot{x}}\right] = 0 \tag{9.62}$$

$$\frac{\partial f}{\partial y} - \frac{d}{dt}\left[\frac{\partial f}{\partial \dot{y}}\right] = 0. \tag{9.63}$$

Similarly for the n-dimensional case (i.e. n-dependent and one independent variables)

$$I = \int_{t_1}^{t_2} f(\mathbf{x}, \dot{\mathbf{x}}, t)dt, \quad \mathbf{x}\epsilon R^n \tag{9.64}$$

the extremum functions have to satisfy

$$\frac{\partial f}{\partial x_i} - \frac{d}{dt}\left[\frac{\partial f}{\partial \dot{x}_i}\right] = 0, \; i = 1, \ldots, n. \tag{9.65}$$

Another extension of the Euler-Lagrange equations is required when f is a function of more than one independent variable but one dependent variable, i.e.

$$I = \iint_D f(x, y, w, w_x, w_y)dA, \quad D\epsilon R^2. \tag{9.66}$$

To prove the analog of Equation (12.32) in this case let

$$W(x, y) = w(x, y) + \epsilon\eta(x, y) \tag{9.67}$$

where $w(x, y)$ is the extremal function we are looking for.

$$0 = \frac{dI}{d\epsilon}\bigg|_{\epsilon=0} = \iint_D \left[\frac{\partial f}{\partial w}\eta + \frac{\partial f}{\partial w_x}\eta_x + \frac{\partial f}{\partial w_y}\eta_y\right] dA. \tag{9.68}$$

By applying Green's theorem in two dimensions (see Appendix) we obtain

$$\iint_D \left[\frac{\partial f}{\partial w_x}\eta_x + \frac{\partial f}{\partial w_y}\eta_y\right] dA = \tag{9.69}$$

$$-\iint_D \eta \left[\frac{\partial}{\partial x}\left[\frac{\partial f}{\partial w_x}\right] + \frac{\partial}{\partial y}\left[\frac{\partial f}{\partial w_y}\right]\right] dA$$

$$+ \int_{\partial D} \eta \left[\frac{\partial f}{\partial w_x} \, dy - \frac{\partial f}{\partial w_y} \, dx\right].$$

Assuming that w is specified on the boundary of the domain ∂D we deduce that $\eta|_{\partial D} = 0$. Hence the extremum function $w(x, y)$ has to satisfy the following partial differential equation:

$$\frac{\partial f}{\partial w} - \frac{\partial}{\partial x}\left[\frac{\partial f}{\partial w_x}\right] - \frac{\partial}{\partial y}\left[\frac{\partial f}{\partial w_y}\right] = 0. \tag{9.70}$$

Exercises

1. Use variations on Equation (12.33) to derive Equation (12.34).

2. Find the functions $x(t), y(t), z(t)$ which minimize the integral

$$I = \int_{t_1}^{t_2} \sqrt{\dot{x}^2 + \dot{y}^2 + \dot{z}^2} \, dt \qquad (9.71)$$

and satisfy the boundary conditions

$$x(t_i) = x_i \quad y(t_i) = y_i \quad z(t_i) = z_i \quad i = 1, 2. \qquad (9.72)$$

3. On a Riemanian manifold the infinitesimal distance is given by

$$(ds)^2 = g_{ij}(\mathbf{x})dx_i dx_j \qquad (9.73)$$

(implicit simulation on i, j). To find the equation of the curve $\mathbf{x}(t)$ of minimum length that connects two points on the manifold (=geodesic curve) it is enough to minimize the square of the distance , i.e. minimize

$$I(\mathbf{x}) = \int_{t_1}^{t_2} g_{ij}(\mathbf{x}) \frac{dx_i}{dt} \frac{dx_j}{dt} \, dt. \qquad (9.74)$$

Derive the explicit form of Euler-Lagrange equation for $\mathbf{x}(t)$.

9.5 APPLICATIONS

A key point in the application of variational principles to mechanical (and other) systems is Hamilton's principle.

Definition: If $V : R^n \to R$ the gradient of V is defined as

$$grad V = \nabla V = \left[\frac{\partial V}{\partial x_1}, \ldots, \frac{\partial V}{\partial x_n} \right].$$

Definition: A system is called conservative if the forces acting on the system can be expressed as the gradient of some scalar function V which is called the potential of the force.

Hamilton Principle: For a conservative system the actual motion (or trajectory) will minimize the variational integral

$$I = \int_{t_1}^{t_2} (T - V)dt \qquad (9.75)$$

where T is the kinetic energy of the system. $(T - V)$ is also called the Lagrangian of the system.

Remark: The kinetic energy of a point particle of mass m is $\frac{1}{2}m\mathbf{v}^2$ where $\mathbf{v}$ is the velocity of the particle.

Example 1: Derive the equations of motion for a system of two masses and three springs as shown in the diagram:

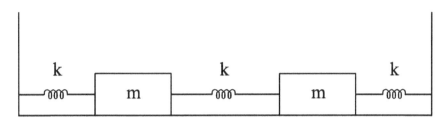

Figure 9.2 System of two masses and three springs

Solution: Let x, y be the displacement of the masses from equilibrium at time t. The kinetic energy is then given by

$$T = \frac{1}{2}m\dot{x}^2 + \frac{1}{2}m\dot{y}^2. \tag{9.76}$$

The potential energy of the springs in this position is given by

$$V = \frac{1}{2}kx^2 + \frac{1}{2}k(y - x)^2 + \frac{1}{2}.ky^2 \tag{9.77}$$

Therefore the variational integral for this system is

$$I = \int_{t_1}^{t_2} \left[\frac{1}{2}m(\dot{x}^2 + \dot{y}^2) - k(x^2 - xy + y^2) \right] dt. \tag{9.78}$$

This is a variational integral with one independent and two dependent variables. Euler-Lagrange Equations (12.31)-(12.32) then lead to

$$m\ddot{x} + 2kx - ky = 0 \tag{9.79}$$

$$m\ddot{y} + 2ky - kx = 0. \tag{9.80}$$

Example 2: Vibrating Membrane (Drum head)

We consider a "flexible membrane" which is stretched over some region D in the $x - y$ plane and bounded by a curve C. Denote the deviation of the membrane from the $x - y$ plane by $w(x, y, t)$ (where we assume that on the boundary $w(x, y, t)|_C = 0$). The membrane is set in motion by an initial disturbance from the equilibrium position, and we want to derive an equation of motion for $w(x, y, t)$.

Solution: The kinetic energy of the membrane is obviously given by

$$T = \frac{1}{2} \int \int_D \rho(x, y) w_t^2 dx \, dy. \tag{9.81}$$

To evaluate the expression for the potential energy we first note that for conservative systems $V(b) - V(a)$ is the amount of work needed to take the system from state a to state b.

For our system we can assume that the amount of work needed to take the membrane from its equilibrium state $(x - y$ plane) to another is proportional to the difference in the surface area (Hooke's law in two dimensions). Hence

$$V(w) = V(w) - V(w = 0) = k \int \int_D \left\{ \sqrt{1 + w_x^2 + w_y^2} - 1 \right\} dx \, dy. \tag{9.82}$$

For small deflections we can use the approximation

$$\sqrt{1 + u} \cong 1 + \frac{1}{2}u, \quad |u| << 1 \tag{9.83}$$

to obtain

$$V(w) = \frac{k}{2} \int \int_D \left(w_x^2 + w_y^2 \right) dx \, dy. \tag{9.84}$$

The variational integral which we have to consider is therefore

$$I = \frac{1}{2} \int_{t_1}^{t_2} \left\{ \int \int_D \left[\rho w_t^2 - k(w_x^2 + w_y^2) \right] dx \, dy \right\} dt. \tag{9.85}$$

Euler-Lagrange equations for this case yield (one dependent variable and three independent ones)

$$k(w_{xx} + w_{yy}) - \rho(x, y) w_{tt} = 0 \tag{9.86}$$

or equivalently

$$k \nabla^2 w = \rho w_{tt}, \tag{9.87}$$

which is the wave equation in two dimensions.

Exercises

1. Modify the discussion above so that gravity is included.

2. Derive the equations of motion for the double pendulum.

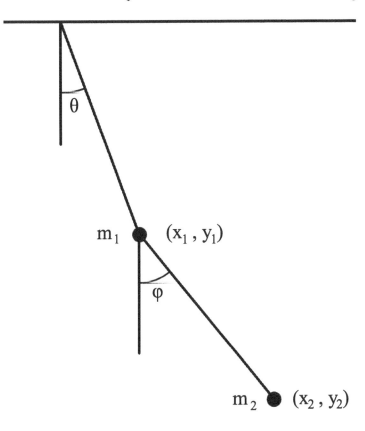

Figure 9.3 Double pendulum

Hint: From the drawing it is obvious that

$$x_1 = \ell_1 \sin \theta, \quad y_1 = \ell_1 \cos \theta \tag{9.88}$$

$$x_2 = \ell_1 \sin \theta + \ell_2 \sin \phi, \quad y_2 = \ell_1 \cos \theta + \ell_2 \cos \phi. \tag{9.89}$$

Hence

$$\dot{x}_1 = \ell_1 \cos \theta \frac{d\theta}{dt}, \quad \dot{y}_1 = -\ell_1 \sin \theta \frac{d\theta}{dt} \tag{9.90}$$

$$\dot{x}_2 = \ell_1 \cos\theta \frac{d\theta}{dt} + \ell_2 \cos\phi \frac{d\phi}{dt} \tag{9.91}$$

$$\dot{y}_2 = -\ell_1 \sin\theta \frac{d\theta}{dt} - \ell_2 \sin\phi \frac{d\phi}{dt}. \tag{9.92}$$

The expressions for the kinetic and potential energy are

$$T = \frac{m_1}{2}\left(\dot{x}_1^2 + \dot{y}_1^2\right) + \frac{m_2}{2}\left(\dot{x}_2^2 + \dot{y}_2^2\right) \tag{9.93}$$

$$V = m_1 g \ell_1 (1 - \cos\theta) + m_2 g(\ell_1 + \ell_2 - \ell_1 \cos\theta - \ell_2 \cos\phi). \tag{9.94}$$

3. Derive the equations of motion of the system shown in Figure 9 where m_1 is constrained to move only in the vertical direction.

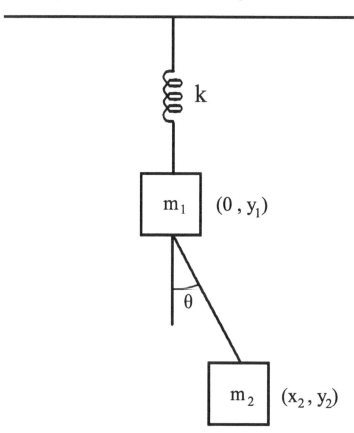

Figure 9.4 Pendulum suspended on a spring mass system

Hint: $x_2 = \ell_2 \sin\theta$

$$y_2 = y_1 + \ell_2 \cos\theta \tag{9.95}$$

$$T = \frac{1}{2}m_1\dot{y}_1^2 + \frac{m_2}{2}\left(\dot{x}_2^2 + \dot{y}_2^2\right) \tag{9.96}$$

$$V = \frac{1}{2}k(y - \ell_0)^2 + m_1g[-y + \ell_0] + m_2g[\ell_o + \ell_2 - (y_1 + \ell_2\cos\theta)] \tag{9.97}$$

where ℓ_0 is the natural length of the spring.

4. Solve the system described by Equations (12.41)-(12.42) with $m_1 \neq m_2$ and $k_1 \neq k_2 \neq k_3$.

9.6 VARIATION WITH CONSTRAINTS

In many applications of variational principles we have to find the extremum of a variational integral subject to constraints. The method used to solve these problems is an extension of the Lagrange multiplier approach for functions. We shall illustrate it with few examples in this section.

Example 1: Find $\psi(x)$ so that

1. (Variational Integral)

$$I(\psi) = \int_{x_1}^{x_2} (\psi_x + \psi_x^2)dx \qquad \text{is extremum} \tag{9.98}$$

2. (Constraint)

$$J(\psi) = \int_{x_1}^{x_2} \psi^2(x)dx = 1. \tag{9.99}$$

3. (Boundary Conditions)

$$\psi(x_1) = \psi(x_2) = 0. \tag{9.100}$$

Solution: Let $\psi(x)$ be the solution. For a small variation near $\psi(x)$ we can write

$$\bar{\psi}(x) = \psi(x) + \epsilon\eta(x). \tag{9.101}$$

Hence

$$I(\epsilon) = \int_{x_1}^{x_2} [\psi_x + \epsilon\eta_x(x) + (\psi_x + \epsilon\eta_x)^2]dx \tag{9.102}$$

$$J(\epsilon) = \int_{x_1}^{x_2} (\psi + \epsilon\eta)^2 dx = 1 \tag{9.103}$$

and $\eta(x_1) = \eta(x_2) = 0$.

Applying the Lagrange multiplier approach to the function $I(\epsilon)$ and the constraint $J(\epsilon)$ we now form

$$K(\epsilon, \lambda) = I(\epsilon) + \lambda J(\epsilon). \tag{9.104}$$

Since by assumption ψ is the solution of our problem we must have

$$\frac{\partial K(\epsilon, \lambda)}{\partial \epsilon}\bigg|_{\epsilon=0} = 0 \tag{9.105}$$

and

$$J(\epsilon = 0) = 1. \tag{9.106}$$

But

$$\frac{\partial K}{\partial \epsilon}\bigg|_{\epsilon=0} = \int_{x_1}^{x_2} [\eta_x(x) + 2\psi_x \eta_x + 2\eta\psi\lambda]dx. \tag{9.107}$$

However,

$$\int_{x_1}^{x_2} (\eta_x + 2\psi_x\eta_x)dx = (\eta + 2\psi_x\eta)|_{x_1}^{x_2} - \int_{x_1}^{x_2} 2\psi_{xx}\eta \, dx = -\int_{x_1}^{x_2} \psi_{xx}\eta \, dx. \tag{9.108}$$

Thus we infer that for all η

$$\int_{x_1}^{x_2} (-2\psi_{xx} + 2\lambda\psi)\eta \, dx = 0; \tag{9.109}$$

i.e. ψ must satisfy

$$\psi_{xx} - \lambda\psi = 0 \tag{9.110}$$

with the constraint

$$\int_{x_1}^{x_2} \psi^2 dx = 1. \tag{9.111}$$

A direct method to obtain the same result is to construct the Lagrange multiplier functional

$$K(\psi, \psi_x) = I(\psi, \psi_x) + \lambda J(\psi, \psi_x) \tag{9.112}$$

and compute the optimal solution as the simultaneous solution to a Euler-Lagrange equation for the variational principle K and the constraint J. In fact,

$$\frac{\partial K}{\partial \psi} - \frac{d}{dx}\left[\frac{\partial K}{\partial \psi_x}\right] = 2\lambda\psi - \frac{d}{dx}[1 + 2\psi_x] = 0, \tag{9.113}$$

i.e.

$$\psi_{xx} - \lambda\psi = 0.$$

Example 2: Find the curve of length 3 which passes through $(0,0)$ and $(1,0)$ for which the area between the curve and the x-axis is a maximum.

Solution: If the equation of the curve is $\psi = \psi(x)$, then we want to find the extremum of

$$I(\psi) = \int_0^1 \psi(x)dx, \quad \psi(0) = 0, \quad \psi(1) = 0$$

subject to the constraint

$$J(\psi) = \int_0^1 \sqrt{1 + \psi'^2} \, dx = 3.$$

Following the Lagrange multiplier approach we obtain the following Euler-Lagrange equation:

$$\frac{d}{dx}\left[\frac{\partial}{\partial\psi'}\left[\psi + \lambda\sqrt{1 + \psi'^2}\right]\right] - \frac{\partial}{\partial\psi}\left[\psi + \lambda\sqrt{1 + \psi'^2}\right] = 0. \tag{9.114}$$

Hence

$$\lambda\frac{d}{dx}\frac{\psi'}{\sqrt{1 + \psi'^2}} - 1 = 0. \tag{9.115}$$

Integrating Equation (9.115) we find (after some algebra) that ψ must satisfy

$$(x - c_1)^2 + (\psi - c_2)^2 = \lambda^2;$$

i.e. ψ is a circle. The three constants c_1, c_2, and λ are to be determined so that $\psi(0) = \psi(1) = 0$ and the requirement that the curve length is 2.

Exercises

1. Find the explicit expression(s) for ψ in Example 1 if $x_1 = 0$ and $x_2 = 2$.

2. Find the extremum of the variational integral

$$I(\psi) = \int_{x_1}^{x_2} \left[\psi^2\psi_x + \psi_x^2\right] dx$$

subject to the constraint

$$J(\psi) = \int_{x_1}^{x_2} \psi^2 dx = 1.$$

3. Show that the extremum of

$$I(\psi) = \int_{x_1}^{x_2} (p(x)(\psi'(x))^2 - q(x)\psi^2)dx$$

subject to the constraint

$$J(\psi) = \int_{x_1}^{x_2} r(x)\psi^2(x)dx = 1$$

$$\psi(x_1) = a \quad \psi(x_2) = b$$

are solutions of the differential equation

$$(p\psi')' + (q + \lambda r)\psi = 0.$$

9.7 AIRPLANE CONTROL; MINIMUM FLIGHT TIME

As is well known, some (sea) fighter planes take off vertically. In this section we consider a simplified model (with no air drag) for the vertical take off of such a plane in order to find out the optimal thrust distribution so that the plane achieves a specific height in minimum time. We assume that the total amount of fuel available is constant.

To formulate a mathematical model for this problem we introduce the following data and approximations.

1. The simplified equation of motion for the plane (when air drag is neglected) is given by

$$\ddot{x} = -g + u(t) \tag{9.116}$$

where $x(t)$ is the plane height at time t from the ground and u is the thrust.

2. The fuel consumption of the plane engine is proportional to the square of the thrust. Since the amount of fuel available is constant, it follows that the total thrust of the engine over the flight time T must satisfy the constraint

$$\int_0^T u(t)^2 dt = \text{constants} = c. \tag{9.117}$$

3. The boundary conditions on the flight of the plane are

(a)

$$x(0) = \dot{x}(0) = 0; \tag{9.118}$$

i.e. the plane starts from the ground with velocity 0.

(b)

$$x(T) = h; \tag{9.119}$$

i.e. the plane must achieve a predetermined height.

4. We want to minimize the vertical flight time T, i.e. minimize

$$I(T) = \int_0^T 1.dt = T. \tag{9.120}$$

Model: In this problem we must minimize Equation (9.120) subject to the constraints given by Equations (9.116), (10.103) and the boundary conditions Equation (9.118), and Equation (9.119). As a first step, however, we must convert the constraint given by the differential equation Equation (9.116) into an "algebraic" relationship so that the problem becomes amenable for treatment by the method of Lagrange multipliers as described in the previous section. Integrating Equation (9.116) over the interval $[0, \tau]$ and using the condition $\ddot{x}(0) = 0$ we obtain

$$\dot{x}(\tau) = -g\tau + \int_0^\tau u(s)ds . \tag{9.121}$$

Integrating this equation over $[0, t]$ and using the condition $x(0) = 0$ yields

$$x(t) = \frac{-gt^2}{2} + \int_0^t \left[\int_0^\tau u(s)ds \right] d\tau. \tag{9.122}$$

We now invert the order of integration of the double integral in Equation (9.122)

$$\int_0^t \left\{ \int_0^\tau u(s)ds \right] d\tau = \int_0^t \int_s^t u(s)d\tau ds$$

$$= \int_0^t (t-s)u(s)ds. \tag{9.123}$$

Hence,

$$x(t) = \frac{-gt^2}{2} + \int_0^t (t-s)u(s)ds. \tag{9.124}$$

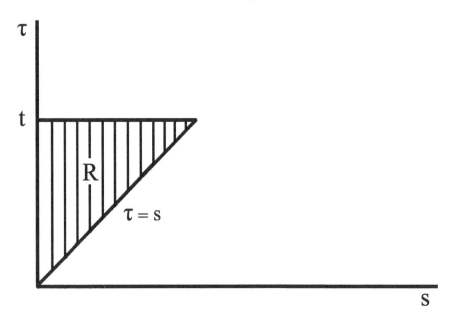

Figure 9.5   Region of integration in Equation (9.122)

Thus the variational problem under consideration is to minimize

$$I = \int_0^T 1\,dt \tag{9.125}$$

subject to the constraints

$$J_1(T, u) = -\int_0^T gt\,dt + \int_0^T (T - t)u(t)\,dt = h \tag{9.126}$$

$$J_2(T, u) = \int_0^T u^2(t)\,dt = c \, . \tag{9.127}$$

We see then that the variational problem under consideration has two dependent variables T, u, one independent variable t, and two constraints.

To compute the optimal solution we now follow the Lagrange multiplier approach and construct the functional

$$K(T, u) = I - \lambda_1 J_1 - \lambda_2 J_2 \, . \tag{9.128}$$

The optimal solution must then satisfy Euler-Lagrange equations for T, u and the constraints given by Equations (9.126)-(9.127). Thus

$$\frac{\partial K}{\partial T} - \frac{d}{dt}\left[\frac{\partial K}{\partial T'}\right] = 1 + \lambda_1 gT - \lambda_2 u^2(T) - \lambda_1 \int_0^T u(t)\,dt = 0 \tag{9.129}$$

$$\frac{\partial K}{\partial u} - \frac{d}{dt}\left[\frac{\partial K}{\partial \dot{u}}\right] = -\lambda_1(T - t) - 2\lambda_2 u(t) = 0. \tag{9.130}$$

Hence

$$u(t) = -\frac{\lambda_1(T - t)}{2\lambda_2} \tag{9.131}$$

(and therefore, as expected $u(T) = 0$).

Substituting Equation (9.131) in Equations (9.126), (9.127) and (10.115) leads to the following three equations in the unknowns T, λ_1, λ_2.

$$T^2(\lambda_1 T + 3\lambda_2 g) = -6\lambda_2 h \tag{9.132}$$

$$\lambda_1^2 T^3 = 12\lambda_2^2 c^2 \tag{9.133}$$

$$\lambda_1 T(\lambda_1 T + 4\lambda_2 g) = -4\lambda_2 . \tag{9.134}$$

From Equation (9.133) we infer

$$\frac{\lambda_1}{\lambda_2} = -\frac{2\sqrt{3}c}{T^{3/2}}; \tag{9.135}$$

hence for the optimal flight time Equation (9.132) yields the equation

$$3gT^2 + 6h = 2\sqrt{3}cT^{3/2} . \tag{9.136}$$

Finally, using Equations (9.131) and (9.135), we obtain an explicit expression for the thrust as a function of time:

$$u(t) = \sqrt{\frac{3}{T}}c\left[1 - \frac{t}{T}\right]. \tag{9.137}$$

Exercises

1. Explain in detail the inversion of the integral in Equation (9.123).

2. Reconsider the minimum lift off time problem when air drag has to be taken into account; i.e. the equation of motion for the plane is given by

$$\ddot{x} = -g - \alpha\dot{x} + u(t)$$

where α is a positive constant.

3. Reconsider the problem of this section when one wants to minimize the fuel consumption (e.g., in lift off from the moon surface).

9.8 APPLICATIONS IN ELASTICITY

In this section we discuss the modeling of transverse vibrations in elastic bars and thin plates using variational principles.

A. Transverse vibrations in an elastic bar.

Consider an elastic bar of constant cross section, density ρ per unit length, and length L. To model the transverse vibrations of such a bar we ignore the possible distortion of the cross sections and assume that any small section of the bar moves as a rigid entity.

If $u(x, t)$ is the displacement from equilibrium, the total kinetic energy of the bar is given by

$$T = \frac{\rho}{2} \int_0^L \dot{u}^2 dx . \tag{9.138}$$

The strain potential energy due to the vibrations is given by

$$J = \iint_D z^2 dydz. \tag{9.139}$$

By Hamilton's principle the motion of the bar will be an extremum of the variational integral

$$I(u) = \frac{1}{2} \int_{t_1}^{t_2} \int_0^L (\rho \dot{u}^2 - EJu_{xx}^2)dxdt \tag{9.140}$$

with the boundary condition

$$u = u_x = 0 \text{ at } x = 0, L \tag{9.141}$$

(clamped rod).

To derive the appropriate Euler-Lagrange equation for $I(u)$ we introduce

$$u = u + \epsilon\eta \tag{9.142}$$

in Equation (10.91) where u is the (sought for) extremal solution. Hence

$$I'(0) = \int_{t_1}^{t_2} \int_0^L \left[\frac{\partial f}{\partial \dot{u}}\dot{\eta} + \frac{\partial f}{\partial u_{xx}}\right] \eta_{xx} dxdt = 0 \tag{9.143}$$

where

$$f = \frac{1}{2} \left[\rho \dot{u}^2 - EJu_{xx}^2 \right] . \tag{9.144}$$

Using integration by parts and the boundary conditions we obtain

$$\int_{t_1}^{t_2} \int_0^L \frac{\partial f}{\partial \dot{u}} \dot{\eta} dx dt = - \int_{t_1}^{t_2} \int_0^L \frac{\partial}{\partial t} \left[\frac{\partial f}{\partial \dot{u}} \right] \eta dx dt \tag{9.145}$$

$$\int_0^L \frac{\partial f}{\partial u_{xx}} \eta_{xx} dx = \int_0^L \frac{\partial^2}{\partial x^2} \left[\frac{\partial f}{\partial u_{xx}} \right] \eta dx . \tag{9.146}$$

Hence

$$\frac{\partial}{\partial t} \left[\frac{\partial f}{\partial \dot{u}} \right] - \frac{\partial^2}{\partial x^2} \left[\frac{\partial f}{\partial u_{xx}} \right] = 0 \tag{9.147}$$

which yields

$$\rho \frac{\partial^2 u}{\partial t^2} + EJ \frac{\partial^4 u}{\partial x^4} = 0. \tag{9.148}$$

To solve this equation we need the boundary conditions at $0, L$ and the initial position and velocity of the bar.

B . Transverse vibrations of thin plate.

Consider the transverse vibrations of a uniform thin plate over a domain R with the boundary conditions

$$u = \frac{\partial u}{\partial \mathbf{n}} \bigg|_{\partial R} = 0 . \tag{9.149}$$

The kinetic energy of the plate is

$$T = \frac{1}{2} \rho \int_R \dot{u}^2 dA, \tag{9.150}$$

and the total strain potential energy is

$$V = \frac{1}{2} D \int\int_R \left[(\nabla^2 u)^2 - 2(1 - \sigma)(u_{xx}y_{yy} - u_{xy}^2) \right] dA . \tag{9.151}$$

Here D is the "flexural rigidity" of the plate and σ is the "Poisson ratio" of the material (σ is related to the relationship between the strain-stress tensors in the material of the plate).

The variational integral for this problem is

$$I(u) = \tag{9.152}$$

$$\frac{1}{2} \int_{t_1}^{t_2} \iint_R \left\{ \rho \dot{u}^2 - D \left[(\nabla^2 u)^2 - 2(1-\sigma)(u_{xx}u_{yy} - u_{xy}^2) \right] \right\} dA dt =$$

$$\frac{1}{2} \int_{t_1}^{t_1} \iint_R f dA dt .$$

Letting

$$u = u + \epsilon \eta \tag{9.153}$$

we obtain

$$I'(0) = \int_{t_1}^{t_2} \iint_R \left[\frac{\partial f}{\partial \dot{u}} \dot{\eta} + \frac{\partial f}{\partial u_{xx}} \eta_{xx} + \frac{\partial f}{\partial u_{yy}} \eta_{yy} + \frac{\partial f}{\partial u_{xy}} \eta_{xy} \right] dA dt = 0 \tag{9.154}$$

Using integration by parts, Green's theorem and the boundary conditions yields:

$$\int_{t_1}^{t_2} \iint_R \frac{\partial f}{\partial \dot{u}} \dot{\eta} dA dt = - \int_{t_1}^{t_2} \iint_R \eta \frac{\partial}{\partial t} \left[\frac{\partial f}{\partial \dot{u}} \right] dA dt \tag{9.155}$$

$$\iint_R \frac{\partial f}{\partial u_{xx}} \eta_{xx} = \iint_R \eta \frac{\partial^2}{\partial x^2} \left[\frac{\partial f}{\partial u_{xx}} \right] dA + \int_C \left[\eta_x \frac{\partial f}{\partial u_{xx}} - \eta \frac{\partial}{\partial x} \left[\frac{\partial f}{\partial u_{xx}} \right] \right] dy \tag{9.156}$$

etc.

Combining all these results together we finally obtain

$$\rho \frac{\partial^2 u}{\partial t^2} + D \nabla^2 (\nabla^2 u) = 0 . \tag{9.157}$$

This equation is referred to as the biharmonic equation.

9.9 RAYLEIGH-RITZ METHOD

Variational formulation of differential equations can be used to obtain approximate solutions for these equations. The basic technique is due to Rayleigh-Ritz, and it was the "precursor" to the current Finite Element Methods that are used for the numerical solution of partial differential equations in various applications.

We present this method through two examples:

Example 1: The bending of an elastic bar under uniform loading.

In the previous section we derived a variational principle for the vibrations of an elastic bar. If we apply on such a bar a uniform static loading, p, the appropriate variational principle for the static shape of this bar will be

$$I(u) = \int_0^L \left[\frac{1}{2} EJ(u'')^2 - pu \right] dx. \tag{9.158}$$

Euler-Lagrange equations then yield

$$EJu_{xxxx} - p = 0. \tag{9.159}$$

If the rod is clamped at $x = 0$, the corresponding boundary conditions are

$$u(0) = u'(0) = 0. \tag{9.160}$$

We can solve Equation (9.159) directly (as an ordinary differential equation) or use the variational principle given by Equation (9.158) to obtain at least an approximation to the solution. To this end we consider (as an example) the function space $\mathcal{S} = \{1, x, x^2, x^3, x^4\}$ and attempt to find the best approximation to the solution in this space; i.e., we seek to find the values of a, b, c, d, e which yield the best approximation to the solution u in the form

$$u = ax^4 + bx^3 + cx^2 + dx + e. \tag{9.161}$$

These values of a, b, c, d, e will minimize $I(u)$ in the function space $\mathcal{S}$ subject to the boundary conditions.

As a first step toward the solution we apply the boundary conditions Equation (9.160) to Equation (9.161). This yields $d = e = 0$, i.e.

$$u(x) = ax^4 + bx^3 + cx^2. \tag{9.162}$$

Substituting Equation (9.162) in Equation (9.158) we obtain after integration:

$$I(u) = -pL^3 \left[\frac{1}{5} aL^2 + \frac{1}{4} bL + \frac{1}{3} c \right] + \tag{9.163}$$

$$EJ \left[\frac{72}{5} a^2 L^5 + 18abL^4 + \frac{1}{3}(24ac + 18b^2))L^3 + 6bcL^2 + 2c^2 L \right].$$

This expression attains its minimum at a point where

$$\frac{\partial I}{\partial a} = \frac{\partial I}{\partial b} = \frac{\partial I}{\partial c} = 0. \tag{9.164}$$

This leads to a system of three linear equations for a, b, c whose solution is

$$a = \frac{p}{24EJ}, \quad b = \frac{-pL}{6EJ}, \quad c = \frac{pL^2}{4EJ}. \tag{9.165}$$

Example 2: Solve

$$\nabla^2 u = c \tag{9.166}$$

on the square $\Omega = [-a, a] \times [-a, a]$ subject to the Dirichlet boundary condition

$$u|_{\partial\Omega} = 0. \tag{9.167}$$

Solution: The solution of Equation (9.166) satisfies the variational principle

$$I(u) = \int_\Omega \left[\frac{1}{2}(u_x^2 + u_y^2) + cu\right] dxdy . \tag{9.168}$$

Let $\{\phi_i\}$ $i = 1 \ldots N$ be a set of functions which satisfy the boundary condition (9.167). We seek an approximate solution of Equations (9.166), (9.167) in the form;

$$u(x, y) = \sum a_i\phi_i(x, y) \tag{9.169}$$

which minimizes the functional in Equation (9.168), i.e.

$$\frac{\partial I(\sum a_i\phi_i(x, y))}{\partial a_k} = 0 \quad k = 1, \ldots, N. \tag{9.170}$$

This yields (after substitution in Equation (9.168))

$$\sum_i a_i \int_\Omega \left[\left(\frac{\partial\phi_i}{\partial x}\right)\left(\frac{\partial\phi_k}{\partial x}\right) + \left(\frac{\partial\phi_i}{\partial y}\right)\left(\frac{\partial\phi_k}{\partial y}\right)\right] dxdy + c\int_\Omega \phi_k dxdy = 0. \tag{9.171}$$

This is a system of N linear equations for the coefficients a_i whose solution yields (through Equation(9.169)) an approximate solution to Equations (9.166), (9.167).

9.10   THE FINITE ELEMENT METHOD IN 2-D

We shall discuss this method within the context of the following boundary value problem

$$\nabla^2 u = 0 \ \text{ in } \ \Omega \qquad\qquad (9.172)$$

$$u = u_D \ \text{ on } \ \Gamma_D$$

$$\frac{\partial u}{\partial \mathbf{n}} = (\nabla u) \cdot \mathbf{n} = u_N \ \text{ on } \ \Gamma_N$$

where Γ_D is the part of the boundary of Ω where Dirichlet boundary conditions are being specified and Γ_N is the part where Neumann boundary conditions are specified ($\mathbf{n}$ is the normal to the boundary). Thus, the boundary of the domain satisfies

$$\partial\Omega = \Gamma_D \cup \Gamma_N = \Gamma.$$

9.10.1   Geometrical Triangulations

The first step towards the application of the Finite Element Method (FEM) is a geometrical one. We have to **triangulate the region**, i.e. divide the region into triangles (not necessarily of equal size).

Remark: In the following we consider only "triangular elements;" however, in many applications other shapes are used.

   In this triangulation process one must be careful to leave no gaps between the elements and there should NOT be any "free nodes;" i.e. nodes located on an element's edges are not allowed (in this context "nodes" is another name for the elements vertices or "grid points"). Also triangles should not overlap with each other.

Guidelines for "good" triangulations:

1. Avoid extremely irregular mesh. Do not use obtuse triangles; i.e. the "sides" of the elements should be "comparable in size."

2. If the physical problem has some sort of symmetry, subdivide the domain according to the symmetry.

3. Use fine mesh in the vicinity of "abrupt changes" (in geometry, boundary conditions, etc.).

The motivation for these "guidelines" is intended to ensure that one obtains a well conditioned system of linear equations for the solution at the vertices of the elements.

Topological Data: We have to number the elements and nodes (without gaps) and prepare a "Dictionary" where

1. The node number and its coordinates (x, y) are given.

2. The element number and its associated nodes (in the counterclockwise order) are given.

This numbering can be done using either row or column schemes (see Figure 9.6 and Figure 9.7). However more general schemes are possible. These numbering schemes attempt to reduce the band of the coefficient matrix for the linear system of equations one has to solve eventually.

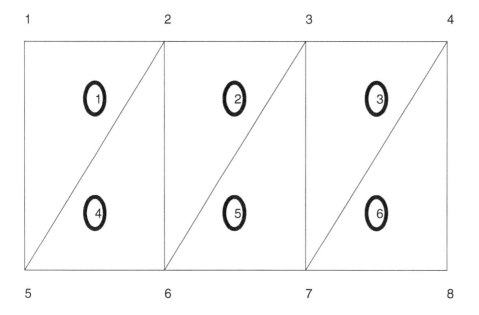

Figure 9.6 Row numbering of the nodes

9.10.2 Linear Interpolation in 2-D

Consider an element where the value of the solution u at the nodes (x_i, y_i), $i = 1, 2, 3$ is given by u_1, u_2, u_3. We want to find a linear interpolating function

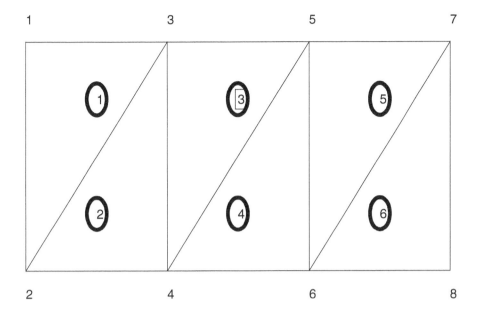

Figure 9.7 Column numbering of the nodes

over the element

$$u^e(x, y) = \alpha + \beta x + \gamma y$$

using the values of the solution at the nodes, viz.

$$u_1 = \alpha + \beta x_1 + \gamma y_1$$

$$u_2 = \alpha + \beta x_2 + \gamma y_2$$

$$u_3 = \alpha + \beta x_3 + \gamma y_3.$$

This is a system of linear equations which can be solved by Cramer's method for α, β, γ. The solution can be rewritten in the form

$$u^e(x, y) = \Sigma \phi_i(x, y) u_i$$

where

$$\phi_i(x, y) = \frac{1}{2\Delta^e}(a_i + b_i x + c_i y)$$

and

$$\Delta^e = \frac{1}{2}(b_2 c_3 - b_3 c_2),$$

$$a_1 = x_2y_3 - x_3y_2 \quad b_1 = y_2 - y_3 \quad c_1 = x_3 - x_2 \qquad (9.173)$$

$$a_2 = x_3y_1 - x_1y_3 \quad b_2 = y_3 - y_1 \quad c_2 = x_1 - x_3 \qquad (9.174)$$

$$a_3 = x_1y_2 - x_2y_1 \quad b_3 = y_1 - y_2 \quad c_3 = x_2 - x_1. \qquad (9.175)$$

Δ^e is the element area up to a sign.

Remark: Observe that $\phi_i(x, y)$ have the nice property that $\phi_i(x_i, y_i) = 1$ and $\phi_i(x_j, y_j) = 0$ $j \neq i$. So this formulation preserves the fact that u_i are the actual values of the solution at the nodes.

It follows then from this discussion that the values of the solution at the nodes, u_i, can be used to determine (by interpolation) the solution over the whole domain Ω.

9.10.3   Galerkin Formulation of FEM

We begin by reminding the reader of the following theorem:

Gauss-Green's Theorem: (Integration by parts in 2-D)

$$\int_\Omega (\nabla^2 u) w \, dA = \int_\Gamma \frac{\partial u}{\partial \mathbf{n}} w \, d\Gamma - \int_\Omega (\nabla u \cdot \nabla w) dA. \qquad (9.176)$$

In particular if u is a solution of the Laplace equation then

$$\iint_\Omega \left(\frac{\partial u}{\partial x} \frac{\partial w}{\partial x} + \frac{\partial u}{\partial y} \frac{\partial w}{\partial y} \right) dA = \int_\Gamma \frac{\partial u}{\partial \mathbf{n}} w \, d\Gamma. \qquad (9.177)$$

In the Galerkin formulation we attempt to write the solution on each element as

$$u^e(x, y) = \sum_{i=1}^3 \phi_i^e(x, y) u_i \qquad (9.178)$$

and let

$$R^e(x, y) = \nabla^2 u^e(x, y)$$

be the residual from the exact solution on each element. We then determine the u_i's by imposing the orthogonality constraint

$$\sum_{e,i} \iint_{\Omega^e} \phi_i^e(x, y) R^e(x, y) dA = 0. \qquad (9.179)$$

If the residuals $R^e(x, y)$ are negligible we can assume that approximately

$\nabla^2 u^e(x, y) \approx 0$ and use Equation (9.177) to rewrite Equation (9.179) on each element as

$$\sum_{j=1}^{3} \left\{ \int\int_{\Omega^e} \left(\frac{\partial \phi_i}{\partial x} \frac{\partial \phi_j}{\partial x} + \frac{\partial \phi_i}{\partial y} \frac{\partial \phi_j}{\partial y} \right) dA \right\} u_j = \int_{\Gamma_N^e} u_N \phi_i d\Gamma \quad i = 1, 2, 3$$

$$(9.180)$$

where Γ_N^e denote the intersection, if any, of Γ_N and the triangle e.

Introducing

$$K_{i,j}^e = \int\int_{\Omega^e} \left(\frac{\partial \phi_i}{\partial x} \frac{\partial \phi_j}{\partial x} + \frac{\partial \phi_i}{\partial y} \frac{\partial \phi_j}{\partial y} \right) dA,$$

and

$$F_i^e = \int_{\Gamma_N^e} u_N \phi_i d\Gamma,$$

Equation (9.180) can be rewritten as

$$\sum_j K_{i,j}^e u_j = F_i^e. \tag{9.181}$$

Using the expression for the the functions $\phi_i(x, y)$, which was derived earlier, we find the following matrix representation for $K_{i,j}^e$

$$K^e = \frac{1}{4(\Delta^e)^2} \begin{pmatrix} b_1^2 + c_1^2 & b_1 b_2 + c_1 c_2 & b_1 b_3 + c_1 c_3 \\ b_1 b_2 + c_1 c_2 & b_2^2 + c_2^2 & b_2 b_3 + c_2 c_3 \\ b_1 b_3 + c_1 c_3 & b_2 b_3 + c_2 c_3 & b_3^2 + c_3^2 \end{pmatrix}. \tag{9.182}$$

Observe that the matrix K^e is symmetric. These equations have to be "assembled" for all the elements (since u_i may appear in several elements) and solved for the nodal values u_i.

Example 3: A Four Element Model.

Consider the following FEM model made of four equal size right angle, isosceles triangular elements where the sides are of length $\sqrt{2}$ (see Figure 9.8)).

Column numbering is used for the nodes. Observe also the counterclockwise numbering of the nodes in each element. The area of each triangle is 1 and the coordinates of the nodes in order are $(0, 0)$, $(0, -\sqrt{2})$, $(\sqrt{2}, 0)$, $(\sqrt{2}, -\sqrt{2}), (2\sqrt{2}, 0), (2\sqrt{2}, -\sqrt{2})$ (the element number is circled). Using Equations (9.173), (9.182) we obtain the following element equations:

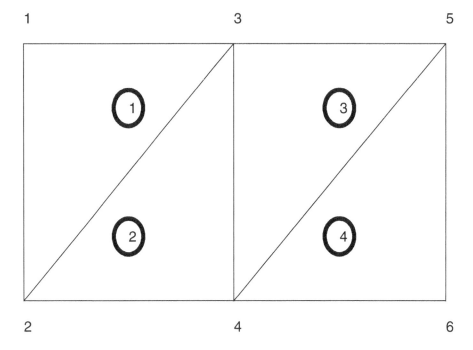

Figure 9.8 Four element FEM model

1. $\dfrac{1}{4} \begin{pmatrix} 4 & -2 & -2 \\ -2 & 2 & 0 \\ -2 & 0 & 2 \end{pmatrix} \begin{pmatrix} u_1 \\ u_2 \\ u_3 \end{pmatrix} = \begin{pmatrix} F_1^1 \\ F_2^1 \\ F_3^1 \end{pmatrix}$

2. $\dfrac{1}{4} \begin{pmatrix} 2 & -2 & 0 \\ -2 & 4 & -2 \\ 0 & -2 & 2 \end{pmatrix} \begin{pmatrix} u_2 \\ u_4 \\ u_3 \end{pmatrix} = \begin{pmatrix} F_2^2 \\ F_4^2 \\ F_3^2 \end{pmatrix}$

3. $\dfrac{1}{4} \begin{pmatrix} 4 & -2 & -2 \\ -2 & 2 & 0 \\ -2 & 0 & 2 \end{pmatrix} \begin{pmatrix} u_3 \\ u_4 \\ u_5 \end{pmatrix} = \begin{pmatrix} F_3^3 \\ F_4^3 \\ F_5^3 \end{pmatrix}$

4. $\dfrac{1}{4} \begin{pmatrix} -2 & -2 & 0 \\ -2 & 4 & -2 \\ 0 & -2 & 2 \end{pmatrix} \begin{pmatrix} u_4 \\ u_6 \\ u_5 \end{pmatrix} = \begin{pmatrix} F_4^4 \\ F_6^4 \\ F_3^4 \end{pmatrix}$

Combining these equations yields:

$$\frac{1}{4} \begin{pmatrix} 4 & -2 & -2 & 0 & 0 & 0 \\ -2 & 4 & 0 & -2 & 0 & 0 \\ -2 & 0 & 8 & -4 & -2 & 0 \\ 0 & -2 & -4 & 8 & 0 & -2 \\ 0 & 0 & -2 & 0 & 4 & -2 \\ 0 & 0 & 0 & -2 & -2 & 4 \end{pmatrix} \begin{pmatrix} u_1 \\ u_2 \\ u_3 \\ u_4 \\ u_5 \\ u_6 \end{pmatrix} = \begin{pmatrix} F_1^1 \\ F_2^1 + F_2^2 \\ F_3^1 + F_3^2 + F_3^3 \\ F_4^2 + F_4^3 + F_4^4 \\ F_5^3 + F_5^4 \\ F_6^4 \end{pmatrix}$$

9.11 APPENDIX

Green's Theorems in Two and Three Dimensions

Green's theorems are the analog of "integration by parts" in higher dimensions. However, some variations of these theorems are useful in other contexts.

Theorem (Green's Theorem in two dimensions) Let D be a region in two dimensions and let ∂D be its boundary (traced in the positive direction), then for any two (smooth) functions P, Q (and proper assumptions on D),

$$\iint_D \left[\frac{\partial P}{\partial x} + \frac{\partial Q}{\partial y} \right] dA = \int_{\partial D} (P\,dy - Q\,dx).$$

Corollary: (Integration by parts)

If $P = G\eta$. $A = F\eta$ then we obtain

$$\iint_D \left[G\frac{\partial \eta}{\partial x} + F\frac{\partial \eta}{\partial y} \right] dA = -\iint_D \eta \left[\frac{\partial G}{\partial x} + \frac{\partial F}{\partial y} \right] dA + \int_{\partial D} \eta(G\,dy - F\,dx).$$

Corollary: (Second order integration by parts)

If we let $Q = 0$ and $P = \left[G\frac{\partial \eta}{\partial x} - \eta\frac{\partial G}{\partial x} \right]$ in Green's theorem we obtain

$$\iint_D G\frac{\partial^2 \eta}{\partial x^2} dA = \iint_D \eta\frac{\partial^2 G}{\partial x^2} dA + \int_{\partial D} \left(G\frac{\partial \eta}{\partial x} - \eta\frac{\partial G}{\partial x} \right) dy.$$

Similar equation holds for differentiation with respect to y.

Theorem: (Green's theorem in three dimensions)

$$\int_V \mathbf{F} \cdot \mathrm{grad}\,\eta\,dV = -\int_V \eta\,\mathrm{div}\mathbf{F}\,dV + \int_{\partial V} \eta\mathbf{F} \cdot \mathbf{n}\,dS.$$

Note that the divergence theorem is obtained as a special case of this theorem with $\eta = 1$.

CHAPTER 10

Modeling Fluid Flow

CONTENTS

10.1   STRAIN AND STRESS

By definition the relative positions of the points in a rigid body cannot change over time. For elastic bodies, on the other hand, these relative positions can change. Strain measures the relative deformation of the points in an elastic body when these deformations are "small."

To analyze this concept, consider two nearby points whose coordinates in the undeformed position are $\mathbf{x} = (x_1, x_2, x_3)$ and $\mathbf{y} = (y_1, y_2, y_3)$. Let these points be displaced now to $\mathbf{x} + \mathbf{d}(\mathbf{x}), \mathbf{y} + \mathbf{d}(\mathbf{y})$ (that is, the deformation is a function of the position).

The distance between the points before and after the deformation respectively is

$$r_0^2 = \sum_{i=1}^{3} (y_i - x_i)^2 \tag{10.1}$$

$$r_1^2 = \sum_{i=1}^{3} (y_i - x_i + d_i(\mathbf{y}) - d_i(\mathbf{x}))^2. \tag{10.2}$$

Since we are assuming that $\| \mathbf{x} - \mathbf{y} \| << 1, \ \| \mathbf{d} \| << 1$ we can write to a first order approximation

$$d_i(\mathbf{y}) - d_i(\mathbf{x}) \cong \sum_{j=1}^{3} \frac{\partial d_i}{\partial x_j}(\mathbf{x})(y_j - x_j). \tag{10.3}$$

Expanding Equation (10.2) using Equation (10.3) and neglecting higher order terms we obtain

$$r_1^2 = r_0^2 + 2 \sum_{i,j=1}^{3} \frac{\partial d_i}{\partial x_j}(\mathbf{x})(y_i - x_i)(y_j - x_j).$$

We now define the extension of the element $\| \mathbf{y} - \mathbf{x} \|$ as

$$e = \lim_{\mathbf{y} \to \mathbf{x}} \frac{r_1 - r_0}{r_0} = \lim_{\mathbf{y} \to \mathbf{x}} \left(\frac{r_1 - r_0}{r_0} \right) \left(\frac{r_1 + r_0}{2r_0} \right) =$$

$$= \frac{1}{2} \lim_{\mathbf{y} \to \mathbf{x}} \frac{r_1^2 - r_0^2}{r_0^2} = \sum_{i,j=1}^{3} \frac{\partial d_i}{\partial x_j}(\mathbf{x}) \cos \theta_i \cos \theta_j.$$

Where we used the fact that

$$\lim_{\mathbf{y} \to \mathbf{x}} \left(\frac{r_1 + r_0}{2r_0} \right) = 1,$$

and defined

$$\cos\theta_i = \lim_{\mathbf{y}\to\mathbf{x}} \frac{y_i - x_i}{r_0}$$

are the cosine direction of the element. We therefore showed that the extension e is a quadratic form in the direction cosines which can be rewritten as

$$e = \sum_{i,j} e_{ij} \cos\theta_i \cos\theta_j \tag{10.4}$$

where

$$e_{ij} = \frac{1}{2}\left(\frac{\partial d_i}{\partial x_j} + \frac{\partial d_j}{\partial x_i}\right). \tag{10.5}$$

Since $e_{ij} = e_{ji}$, this quadratic form is symmetric and e_{ij} are called the components of the strain [the symmetric strain matrix (e_{ij}) forms a second rank tensor].

Strains in elastic bodies are the result of forces acting on the boundary of each volume element. The force per unit area is called the stress $\mathbf{P}$. Observe however that in general $\mathbf{P}$ is not normal to the surface element. The normal component of the stress is called the pressure (or tension), while the tangential component is called the "shearing stress." To analyze the stress $\mathbf{P}$ acting on a surface element dS with normal $\mathbf{n}$ we consider a tetrahedron of which three faces dS_x, dS_y, dS_z lie in the coordinate plane (see Figure. 10.1).

Let $\boldsymbol{\tau}_i, i = 1,2,3$, be the stresses on each face of the tetrahedron. By Newton's third law the condition of equilibrium for the volume enclosed by the tetrahedron is

$$\mathbf{P}dS - \boldsymbol{\tau}_1 dS_x - \boldsymbol{\tau}_2 dS_y - \boldsymbol{\tau}_3 dS_z = 0. \tag{10.6}$$

But

$$d\mathbf{S} = \mathbf{n}dS = dS_x\mathbf{i} + dS_y\mathbf{j} + dS_z\mathbf{k},$$

i.e.

$$dS_x = n_1 dS, \quad dS_y = n_2 dS, \quad dS_z = n_3 dS.$$

Therefore

$$(\mathbf{P} - \boldsymbol{\tau}_1 n_1 - \boldsymbol{\tau}_2 n_2 - \boldsymbol{\tau}_3 n_3)dS = 0.$$

In component form this becomes

$$P_1 = \tau_{11} n_1 + \tau_{21} n_2 + \tau_{31} n_3, \quad etc.$$

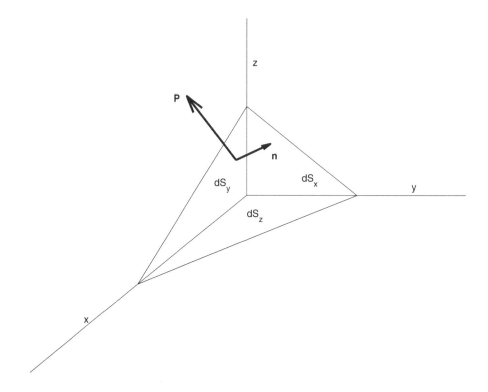

Figure 10.1   Wolume element with stress **P**

Hence we can rewrite Equation (10.6) as

$$P_i = \sum_{j=1}^{3} \tau_{ji} n_j. \tag{10.7}$$

It can be shown that (τ_{ij}) is a symmetric (i.e. $\tau_{ij} = \tau_{ji}$) second rank tensor which is called the "stress tensor."

Using the symmetry of (τ_{ij}) Equation (10.7) can be written as

$$P_i = \sum_{j=1}^{3} \tau_{ij} n_j. \tag{10.8}$$

The basic assumption of **linear elasticity** is that for small deformations the strain and stress tensors are linearly related. Thus

$$\tau_{ij} = \sum_{m,n} \alpha_{ijmn} e_{mn} \tag{10.9}$$

where α_{ijmn} depends on the elastic material.

Since e_{mn}, τ_{ij} are symmetric, it is easy to see that

$$\alpha_{ijmn} = \alpha_{ijnm}, \quad \alpha_{ijmn} = \alpha_{jimn}. \qquad (10.10)$$

Furthermore, it can be shown (using a thermodynamic argument) that

$$\alpha_{ijmn} = \alpha_{mnij}.$$

These constraints reduce the number of independent components in this tensor to 21. When the material is isotropic (that is there is no preferred direction) the most general form of α_{ijmn} is

$$\alpha_{ijmn} = \lambda \delta_{ij} \delta_{mn} + \mu(\delta_{im}\delta_{jn} + \delta_{in}\delta_{jm}) \qquad (10.11)$$

where λ is the Lame's constant and μ is the rigidity. Using Equations (10.5), (10.11) Equation (10.9) now becomes:

$$\tau_{ij} = \lambda \delta_{ij} \sum_k \frac{\partial d_k}{\partial x_k} + \mu \left(\frac{\partial d_i}{\partial x_j} + \frac{\partial d_j}{\partial x_i} \right). \qquad (10.12)$$

10.2   EQUATIONS OF MOTION FOR IDEAL FLUID

The equations which govern the motion of "ideal" fluid with zero viscosity (and constant temperature) consist of one equation which is an expression of the law of "mass conservation" (the "continuity equation") and three equations which express the law of momentum conservation. These are called Euler's equations.

10.2.1   Continuity equation

To derive this equation consider a fluid flow with mass density $\rho(\mathbf{x}, t)$. Let V be a volume contained in the flow. The mass of the fluid in this volume V at time t is given by

$$m(t, v) = \int_V \rho(x, t) d\mathbf{x}. \qquad (10.13)$$

Hence the rate of change of the mass in V is

$$\frac{dm}{dt} = \int_V \frac{\partial \rho}{\partial t}(\mathbf{x}, t) d\mathbf{x}. \qquad (10.14)$$

Now let ∂V denote the boundary of V and $\mathbf{n}$ the unit outward normal to ∂V. The total mass flow rate of the fluid across ∂V in the *outward* direction is

$$\int_{\partial V} \mathbf{q} \cdot \mathbf{n} dS = \int_{\partial V} (\rho \mathbf{u}) \cdot \mathbf{n} dS. \tag{10.15}$$

Mass conservation principle implies, however, that a (positive) rate of change of the mass in V must equal the rate at which the mass is crossing ∂V in the *inward* direction. Therefore

$$\int_V \frac{\partial \rho}{\partial t}(\mathbf{x}, t) dV = -\int_{\partial V} \rho \mathbf{u} \cdot \mathbf{n} dS. \tag{10.16}$$

To convert the right hand side of Equation (10.16) into a volume integral we now invoke the divergence theorem which states that for any "smooth" vector field $\mathbf{F}$ in V

$$\int_{\partial V} \mathbf{F} \cdot \mathbf{n} dS = \int_V div\mathbf{F} dV.$$

Then Equation (10.16) yields

$$\int_V \left\{ \frac{\partial \rho}{\partial t} + div(\rho \mathbf{u}) \right\} dV = 0. \tag{10.17}$$

Since V is arbitrary we infer that the integrand in equation (10.17) is zero, i.e.

$$\frac{\partial \rho}{\partial t} + div(\rho \mathbf{u}) = 0. \tag{10.18}$$

Equation (10.18) is called the continuity equation.

10.2.2 Euler's equations:

To derive these equations we consider a volume of fluid V and a coordinate system that is moving with it (this is referred to as the "Lagrangian picture" for the fluid flow).

At time t the momentum of the fluid in V is

$$\int_V (\rho \mathbf{u}) dV.$$

Hence by Newton's second law and Equation (10.7)

$$\frac{d}{dt} \int_V (\rho u_i) dV = -\int_{\partial V} \sum_j \tau_{ij} n_j dS.$$

The minus sign is due to the fact that $d\mathbf{S} = \mathbf{n}dS$ where $\mathbf{n}$ is the outward normal to ∂V. Applying the divergence theorem yields

$$\frac{d}{dt}\int_V (\rho u_i)dV = -\int_V \sum_j \frac{\partial \tau_{ij}}{\partial x_j}dV. \tag{10.19}$$

Since we are considering a fluid with zero viscosity, there are no shearing stresses and the stress tensor τ_{ij} has only a normal components, viz. the pressure p. Thus

$$\tau_{ij} = p\delta_{ij}.$$

Substituting this expression of τ in Equation (10.19) yields (in vector form)

$$\frac{d}{dt}\int_V (\rho\mathbf{u})dV = -\int_V \nabla p \, dV.$$

Since the volume V is arbitrary we infer therefore that

$$\frac{d}{dt}(\rho\mathbf{u}) = -\nabla p.$$

To convert this equation to one in a fixed (inertial) coordinate system (Eurelian Picture) we use the fact that in a fixed coordinate system $x = x(t)$, $y = y(t)$, $z = z(t)$ and hence for a general vector field $\mathbf{F}(\mathbf{x})$ we have

$$\left.\frac{d\mathbf{F}}{dt}\right|_{\text{Lagrange}} = \frac{\partial F}{\partial t} + \frac{\partial \mathbf{F}}{\partial x}\frac{dx}{dt} + \frac{\partial \mathbf{F}}{\partial y}\frac{dy}{dt} + \frac{\partial \mathbf{F}}{\partial z}\cdot\frac{dz}{dt} = \frac{\partial \mathbf{F}}{\partial t} + (\mathbf{u}\cdot\nabla)\mathbf{F}.$$

Therefore

$$\frac{\partial(\rho\mathbf{u})}{\partial t} + (\mathbf{u}\cdot\nabla)(\rho\mathbf{u}) = -\nabla p. \tag{10.20}$$

When $\rho = $ constant (i.e. the fluid is homogeneous and incompressible) Equations (10.18), (10.20) reduce to

$$\nabla\cdot\mathbf{u} = 0 \tag{10.21}$$

$$\frac{\partial\mathbf{u}}{\partial t} + (\mathbf{u}\cdot\nabla\mathbf{u}) = -\frac{1}{\rho}\nabla p \tag{10.22}$$

or in component form (where $\mathbf{u} = (u, v, w)$)

$$\frac{\partial u}{\partial x} + \frac{\partial v}{\partial y} + \frac{\partial w}{\partial z} = 0 \tag{10.23}$$

$$\frac{\partial u}{\partial t} + u\frac{\partial u}{\partial x} + v\frac{\partial u}{\partial y} + w\frac{\partial u}{\partial z} = -\frac{1}{\rho}\frac{\partial p}{\partial x} \tag{10.24}$$

$$\frac{\partial v}{\partial t} + u\frac{\partial v}{\partial x} + v\frac{\partial v}{\partial y} + w\frac{\partial v}{\partial z} = -\frac{1}{\rho}\frac{\partial p}{\partial y} \tag{10.25}$$

$$\frac{\partial w}{\partial t} + u\frac{\partial w}{\partial x} + v\frac{\partial w}{\partial y} + w\frac{\partial w}{\partial z} = -\frac{1}{\rho}\frac{\partial p}{\partial z}. \tag{10.26}$$

10.3   NAVIER-STOKES EQUATIONS

The equations that govern the motion of viscous fluid consist of the continuity equation and a set of three equations which express the law of momentum conservation. These equations are called Navier-Stokes equations.

To derive these equations we consider again a fluid volume V. The resultant force in the ith-direction due to the stresses acting on this volume (see Equation (10.8)) is

$$\int_{\partial V} P_i dS = \int_{\partial V} \sum_j \tau_{ij} n_j dS. \tag{10.27}$$

Hence momentum conservation yields

$$\frac{d}{dt}\int_V (\rho u_i) dV = \int_{\partial V} \left(\sum_j \tau_{ij} n_j \right) dS + \int_V \rho F_i dV \tag{10.28}$$

where $\mathbf{F}$ represents "body force(s)" per unit mass (such as gravity) acting on the fluid. Using the divergence theorem this converts to

$$\int_V \left[\frac{d(\rho u_i)}{dt} - \sum_j \frac{\partial \tau_{ij}}{\partial x_j} - \rho F_i \right] dV. \tag{10.29}$$

But since V is arbitrary, it follows that

$$\frac{d(\rho u_i)}{dt} = \sum_j \frac{\partial \tau_{ij}}{\partial x_j} + \rho F_i. \tag{10.30}$$

In the following discussion we neglect the body forces.

To motivate the choice of the stress tensor which is appropriate for fluid motion we shall use both formal and physical arguments. From a formal point of view we shall assume that there exists in the fluid a linear relationship between the stress and strain

$$\tau_{ij} = s_{ij} + \sum_{m,n} \alpha_{ijmn} e_{mn} \tag{10.31}$$

where s_{ij} is the "static stress tensor." This motivates us to write e_{ij} in analogy with (10.5) as

$$e_{ij} = \frac{1}{2} \left(\frac{\partial u_i}{\partial x_j} + \frac{\partial u_j}{\partial x_i} \right). \tag{10.32}$$

Note, however, *that in these equations the velocities* u_i *replaced the displacements in Equation (10.5). This is justified by the observation that the relation given by Equation (12.27) implicitly assumes that the strains* e_{ij} *are due to the fluid motion. We observe that in the literature there exist several other models for the relationship between stress and strain especially in the context of fluid turbulence.*

For an isotropic fluid the tensors s_{ij}, α_{ijmn} are also isotropic and we must have

$$s_{ij} = -p_0 \delta_{ij} \tag{10.33}$$

$$\alpha_{ijmn} = \lambda \delta_{ij} \delta_{mn} + \tag{10.34}$$
$$\mu(\delta_{im}\delta_{jn} + \delta_{in}\delta_{jm})$$

where p_0 is independent of e_{mn} and is called the static pressure, λ is called the "bulk viscosity" of the fluid, while μ is called the "dynamic viscosity." Hence

$$\sum_{m,n} \alpha_{ijmn} e_{mn} = \sum_{m,n} [\lambda \delta_{ij}\delta_{mn} + \mu(\delta_{im}\delta_{jn} + \delta_{in}\delta_{jm})]e_{mn} = \lambda \delta_{ij} \sum_m e_{mm} + 2\mu e_{ij}.$$

Therefore

$$\tau_{ij} = -p_0 \delta_{ij} + \lambda \delta_{ij} \sum_m e_{mm} + 2\mu e_{ij}. \tag{10.35}$$

We now define the "dynamic pressure" p as the negative mean of the principal stresses

$$p = \frac{-1}{3} \sum_i \tau_{ii} = p_0 - \left(\lambda + \frac{2}{3}\mu\right) \nabla \cdot \mathbf{u}. \tag{10.36}$$

Substituting Equation (10.36) in Equation (10.35) we obtain

$$\tau_{ij} = -\left[p + \frac{2}{3}\mu\nabla \cdot \mathbf{u}\right]\delta_{ij} + 2\mu e_{ij}. \tag{10.37}$$

Substituting this expression of τ_{ij} in (12.26) we obtain after some algebra

$$\frac{d(\rho\mathbf{u})}{dt} = -\nabla p + \frac{\mu}{3}\nabla(\nabla \cdot \mathbf{u}) + \mu\nabla^2\mathbf{u}, \tag{10.38}$$

where we dropped the body force term $\rho\mathbf{F}$. If the fluid is incompressible, then $\rho = $ constant and $\nabla \cdot \mathbf{u} = 0$ (continuity equation) which leads to

$$\frac{d\mathbf{u}}{dt} = -\frac{1}{\rho}\nabla p + \nu\nabla^2\mathbf{u}, \tag{10.39}$$

where $\nu = \mu/\rho$ is called the "kinematic viscosity" of the fluid. Replacing $\left(\frac{d\mathbf{u}}{dt}\right)$ by its Eurelian equivalent, Equation (10.39) takes the form

$$\frac{\partial\mathbf{u}}{\partial t} + (\mathbf{u} \cdot \nabla)\mathbf{u} = -\frac{1}{\rho}\nabla p + \nu\nabla^2\mathbf{u} \tag{10.40}$$

Thus the motion of incompressibe viscous fluids is governed by Equations (10.21), (10.40). They are referred to as "Navier-Stokes equations (NSE)." In component form these equations are:

$$\frac{\partial u}{\partial x} + \frac{\partial v}{\partial y} + \frac{\partial w}{\partial z} = 0 \tag{10.41}$$

$$\frac{\partial u}{\partial t} + u\frac{\partial u}{\partial x} + v\frac{\partial u}{\partial y} + w\frac{\partial u}{\partial z} = -\frac{1}{\rho}\frac{\partial p}{\partial x} + \nu\nabla^2 u \tag{10.42}$$

$$\frac{\partial v}{\partial t} + u\frac{\partial v}{\partial x} + v\frac{\partial u}{\partial y} + w\frac{\partial v}{\partial z} = -\frac{1}{\rho}\frac{\partial p}{\partial y} + \nu\nabla^2 v \tag{10.43}$$

$$\frac{\partial w}{\partial t} + u\frac{\partial w}{\partial x} + v\frac{\partial w}{\partial y} + w\frac{\partial w}{\partial z} = -\frac{1}{\rho}\frac{\partial p}{\partial z} + \nu\nabla^2 w. \tag{10.44}$$

In the following we consider only incompressible flows.

EXAMPLE - Poiseuille Flow:

This is a "uniform" steady state flow between two infinite parallel plates located at $z = -h$ and $z = h$ which satisfies

$$u = u(z), \quad v = 0, \quad w = 0, \quad \frac{dp}{dx} = const. \tag{10.45}$$

Substituting these assumptions NSE we obtain for the steady state

$$\nu \frac{\partial^2 u}{\partial z^2} = \frac{1}{\rho} \frac{\partial p}{\partial x}. \tag{10.46}$$

Integrating with respect to z leads to

$$u = \frac{z^2}{2\mu} \frac{dp}{dx} + Az + B, \quad \mu = \nu\rho. \tag{10.47}$$

Applying the no slip boundary conditions

$$u = 0 \text{ at } z = \pm h \tag{10.48}$$

yields

$$u = -\frac{1}{2\mu} \frac{dp}{dx} (h^2 - z^2). \tag{10.49}$$

To find p we use

$$\bar{u} = \frac{1}{2h} \int_{-h}^{h} u \, dz = -\frac{2h^3}{3\mu} \frac{dp}{dx}. \tag{10.50}$$

Integrating with respect to x we have

$$p = \frac{3\mu\bar{u}}{h^2} x + p_0 \tag{10.51}$$

Exercises

1. Prove Archimedes law : "When a body is immersed in a (static) fluid a force acts on it in direction opposite to the gravity force. The magnitude of this "buoyancy force," is equal to the weight of the displaced fluid by the immersed body."

   Hint: The hydrostatic pressure at depth z is $-\rho g z \mathbf{k}$ where ρ is the fluid density.

2. Show that if a flow $\mathbf{u}$ can be expressed in the form

$$\mathbf{u} = \nabla\phi \times \nabla\psi$$

   then it admits a **vector potetial** $\mathbf{A} = \phi\nabla\psi$ so that

$$\mathbf{u} = \nabla \times \mathbf{A}.$$

3. Fluid flow is irrotational if $\nabla \times \mathbf{u} = 0$. Show that under this assumption there exists a "**potential function** ϕ so that $\mathbf{u} = \nabla\phi$.

4. Show that for incompressible flow in two dimensions one can define a **Stream function** ψ which is defined as

$$u = \frac{\partial\psi}{\partial y}, \quad v = -\frac{\partial\psi}{\partial x}.$$

5. Show that for irrotational and incompressible flow in two dimensions the function

$$f(z) = \psi(x,y) + i\phi(x,y), \quad z = x + iy$$

is analytic. Hint: Show that $f(z)$ satisfies Cauchy-Reimann equations.

6. Sphere moving uniformly in a stationary viscous fluid.

Since motion is relative, we can consider the equivalent system where the sphere is fixed and the fluid moving (in the reverse direction). Assuming that the fluid velocity is "small", we can neglect the convective term $\mathbf{u} \cdot \nabla\mathbf{u}$ in NSE. Furthermore we neglect the gravitational force. In the steady state of the flow NSE reduce then to

$$\nabla \cdot \mathbf{u} = 0, \quad \nabla p = \nu \nabla^2 \mathbf{u}.$$

The boundary conditions are that $\mathbf{u}_{|R} = \mathbf{0}$ where R is the radius of the sphere.

Show that the solution for $\mathbf{u} = (u, v, w)$ and p of these equations is

$$u = -\frac{3w_0 R}{r^3}xz\left(1 - \frac{R^2}{r^2}\right),$$

$$v = -\frac{3w_0 R}{r^3}yz\left(1 - \frac{R^2}{r^2}\right),$$

$$w = -\frac{3w_0 R}{r^3}z^2\left(1 - \frac{R^2}{r^2}\right) + w_0\left(1 - \frac{R^3}{4r^3} - \frac{3R}{4r^3}\right),$$

$$p = -\frac{3\nu w_0 Rz}{2r^2},$$

where w_0 represents the uniform velocity of the sphere in the fluid (in the z-direction).

10.4 SIMILARITY AND REYNOLDS' NUMBER

Suppose that we wish to solve Navier-Stokes equations for a flow which involves a "representative length" L and a "representative velocity" U. Furthermore, assume that the initial and boundary conditions can be expressed in non-dimensional form using

$$\mathbf{u}' = \frac{\mathbf{u}}{U}, \quad t' = \frac{tU}{L}, \quad x' = \frac{\mathbf{x}}{L}. \tag{10.52}$$

Using these new variables and

$$p' = \frac{p - p_0}{\rho U^2} \tag{10.53}$$

where p_0 is some "representative pressure" we can rewrite Navier-Stokes equations as

$$\sum \frac{\partial u_i'}{\partial x_i'} = 0 \tag{10.54}$$

$$\frac{\partial u_i'}{\partial t'} + \sum_j u_j' \frac{\partial u_i'}{\partial x_j'} = -\frac{\partial p'}{\partial x_i'} + \frac{1}{Re} \sum_j \frac{\partial^2 u_i'}{\partial x_j' \partial x_j'}, \tag{10.55}$$

where

$$Re = \frac{\rho L U}{\mu} = \frac{L U}{\nu} \tag{10.56}$$

is called "Reynold's number."

Navier-Stokes Equations (12.32),(12.33) contain only the Reynold's number as a parameter. The solution $\mathbf{u}', p'$ depends only on $\mathbf{x}', t', Re$ and the dimensionless ratios which are needed to specify the initial and boundary conditions.

This shows that all flows that satisfy the same initial and boundary conditions (when expressed in non-dimensional form) and have the same Reynold's number (i.e. the ratio $\rho L U/\mu$ is the same although ρ, L, u, μ might be different) are described by the same nondimensional solution. Such flows are called *dynamically similar*. In the following we shall always use the nondimensional form of Navier-Stokes equations viz.

$$\nabla \cdot \mathbf{u} = 0 \tag{10.57}$$

$$\frac{\partial \mathbf{u}}{\partial t} + (\mathbf{u} \cdot \nabla)\mathbf{u} = -\nabla p + \frac{1}{Re}\nabla^2 \mathbf{u}. \tag{10.58}$$

This similarity principle is behind the use of a "wind tunnel" where appropriately scaled down models of airplanes (and other systems) can be tested in flows which have the same Reynold's number to which an actual airplane will be subjected.

10.5 DIFFERENT FORMULATIONS OF NAVIER-STOKES EQUATIONS

The variables $\mathbf{u}, p$ which characterize the fluid flow are usually called "primitive variables." Equations (12.35), (12.36), which are formulated in terms of these variables, constitute a system of coupled nonlinear partial differential equations. The first of these equations, the continuity equation, is a first order PDE which does not include the pressure. This fact can lead to instabilities when one attempts to solve Navier-Stokes equations numerically, and several other formulations are used to alleviate this problem.

We present here three such formulations for incompressible flow in two dimensions.

A. Navier-Stokes equations in conservative form.

The explicit form of Navier-Stokes equations in two dimensions is

$$\frac{\partial u}{\partial x} + \frac{\partial u}{\partial y} = 0 \tag{10.59}$$

$$\frac{\partial u}{\partial t} + u\frac{\partial u}{\partial x} + v\frac{\partial u}{\partial y} = -\frac{\partial p}{\partial x} + \frac{1}{Re}\nabla^2 u \tag{10.60}$$

$$\frac{\partial v}{\partial t} + u\frac{\partial v}{\partial x} + v\frac{\partial v}{\partial y} = -\frac{\partial p}{\partial y} + \frac{1}{Re}\nabla^2 v. \tag{10.61}$$

Using the continuity equation we can write

$$u\frac{\partial u}{\partial x} + v\frac{\partial u}{\partial y} = \frac{\partial}{\partial x}(u^2) + \frac{\partial}{\partial y}(uv)$$

$$u\frac{\partial v}{\partial x} + v\frac{\partial v}{\partial y} = \frac{\partial}{\partial x}(uv) + \frac{\partial}{\partial y}(v^2).$$

Hence Equations (12.39)-(12.40) can be rewritten in "conservative form" as:

$$\frac{\partial \mathbf{u}}{\partial t} + \frac{\partial F(\mathbf{u})}{\partial x} + \frac{\partial G(\mathbf{u})}{\partial y} = \frac{1}{Re}\left(\frac{\partial R(\mathbf{u})}{\partial x} + \frac{\partial S(\mathbf{u})}{\partial y}\right) \tag{10.62}$$

where

$$\mathbf{u} = \begin{pmatrix} 1 \\ u \\ v \end{pmatrix}, \quad F(\mathbf{u}) = \begin{pmatrix} u \\ u^2 + p \\ uv \end{pmatrix}, \quad G(\mathbf{u}) = \begin{pmatrix} v \\ uv \\ v^2 + p \end{pmatrix}$$

$$R(\mathbf{u}) = \begin{pmatrix} 0 \\ \tau_{xx} \\ \tau_{xy} \end{pmatrix}, \quad S(u) = \begin{pmatrix} 0 \\ \tau_{xy} \\ \tau_{yy} \end{pmatrix}$$

and

$$\tau_{xx} = \left(\frac{\partial u}{\partial x} - \frac{\partial v}{\partial y} \right), \quad \tau_{xy} = \left(\frac{\partial u}{\partial y} + \frac{\partial v}{\partial x} \right), \quad \tau_{yy} = \left(\frac{\partial v}{\partial y} - \frac{\partial u}{\partial x} \right).$$

B. Elliptic Pressure Equation

For steady state flows (in two dimensions) the continuity equation can be replaced by an elliptic equation for the pressure. To derive this equation we differentiate equations (12.39), (12.40) with respect to x, y respectively and add. Using the continuity equation this leads to

$$\nabla^2 p = 2 \left(\frac{\partial u}{\partial x} \frac{\partial v}{\partial y} - \frac{\partial u}{\partial y} \frac{\partial v}{\partial x} \right). \tag{10.63}$$

C. Vorticity-Stream Function Formulations

The vorticity of the flow $\mathbf{u}$ is defined as

$$\boldsymbol{\omega} = \nabla \times \mathbf{u}. \tag{10.64}$$

In two dimensions, $\boldsymbol{\omega} = (0, 0, \omega)$ where

$$\omega = \frac{\partial v}{\partial x} - \frac{\partial u}{\partial y}. \tag{10.65}$$

Introducing the stream function ψ, which is defined as

$$u = \frac{\partial \psi}{\partial y}, \quad v = -\frac{\partial \psi}{\partial x}, \tag{10.66}$$

we find that the continuity equation is satisfied automatically and

$$\nabla^2 \psi = -\omega. \tag{10.67}$$

Furthermore, if we differentiate (12.39),(12.40) with respect to y, x respectively and subtract we obtain

$$\frac{\partial \omega}{\partial t} + \frac{\partial \psi}{\partial y}\frac{\partial \omega}{\partial x} - \frac{\partial \psi}{\partial x}\frac{\partial \omega}{\partial y} = \frac{1}{Re}\nabla^2 \omega. \tag{10.68}$$

Equations (12.45)-(12.46) give a complete description of the flow in two dimensions) in terms of the non-primitive variables ψ, ω.

Although this form of Navier-Stokes equations is robust numerically, it is usually more difficult to find and apply boundary and initial conditions on the flow in this formulation than in the primitive one.

D. Beltrami Flow

Definition: A flow for which $\boldsymbol{\omega} \times \mathbf{u} = 0$ is called a Beltrami flow. A flow for which $\mathrm{curl}(\boldsymbol{\omega} \times \mathbf{u}) = 0$ is called a generalized Beltrami flow.

To see the implications of these definitions we consider the vorticity equation in three dimensions

$$\frac{\partial \boldsymbol{\omega}}{\partial t} + (\mathbf{u} \cdot \nabla)\boldsymbol{\omega} - (\boldsymbol{\omega} \cdot \nabla)\mathbf{u} = \nu\nabla^2 \boldsymbol{\omega}. \tag{10.69}$$

This equation is obtained by applying the **curl** operator to Navier-Stokes equations and using the continuity equation to simplify the result. Equation (12.47) can be rewritten as

$$\frac{\partial \boldsymbol{\omega}}{\partial t} + \mathrm{curl}(\boldsymbol{\omega} \times \mathbf{u}) = \nu\nabla^2 \boldsymbol{\omega}. \tag{10.70}$$

Thus for a Beltrami (or generalized Beltrami) flow the vorticity equation reduces to the heat equation.

We note that for steady two dimensional motion the condition $\mathrm{curl}(\boldsymbol{\omega} \times \mathbf{u}) = 0$ is equivalent to

$$\frac{\partial(\omega, \psi)}{\partial(x, y)} = 0. \tag{10.71}$$

Therefore any functional relation in the form $\omega = f(\psi)$ is a solution of this equation.

E. Universal Solutions

There is a class of solutions to the NSE for which the viscous term vanishes i.e. the solutions are independent of ν. These are called universal solutions. To derive these solutions we reconsider Equation (12.45) and observe that

$$\nabla^2 \boldsymbol{\omega} = -\nabla \times (\nabla \times \boldsymbol{\omega})$$

(Since $\nabla \cdot \boldsymbol{\omega} = 0$). Hence Equation (12.47) can be rewritten as

$$\nabla \times \left(\frac{\partial \mathbf{u}}{\partial t} + \boldsymbol{\omega} \times \mathbf{u} + \nu \nabla \times \boldsymbol{\omega}\right) = 0.$$

It is obvious therefore that the condition

$$\nabla \times (\nabla \times \boldsymbol{\omega}) = 0,$$

which is equivalent to

$$\frac{\partial \boldsymbol{\omega}}{\partial t} + \nabla \times (\boldsymbol{\omega} \times \mathbf{u}) = 0,$$

leads to a universal solution. It follows then that steady Beltrami flows are universal solutions of NSE.

10.6 CONVECTION AND BOUSSINESQ APPROXIMATION

In many fluid problems one must consider the effects of temperature gradients on the flow. The motion of a fluid subject to such temperature gradients is called convection.

To derive a model for these flows we shall use the "Boussinesq Approximation."

In this approximation all the physical properties of the fluid are assumed to be independent of the temperature except for the density. These density variations lead to a "rbBuoyancy term" in the Navier Stokes equations.

To obtain an expression for this buoyancy term we consider a parcel of fluid which has a temperature and density T, ρ respectively while the ambient temperature and density is T_0, ρ_0. Under these conditions a buoyancy force will act on this parcel (Archimedes law)

$$\mathbf{F} = -\frac{(\rho_0 - \rho)}{\rho_0}\mathbf{g} \tag{10.72}$$

($\mathbf{F}$ is force per unit mass).

If the density variations are small, then from the equation of state $\rho = \rho(T)$ we obtain

$$\frac{\rho}{\rho_0} \cong 1 - \beta(T - T_0) \tag{10.73}$$

where $\beta > 0$ is the thermal expansion coefficient. Hence

$$\mathbf{F} = -\beta(T - T_0)\mathbf{g}. \tag{10.74}$$

Navier-Stokes equations become then

$$\nabla \cdot \mathbf{u} = 0 \tag{10.75}$$

$$\frac{\partial \mathbf{u}}{\partial t} + (\mathbf{u} \cdot \nabla)\mathbf{u} = -\nabla p + \nu \nabla^2 \mathbf{u} - \beta(T - T_0)\mathbf{g} \tag{10.76}$$

(observe that we are still using the incompressibility assumptions in these equations).

Since the temperature is part of Equations (10.75)-(10.76) we need an additional equation to close this set of equations.

To do so we consider again a parcel of fluid (Lagrangian picture) and apply to it the law of energy conservation.

Let $Q(t)$ be the amount of heat in the parcel at time t.

$$Q = \int_V c\rho T(x, t)dV \tag{10.77}$$

where c is the specific heat of the fluid. Then

$$\frac{dQ}{dt} = \int_V c\rho \frac{dT}{dt}dV = -\int_{\partial V} \mathbf{q}(x, t) \cdot d\mathbf{S} = -\int_V div\mathbf{q}dV \tag{10.78}$$

where $\mathbf{q}$ is the heat flux. Using Fourier law of heat conduction $\mathbf{q} = -\kappa \nabla T$ this yields

$$\frac{dT}{dt} = k\nabla^2 T \tag{10.79}$$

where $k = \frac{\kappa}{c\rho}$ is the thermodynamic conductivity. Rewriting Equation (10.79) in the Eulerian picture we therefore have

$$\frac{\partial T}{\partial t} + (\mathbf{u} \cdot \nabla)T = k\nabla^2 T. \tag{10.80}$$

Equations (10.75), (10.76), (10.80) are referred to as Boussinesq equations.

Boussinesq equations can be applied to situations where there is no "reference" velocity unit, e.g. cavity flow. These flows are referred to as "natural" or "free convection." In these cases we use ν/L to represent the "reference velocity" when we perform the transformation to the nondimensional form of the equations.

On the other hand, when a characteristic external velocity U exists, we refer to the flow as "forced convection." The nondimensional form of the equations is obtained by the transformation

$$\bar{t} = \frac{tU}{L}, \quad \bar{\mathbf{x}} = \frac{\mathbf{x}}{L}, \quad \bar{\mathbf{u}} = \frac{\mathbf{u}}{U}, \quad \bar{T} = \frac{T - T_0}{\Delta T}, \quad \bar{p} = \frac{p - p_0}{\rho U^2} \qquad (10.81)$$

where T_0 is the characteristic temperature and ΔT is "the characteristic temperature difference."

With these scalings, Boussinesq equations for forced convection take the following form:

$$\nabla \cdot \mathbf{u} = 0 \qquad (10.82)$$

$$\frac{\partial \mathbf{u}}{\partial t} + (u \cdot \nabla)\mathbf{u} = -\nabla p + \frac{1}{Re}\nabla^2 \mathbf{u} - \frac{1}{Fr}T\mathbf{k} \qquad (10.83)$$

$$\frac{\partial T}{\partial t} + (\mathbf{u} \cdot \nabla)T = \frac{1}{Re\,Pr}\nabla^2 T \qquad (10.84)$$

(where we dropped the bars over the scaled quantities).

In these equations we introduced the Prandtl and Froude numbers Pr, Fr which are defined as

$$\frac{1}{Fr} = \frac{g\beta\Delta T L}{U^2}, \quad Pr = \frac{\nu}{k}. \qquad (10.85)$$

Other "nondimensional numbers" that appear frequently in the literature are the Grashof number

$$Gr = \frac{\beta g \Delta T L^3}{\nu^2} \qquad (10.86)$$

and the Rayleigh number

$$Ra = \frac{\beta g \Delta T L^3}{\nu k} = Gr\,Pr. \qquad (10.87)$$

Observe also that for forced convection we have the relation

$$Gr = Re^2 Fr. \qquad (10.88)$$

10.7   COMPLEX VARIABLES IN 2-D HYDRODYNAMICS

If a 2-D flow is irrotational, i.e. $\nabla \times \mathbf{u} = 0$, then there exists a potential function ϕ so that

$$u = \frac{\partial \phi}{\partial x}, \quad v = \frac{\partial \phi}{\partial y} \tag{10.89}$$

(or equivalently $u = -\frac{\partial \phi}{\partial x}, \quad v = -\frac{\partial \phi}{\partial y}$). The continuity equation in the Navier-Stokes equation will be satisfied if

$$\nabla^2 \phi = 0. \tag{10.90}$$

We observe, however, that the continuity equation $\nabla \cdot \mathbf{u} = 0$ implies that there exists a vector potential $\mathbf{A}$ so that

$$\mathbf{u} = \nabla \times \mathbf{A}. \tag{10.91}$$

In 2-D $\mathbf{A}$ has only one component

$$\mathbf{A} = \psi \mathbf{k} \tag{10.92}$$

which implies that

$$u = \frac{\partial \psi}{\partial y}, \quad v = -\frac{\partial \psi}{\partial x} \ . \tag{10.93}$$

Now the continuity equation is satisfied automatically, and if the flow is irrotational

$$\nabla \times \mathbf{u} = \left(\frac{\partial v}{\partial x} - \frac{\partial u}{\partial y} \right) \mathbf{k} = 0 \tag{10.94}$$

then ψ must satisfy

$$\nabla^2 \psi = 0 \ . \tag{10.95}$$

We stress that a stream function for incompressible flow in 2-D always exists. A potential function on the other hand exists only for irrotational flow.

We conclude that when the flow is two dimensional and irrotational both ϕ, ψ exists and each must satisfy Laplace Equations (10.90), (10.95). They satisfy also Cauchy-Riemann equations, viz.

$$\frac{\partial \phi}{\partial x} = \frac{\partial \psi}{\partial y}, \quad \frac{\partial \phi}{\partial y} = -\frac{\partial \psi}{\partial x} \ . \tag{10.96}$$

Therefore we can define a complex analytic function W

$$W = \phi + i\psi. \tag{10.97}$$

This allows for the use of complex variables methods for the solution of fluid flow problems under these assumptions.

We also have

$$w = \frac{dW}{dz} = \frac{dW/dx}{dz/dx} = \frac{dW}{dx} = \frac{\partial\phi}{\partial x} + i\frac{\partial\psi}{\partial x} = u - iv, \tag{10.98}$$

i.e. w represents the "conjugate velocity" and $\bar{w} = u + iv$.

Example: Laminar parallel flow around a sphere of radius 1.

At $x = \pm\infty$ the flow remains parallel; therefore, we shall assume that $u_\infty = 1, \ v_\infty = 0$. Furthermore, no fluid can penetrate the sphere and hence we must have in polar coordinates that $u_r(1, \theta) = 0$.

Since the relationship between the velocities (u, v) in Cartesian and (u_r, u_θ) in polar coordinates is given by

$$u = u_r \cos\theta - u_\theta \sin\theta, \ \ v = u_r \sin\theta + u_\theta \cos\theta,$$

we can satisfy these boundary conditions by a complex potential of the form

$$W = z + \frac{1}{z}. \tag{10.99}$$

Hence

$$\frac{dW}{dz} = 1 - \frac{1}{z^2} = u - iv, \tag{10.100}$$

which yields

$$u = 1 - \frac{x^2 - y^2}{r^4}, \ \ v = -\frac{2xy}{r^4}. \tag{10.101}$$

10.8   BLASIUS BOUNDARY LAYER EQUATION

To discuss the effects of viscosity on fluid motion near walls we consider in this section the steady state solution of NSE over infinite plate in the $x - z$ plane, $0 < x < \infty, -\infty < z < \infty$ subject to a uniform flow

$$\mathbf{u} = U\mathbf{i} \text{ as } x \to -\infty \text{ for all } y, z \tag{10.102}$$

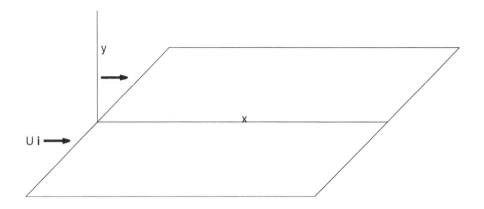

Figure 10.2   Unidirectional flow over a plate in the $x - z$ plane

(see Figure 10.2). Due to the symmetry of the problem with respect to z, the governing equations for this flow are the 2-D NSE, i.e., $\mathbf{u} = (u(x, y), \ v(x, y), 0)$, and

$$\frac{\partial u}{\partial x} + \frac{\partial v}{\partial y} = 0 \tag{10.103}$$

$$u\frac{\partial u}{\partial x} + v\frac{\partial u}{\partial y} = -\frac{1}{\rho}\frac{\partial p}{\partial x} + \nu\nabla^2 u \tag{10.104}$$

$$v\frac{\partial v}{\partial x} + v\frac{\partial v}{\partial y} = -\frac{1}{\rho}\frac{\partial p}{\partial y} + \nu\nabla^2 v \tag{10.105}$$

subject to the upstream boundary conditions Equation(10.102) and

$$u = v = 0 \ \ for \ \ 0 < x < \infty \ \ at \ y = 0. \tag{10.106}$$

We assume also that

$$\lim_{y\to\infty} p = constant, \quad \lim_{y\to\infty} v = 0.$$

We now seek (following Prandtl) to find a solution to this system near $y = 0$ as the viscosity $\nu \to 0$.

From Equation (10.103) we deduce the existence of a stream function $\psi(x, y)$ so that

$$u = \frac{\partial \psi}{\partial y} \quad v = -\frac{\partial \psi}{\partial x}. \tag{10.107}$$

However, as we expect the solution to develop "a singularity" near $y = 0$ (i.e.

we expect that there will exist a region near $y = 0$ where the solution (u, v) changes very rapidly although ψ is very small). To study this behavior we introduce a stretching transformation

$$x' = x, \quad \zeta = \alpha^{-1}y, \quad \psi'(x', \zeta) = \alpha^{-1}\psi(x, y) \tag{10.108}$$

(where α will be determined later).

Then

$$u'(x', \zeta) = \frac{\partial \psi'}{\partial \zeta} = \frac{\partial \psi}{\partial y} = u(x, y) \tag{10.109}$$

$$v'(x', \zeta) = -\frac{\partial \psi'}{\partial x'} = -\alpha^{-1}\frac{\partial \psi}{\partial x} = \alpha^{-1}v(x, y). \tag{10.110}$$

Dropping the primes from x, u Equations (10.104), (10.105) transform into

$$u\frac{\partial u}{\partial x} + v'\frac{\partial u}{\partial \zeta} = \frac{-1}{\rho}\frac{\partial p}{\partial x} + \nu\left(\frac{\partial^2 u}{\partial x^2} + \frac{1}{\alpha^2}\frac{\partial^2 u}{\partial \zeta^2}\right) \tag{10.111}$$

$$\alpha^2\left(u\frac{\partial v'}{\partial x} + v'\frac{\partial v'}{\partial \zeta}\right) = -\frac{1}{\rho}\frac{\partial p}{\partial \zeta} + \nu\left(\alpha^2\frac{\partial^2 v'}{\partial x^2} + \frac{\partial^2 v'}{\partial \zeta^2}\right) \tag{10.112}$$

with the boundary conditions

$$\lim_{\zeta \to \infty}\lim_{\nu \to 0} u(v, \zeta) = U, \quad \lim_{\zeta \to \infty}\lim_{\nu \to 0} v(x, \zeta) = 0 \tag{10.113}$$

$$\lim_{\zeta \to \infty}\lim_{\nu \to 0} p(x, \zeta) = p_0 = \text{ const}. \tag{10.114}$$

In the limit $\alpha \to 0$, $\nu \to 0$ we infer from Equation (10.112) that $\frac{\partial p}{\partial \zeta} = 0$ and hence from Equation (10.114) that $p = p_0 = $ constant. Letting $\alpha^2 = \nu$ we now see that as $\nu \to 0$, the equations governing the flow reduce to

$$u\frac{\partial u}{\partial x} + v'\frac{\partial u}{\partial \zeta} = \frac{\partial^2 u}{\partial \zeta^2} \tag{10.115}$$

$$u = \frac{\partial \psi'}{\partial \zeta}, \quad v' = \frac{-\partial \psi'}{\partial x} \tag{10.116}$$

with the boundary conditions

$$u(x, 0) = v'(x, 0) = 0 \qquad 0 < x < \infty$$

$$\lim_{\zeta \to \infty} u(x, \zeta) = U \qquad 0 < x < \infty.$$

Prandtl now observed that Equation (10.115) is invariant under the transformation

$$(x, \zeta, \psi') \to (\beta^2 x, \beta\zeta, \beta\psi'), \;\; \beta > 0. \tag{10.117}$$

(i.e. if $u(x, \zeta)$ is a solution, then $u(\beta^2 x, \beta\zeta)$ is also a solution of the equation). This suggests that we introduce a similarity variable (which remains unchanged by this transformation)

$$w = \zeta \left(\frac{U}{x} \right)^{1/2} = y \left(\frac{U}{\nu x} \right)^{1/2} \tag{10.118}$$

and look for a solution ψ' in the form

$$\psi'(x, \zeta) = (Ux)^{1/2} f(w) \tag{10.119}$$

i.e.

$$u = U f'(\eta), \;\; v' = \frac{1}{2} \left(\frac{U}{x} \right)^{1/2} (wf' - f). \tag{10.120}$$

Substituting these expressions in Equation (10.115) we find that f has to satisfy Blasius equation

$$f'''(w) + \frac{1}{2} f(w) f''(w) = 1 \tag{10.121}$$

with the boundary conditions $f(0) = f'(0) = 0$ and

$$\lim_{w \to \infty} f'(w) = 1.$$

Finally we observe that throughout our discussion we neglected the edge effects near $x = 0$. Thus the solution derived above is valid only for $x \gg 0$.

10.9 INTRODUCTION TO TURBULENCE MODELING

Although turbulence is a rather familiar phenomenon, there is no formal definition which covers all of its aspects. It is generally accepted however that "turbulence" refers to irregular fluid motion which seems to have "random variations" in space-time so that a statistical treatment of the flow is justified. This is in spite of the fact that the equations which govern the motion of the fluid are well known to be deterministic in nature.

From a less formal point of view we note that in general "turbulence" is generated by the viscous effects at fixed walls or by the flow of layers of fluids with different velocities over one another. Furthermore, turbulence is said to consist of a (nonlinear) superposition of eddies. However, the size of these eddies cannot go to zero since the smaller the eddy, the greater is the associated velocity gradient. These eddies have kinetic energy which is determined by their vorticity. Thus a prominent role in the description of turbulent motion is given to the "vortical structure" and the energy spectrum of the flow.

In view of the "definition" given above it is natural to decompose the (Eulerian) velocity of the fluid $\mathbf{u}$ as

$$\mathbf{u} = \bar{\mathbf{u}} + \mathbf{u}' \tag{10.122}$$

where $\bar{\mathbf{u}}$ is the "average value of the velocity" (or the "large scale component of the flow") and u' are the fluctuations (or "turbulent residuals") around this mean, i.e. $\bar{\mathbf{u}}' = 0$. However, from a mathematical point of view there are (at least) three ways to obtain $\bar{\mathbf{u}}$:

1. Space average (homogeneous turbulence)

$$\bar{\mathbf{u}}^s(t) = \lim_{V \to \infty} \frac{1}{V} \int_V \mathbf{u}(\mathbf{x}, t) d\mathbf{x} \tag{10.123}$$

2. Time average (stationary turbulence)

$$\bar{\mathbf{u}}^t(x) = \lim_{T \to \infty} \int_{-T}^{T} \mathbf{u}(x, t) dt \tag{10.124}$$

3. Ensemble average of N identical systems

$$\bar{\mathbf{u}}^E(x, t) = \frac{1}{N} \sum_{n=1}^{N} \mathbf{u}_n(x, t). \tag{10.125}$$

For stationary and homogeneous turbulence the "ergodic hypothesis" asserts that these three averages are identical:

$$\bar{\mathbf{u}}^t = \bar{\mathbf{u}}^s = \bar{\mathbf{u}}^E. \tag{10.126}$$

Obviously actual turbulence does not satisfy the conditions needed to satisfy the ergodic hypothesis. Moreover, in actual experiments the averaging must be carried over finite time intervals or spatial extensions.

Whatever averaging procedure is used, we shall assume that the averaging process is linear i.e. if f, g are two independent "flow variables" then

$$\overline{f + g} = \bar{f} + \bar{g}, \quad \bar{a} = a \ \text{(a constant)} \tag{10.127}$$

$$\overline{af} = a\bar{f}. \tag{10.128}$$

It follows then that if

$$f = \bar{f} + f', \quad g = \bar{g} + g' \tag{10.129}$$

then

$$\bar{\bar{f}} = \bar{f} \tag{10.130}$$

$$\bar{f} = \overline{\bar{f} + f'} = \bar{\bar{f}} + \bar{f'} = \bar{f} + \bar{f'}, \quad \text{i.e.} \ \ \bar{f'} = 0 \tag{10.131}$$

$$\overline{fg} = \overline{(\bar{f} + f')(\bar{g} + g')} = \bar{f}\bar{g} + \overline{\bar{f}g'} + \overline{f'\bar{g}} + \overline{f'g'} = \bar{f}\bar{g} + \overline{f'g'}. \tag{10.132}$$

Also observe that

$$\frac{\partial \bar{f}}{\partial s} = \overline{\frac{\partial f}{\partial s}}. \tag{10.133}$$

10.9.1 Incompressible Turbulent Flow

Substituting the decomposition

$$\mathbf{u} = \bar{\mathbf{u}} + \mathbf{u}', \quad p = \bar{p} + p' \tag{10.134}$$

in Navier-Stokes equations and averaging with respect to time we obtain (using summation over repeated indices)

$$\frac{\partial \bar{u}_i}{\partial t} + u_j \frac{\partial \bar{u}_i}{\partial x_j} = -\frac{\partial \bar{p}}{\partial x_i} + \frac{1}{Re} \nabla^2 \bar{u}_i - \frac{\partial}{\partial x_j}(\overline{u_i' u_j'}) \tag{10.135}$$

$$\frac{\partial \bar{u}_i}{\partial x_i} = 0 \tag{10.136}$$

where we have used the continuity equation to simplify some of the resulting expressions. It follows then that in order to solve Navier-Stokes equations for the mean flow we need to model the residual term $\frac{\partial}{\partial x_j}(\overline{u_i' u_j'})$. In general this is referred to as the "closure problem."

Rewriting the right hand side of Equation (10.135) as

$$-\frac{\partial \bar{p}}{\partial x_i} + \frac{\partial}{\partial x_j}\left(\frac{1}{Re}\frac{\partial \bar{u}_i}{\partial x_j} - \overline{u_i' u_j'}\right) \tag{10.137}$$

we see that a natural interpretation of the term $u_i' u_j'$ is as a "new" "turbulent stress tensor" in addition to the viscous stresses. This is referred to as "Reynold's stress tensor." In view of the equal footings in which the viscous and the Reynold's tensors appear in Equation(10.135) it is natural to assume that the effect of the new tensor is also proportional to the velocity gradients. This is called the Boussinesq approximation.

$$\overline{u_i' u_j'} = \frac{2}{3}k\delta_{ij} - 2\nu_T\left(\frac{\partial \bar{u}_i}{\partial x_j} + \frac{\partial \bar{u}_j}{\partial x_i}\right) \tag{10.138}$$

where

$$k = \frac{1}{2}\overline{u_i' u_i'} \tag{10.139}$$

is the turbulent kinetic energy and ν_T the "turbulent eddy viscosity." In tensor notation equation (10.138) can be written as

$$\bar{\tau} = \overline{\mathbf{u}'\mathbf{u}'} = \frac{2}{3}k\mathbf{I} - \nu_T[\text{grad}\bar{\mathbf{u}} + (\text{grad}\bar{\mathbf{u}})^T]. \tag{10.140}$$

From an intuitive point of view this treatment of the tensor $\bar{\tau}_{ij}$ corresponds to a physical picture where turbulence eddies are considered as "lumps" of fluids. These lumps collide and exchange momentum like molecules in Maxwell's kinetic theory of gases.

In the literature there have been many attempts to generalize the relation in Equation(10.140). For example, attempts were made to replace Equation (10.140) by

$$\bar{\tau}_{ij} = \overline{u_i' u_j'} = \nu_T^{(0)}\delta_{ij} + \nu_T^{(1)}\bar{D}_{ij} + \nu_T^{(2)}\bar{D}_{ik}\bar{D}_{kj}$$

where

$$\bar{D}_{ij} = \frac{\partial \bar{u}_i}{\partial x_j} + \frac{\partial \bar{u}_j}{\partial x_i}$$

and $\nu_T^{(0)}$, $\nu_T^{(1)}$, $\nu_T^{(2)}$ are scalars which might depend only on the invariants that can be formed by the tensor $\bar{D}_{ij}$.

In principle, however, there is no reason why $\nu_T^{(i)}$ should be scalars rather than tensors. In fact, models of the form

$$\bar{\tau}_{ij} = (\nu_T)_{ik}\bar{D}_{jk}$$

or

$$\bar{\tau}_{ij} = (\nu_T)_{ijkl} \bar{D}_{kl}$$

(i.e. where ν_T is a second or fourth order tensor) were also suggested. When aquation (10.138) is used to model τ_{ij}, the resulting Navier-Stokes equations for the mean flow contain a new parameter ν_T which is a function of space-time. It is the objective of all (statistical) turbulent models to calculate the correct value of the eddy viscosity ν_T.

A completely different approach to model the Reynold's stress tensor is offered by the observation that the Navier-Stokes equations describe the behavior of a Newtonian fluid (i.e. where strain and stress are linearly related). However, it may be argued that in turbulent flow the fluid behavior is not Newtonian. As a result, a **nonlinear** stress-strain relationship should be assumed. Another possibility is to consider turbulent fluid as "viscoelastic" with memory effects (i.e. the state of the fluid at t depends on its history in the time interval $[t - \tilde{\tau}, t]$, where $\tilde{\tau} > 0$).

10.9.2 Modeling Eddy Viscosity

Even though the eddy viscosity hypothesis Equation (10.138), has some conceptual difficulties, it proved successful in many flow simulations. Models for the computation ν_T are usually classified as zero, one, and two equations models.

- Zero equation models.

  Models in this category stipulate an algebraic relationship to compute ν_T. We mention two of these.

  – Prandtl mixing length
    This model was proposed for two dimensional boundary layers (near "walls")

    $$\nu_T = \ell^2 \left| \frac{\partial \bar{u}}{\partial y} \right| \qquad (10.141)$$

    where y is in the direction normal to the boundary and ℓ is determined experimentally.

– Smagorinski model

$$\nu_T = C(2\bar{D}_{ij}\bar{D}_{ij})^{1/2}. \tag{10.142}$$

– This model usually is used in a "large eddy simulation" of atmospheric flow where the grid step is large. Under these circumstances this model gives a computationally reasonable representation of the subgrid-scale-turbulence.

- One equation model.

Based on conjectures by Prandtl and Kolmogorov, ν_T is assumed to be a function of the turbulent kinetic energy k

$$\nu_T = c\ell k^{1/2} \tag{10.143}$$

where c is a constant and ℓ is the "turbulence scale" (eddy size). As we shall see in the next section one can derive a partial differential equation for k. However, as the relation (10.143) requires the specification of ℓ (for which no equations are given), the model is not complete. Due to this deficiency most current applications today use two equation models. This is discussed in the next section.

10.9.3 $k - \epsilon$ Model

As we saw in the previous section, zero and one equation models require the use of an empirical "length scale." As this quantity depends on the geometry and the boundary conditions, it is apparent that a second equation either for this quantity or its "equivalent" is needed for a complete specification of ν_T. Two equation models, in spite of their deficiencies, are utilized today for most turbulence research and applications. They can be used to compute properties of turbulent flow with no apriori assumptions on the structure of the flow.

The most prominent among these two equation models is the $k - \epsilon$ model and its variants. Here k, ϵ represent respectively the turbulent kinetic energy and dissipation per unit mass.

$$k = \frac{1}{2}[(u_1')^2 + (u_2')^2 + (u_3^1)^3] \tag{10.144}$$

$$\epsilon = \nu_T \overline{\frac{\partial u_i'}{\partial x_k} \frac{\partial u_i'}{\partial x_k}} \tag{10.145}$$

(summation over i, k) and ν_T is given by

$$\nu_T = \mu k^2 / \epsilon, \quad c = 0.09. \tag{10.146}$$

The derivation of the equations for the evolution of these quantities is rather involved algebraically and requires the modeling of terms containing various double and triple averages of $\dfrac{\partial u_i'}{\partial x_j}$, e.g.

$$\overline{\frac{\partial u_i'}{\partial x_k} \frac{\partial u_i'}{\partial x_m} \frac{\partial u_k'}{\partial x_m}}.$$

The resulting evolution equations contain several constants which have to be determined experimentally (and adjusted for different applications). Overall, the resulting model should be considered "phenomenological." Yet the model has been successful in many applications and is currently the "industry standard" (we do not present this derivation here).

We should mention at this juncture that there exists an ongoing research effort to model the evolution of turbulent flow using the Reynold stress tensor Equation (10.140). This approach lessens to some extent the need for the various modeling approximations which are made in the $k - \epsilon$ model. However, in this approach six coupled partial differential equations have to be solved in addition to the Navier-Stokes equations. Such a scheme requires heavy computational efforts. Some attempts for reductions based on some algebraic relations were suggested in the literature.

10.9.4   The Turbulent Energy Spectrum

One of the major characteristics of turbulence is the existence of eddies in the flow. These eddies can be viewed as a tangle of vortex elements. These vortex elements undergo "vortex stretching" which leads to the breakup of the large eddies into smaller ones. Thus in turbulence we speak about "energy cascade" from the large scale eddies to the smaller ones across a continuous spectrum of scales. At the smallest scales eddies lose energy due to viscous stresses which convert energy to heat. This suggests that for high Reynolds numbers (where we expect strong turbulence effects) the flow can be decomposed into three "levels." These are the mean flow, the large scale motion, or eddies, and the small scale motion.

The large eddies determine the rate at which energy is fed to the turbulent motion. Due to their dimension these large eddies depend strongly on the geometry and the boundary conditions of the problem. On the other hand at the small scales ($\approx$ large wave numbers) the character of the turbulent motion is determined by the energy flux, and the rate of dissipation must equal to the energy supply in this range. This led Kolmogorov to make the following conjecture:

"At sufficiently high Reynolds numbers there exists a range of (high) wave numbers where turbulence is statistically in equilibrium and uniquely determined by the dissipation ϵ and the viscosity ν. This state of equilibrium is universal."

To give this conjecture a more quantitative representation we note (following Kolmogorov) the following dimensional analysis.

$$[E] = \text{energy} = [L^3 T^{-2}]$$

$$[\nu] = \text{viscosity} = [L^2 T^{-1}]$$

$$[\epsilon] = \text{dissipation} = [L^2 T^{-3}]$$

$$[k] = \text{wave number} = [L^{-1}].$$

It is then "easy" to see that proper dimensional relationship between E and ϵ, k can be obtained by a formula of the form

$$E(k, t) = A\epsilon^{2/3} k^{-5/3} \tag{10.147}$$

(where ν has been eliminated in favor of the wave number k). This is the famous "Kolmogorov 5/3 rule", and the range of wave numbers for which it is assumed to be true is called "the inertial range."

Since its inception, the Kolmogorov hypothesis has been the subject of spirited debate, experimentation (which yielded conflicting data in some cases), and refinements. As Equation (10.147) was derived by dimensional analysis, there have been many attempts to derive it from first principles. So far, however, this remains as one of the fundamental open problems in fluid dynamics.

A typical energy spectrum of turbulence motion is presented in Fig. 8.3.

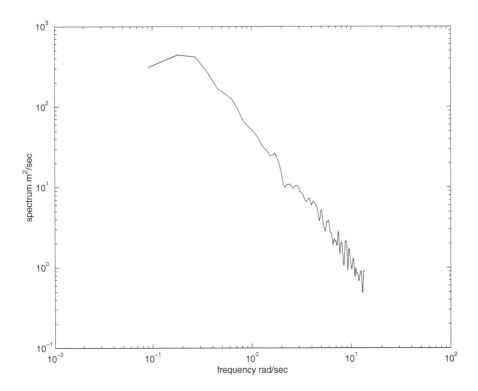

Figure 10.3 Spectrum of a turbulent flow demonstrating Kolmogorov 5/3 law

10.10 STABILITY OF FLUID FLOW

Notation: In this section indices which appear twice are summed over.

Navier-Stokes equations for incompressible flow are

$$\frac{\partial u_k}{\partial x_k} = 0 \tag{10.148}$$

$$\frac{\partial u_j}{\partial t} + u_k \frac{\partial u_j}{\partial x_k} = -\frac{\partial p}{\partial x_j} + \frac{1}{Re}\nabla^2 u_j. \tag{10.149}$$

Let (U_j, P) be the steady state solution of these equation on a domain Ω

$$\frac{\partial U_k}{\partial x_k} = 0 \tag{10.150}$$

$$U_k \frac{\partial U_j}{\partial X_k} = -\frac{\partial P}{\partial x_j} + \frac{1}{Re}\nabla^2 U_j \tag{10.151}$$

which satisfies the following boundary conditions on $\partial\Omega$,

$$U_j \mid_{\partial\Omega} = F_j, \quad P \mid_{\partial\Omega} = g \tag{10.152}$$

(these are "generic" boundary conditions. They might be more general).

To investigate whether this steady state solution is stable we consider a small perturbation to the flow (u_j', p')

$$u_j' = U_j + u_j', \quad p = P + p'. \tag{10.153}$$

Substituting in NSE and using Equations (11.22), (10.150) we obtain

$$\frac{\partial u_k'}{\partial x_k} = 0 \tag{10.154}$$

$$\frac{\partial u_j'}{\partial t} + U_k \frac{\partial u_j'}{\partial x_k} + u_k' \frac{\partial U_j}{\partial X_k} = -\frac{\partial p'}{\partial x_j} + \frac{1}{Re} \nabla^2 u_j'. \tag{10.155}$$

The boundary conditions on (u_j', p') are homogeneous (viz. zero) since (U_j, P) already satisfy the boundary conditions Equation (11.19).

Equations (10.154), (10.155) form a system of linear partial differential equations for (u_j', p'). Therefore we can use the superposition principle to solve these equations in order to determine the impact of these perturbations on the steady state solution. To this end we use separation of variables to find "elementary solutions" of these equations in the form

$$u_j' = e^{-i\omega t} \phi_j(x, y, z), \tag{10.156}$$

$$p' = e^{-i\omega t} \chi(x, y, z). \tag{10.157}$$

These yield

$$\frac{\partial \phi_k}{\partial x_k} = 0 \tag{10.158}$$

$$-i\omega \phi_j + U_k \frac{\partial \phi_j}{\partial x_k} + \phi_k \frac{\partial U_j}{\partial x_k} = -\frac{\partial \chi}{\partial x_j} + \frac{1}{Re} \nabla^2 \phi_j. \tag{10.159}$$

We now specialize and consider the case of a flow in a channel $[-\infty, \infty] \times [-h, h]$ where (e.g. Poiseuille flow)

$$U = (U(y), 0, 0). \tag{10.160}$$

and a perturbation in the form of (oblique) plane waves ($\equiv$ Tollmien - Schlichting waves)

$$u_j' = \phi_j(y) e^{i(\alpha x + \beta z - \omega t)} \tag{10.161}$$

$$pi' = \chi(y)e^{i(\alpha x + \beta z - \omega t)}, \quad \alpha, \beta \in R, \; \omega \in C. \tag{10.162}$$

By a proper transformation in the $x - z$ plane we can set $\beta = 0$ (i.e. let the x-axis coincide with the wave front). Substituting in Equations (10.154), (10.155) we obtain

$$i\alpha\phi_1 + \frac{d\phi_2}{dy} = 0 \tag{10.163}$$

$$-i\omega\phi_1 + iU\alpha\phi_1 + \phi_2\frac{dU}{dy} = -i\alpha\chi + \frac{1}{Re}\left(\frac{d^2}{dy^2} - \alpha^2\right)\phi_1 \tag{10.164}$$

$$-i\omega\phi_2 + i\alpha U\phi_2 = -\frac{d\chi}{dy} + \frac{1}{Re}\left(\frac{d^2}{dy^2} - \alpha^2\right)\phi_2 \tag{10.165}$$

$$-i\omega\phi_3 + i\alpha U\phi_3 = -\frac{1}{Re}\left(\frac{d^2}{dy^2} - \alpha^2\right)\phi_3. \tag{10.166}$$

Equation (10.163) can be satisfied by introducing ϕ so that

$$\phi_2 = i\alpha\phi, \quad \phi_1 = \frac{d\phi}{dy}.$$

Also note that Equation (10.166) is an "independent equation."

Substituting ϕ in Equations(10.164), (10.165) we obtain

$$i\phi'(-\omega + \alpha U) - i\alpha\phi\frac{dU}{dy} = -i\alpha\chi + \frac{1}{Re}\left(\frac{d^2}{dy^2} - \alpha^2\right)\phi' \tag{10.167}$$

$$-\alpha\phi(-\omega + U) = -\chi' + \frac{1}{Re}\left(\frac{d^2}{dy^2} - \alpha^2\right)(-i\alpha\phi). \tag{10.168}$$

Differentiating Equation (10.167) with respect to y and substituting inEquation (10.168) for χ' we obtain

$$i(U - c)(\phi'' - \alpha^2\phi) - \phi U'' = \frac{1}{\alpha Re}(\phi^{(4)} - 2\alpha^2\phi'' + \alpha^2\phi) \tag{10.169}$$

(where $c = \omega/\alpha$ is the "phase velocity").

Equation (10.169) is called "Orr-Somerfeld equation." It is a 4th order differential equation for ϕ with the boundary conditions

$$\phi(-h) = \phi(h) = \phi'(h) = \phi'(-h) = 0.$$

The stability of the solution to Tollmien-Schlichting wave is controlled by the value of ω. When $Im\,\omega > 0$, the perturbation "explodes" and the solution is unstable. Neutral stability is obtained when $Im\,\omega = 0$.

$$Im\,\omega = \psi(\alpha, Re).$$

The curve $\psi(\alpha, Re) = 0$ in the $\alpha - Re$ plane separates the domain of stability from the unstable domain. It follows that there is a critical value of the Reynold's number where the solution U becomes unstable. At this point waves start to appear in the flow. **THEOREM (Rayleigh):**

At $Re = \infty$, $U(y)$ must have a point of inflection for flow instability.

Proof When $Re = \infty$, the right hand side Orr-Somerfeld equation is zero, and we have

$$\phi'' - \alpha^2\phi = \frac{U''}{U - c}\phi. \qquad (10.170)$$

Taking the complex-conjugate of this equation yields

$$\phi'' - \alpha^2\phi^* = \frac{U''}{U - c^*}\phi^*. \qquad (10.171)$$

Multiplying Equation (10.170) by ϕ^* and Equation (10.171) by ϕ and subtracting leads to

$$\frac{d}{dy}(\phi'\phi^* - \phi\phi^{*\prime}) = \frac{2i\,Im(c)U''}{|U - c|^2}\,|\phi|^2\,.$$

Integrating this equation on $[-h, h]$ and using the boundary conditions on ϕ we obtain

$$2i\,Im(c)\int_{-h}^{h}\frac{U''\,|\phi|^2}{|U - c|^2}dy = 0.$$

Since all the terms in the integrand except U'' are positive, it follows that U'' must be 0 for some $y \in [-h, h]$.

10.11 ASTROPHYSICAL APPLICATIONS

Recent astronomical discoveries lead us to believe that our solar system is not unique. In fact the reverse is true, viz. a large number of stars have planets orbiting around them and the number of known exoplanets at the present time is over four thousand. This data leads to the hypothesis that there is a fundamental physical process (which we do not understand fully as of yet) that leads to the formation of planetary systems.

Many theories were put forward in the past about the origin of the solar system. Originally it was Laplace in 1796 who put forward the hypothesis that planetary systems evolve from a family of isolated rings that were formed from a primitive interstellar gas cloud. Actually such a system of rings around a pro-tostar was observed in 2014 by the Atacama Large Millimeter/submillimeter Array in the constellation Taurus.

Currently the leading theory about the formation of planetary systems is the "nebula theory" whereby a cloud of interstellar gas accreted under it own gravitation, to form in stages the protostar and the planets. Many of the results related to this theory were obtained through elaborate modeling and large scale numerical simulations. These involve, in general, thermodynamic considerations, magnetohydrodynamics modeling, and turbulence. However, some questions about planet formation still persist.

In this section, we present an idealized steady state hydrodynamic model which captures the formation of ring structure in a self gravitating disk of stratified gas. Using this model we show that matter density within the nebular cloud can exhibit oscillations whose peaks are separated by an almost "empty space", viz. a ring structure as hypothesized by Laplace.

The basic assumptions of this model are that the interstellar cloud can be treated as a two-dimensional self gravitating, incompressible, and stratified (viz. non-constant density) gas in which the particle velocities $|\mathbf{u}|$ are non-relativistic (i.e $|\mathbf{u}| \ll \bar{c}$ where $\bar{c}$ is the velocity of sound). The justification for the reduction from three to two dimensions was discussed by many authors.

10.11.1 Derivation of the Model Equations

To model the time dependent non-relativistic flow of a rotating incompressible fluid in two dimensions (x, y) we use the hydrodynamic equations of inviscid and incompressible stratified fluid

$$u_x + v_y = 0 \tag{10.172}$$

$$\rho_t + u\rho_x + v\rho_y = 0 \tag{10.173}$$

$$\rho u_t + \rho(u u_x + v u_y) = -p_x - \rho \phi_x + \rho \omega^2 x \qquad (10.174)$$

$$\rho v_t + \rho(u v_x + v v_y) = -p_y - \rho \phi_y + \rho \omega^2 y \qquad (10.175)$$

$$\nabla^2 \phi = 4\pi G \rho \qquad (10.176)$$

where subscripts indicate differentiation with respect to the indicated variable, $\mathbf{u} = (u, v)$ is the fluid velocity, ρ is its density, p is the pressure, ϕ is the gravitational field and G is the gravitational constant. The terms $\rho \omega^2 x$, $\rho \omega^2 y$ represent the components of the apparent centrifugal force due to the rotation of the gas cloud with angular velocity ω.

We can nondimensionalize these equations by introducing the following scalings

$$t = \frac{L\tilde{t}}{U_0}, \quad x = L\tilde{x}, \quad y = L\tilde{y}, \quad u = U_0 \tilde{u}, \quad v = U_0 \tilde{v}, \qquad (10.177)$$

$$\rho = \rho_0 \tilde{\rho}, \quad p = \rho_0 U_0^2 \tilde{p}, \quad \phi = U_0^2 \tilde{\phi}, \quad \omega = \frac{U_0}{L} \tilde{\omega}.$$

where L, U_0, ρ_0 are some characteristic length, velocity and mass density respectively that characterize the problem at hand. Substituting these scalings in Equations (10.172)-(10.176) and dropping the tildes, these equations remain unchanged (but the quantities that appear in these equations become nondimensional) while G is replaced by $\tilde{G} = \frac{G\rho_0 L^2}{U_0^2}$ (once again we drop the tilde).

In view of eq. (10.172) we can introduce a stream function ψ so that

$$u = \psi_y, \quad v = -\psi_x . \qquad (10.178)$$

Using this stream function we can rewrite Equation (10.173) as

$$\rho_t + J\{\rho, \psi\} = 0 \qquad (10.179)$$

where for any two (smooth) functions f, g

$$J\{f, g\} = \frac{\partial f}{\partial x} \frac{\partial g}{\partial y} - \frac{\partial f}{\partial y} \frac{\partial g}{\partial x}. \qquad (10.180)$$

Using ψ the momentum Equations (10.174),(10.175) become

$$\rho(\psi_{yt} + \psi_y \psi_{yx} - \psi_x \psi_{yy}) = -p_x - \rho \phi_x + \rho \omega^2 x \qquad (10.181)$$

$$\rho(-\psi_{xt} - \psi_y\psi_{xx} + \psi_x\psi_{xy}) = -p_y - \rho\phi_y + \rho\omega^2 y. \tag{10.182}$$

To eliminate p from these equations we differentiate Equations (10.181), (10.182) with respect to y, x respectively and subtract. This leads to

$$\rho_y(\psi_{yt} + \psi_y\psi_{yx} - \psi_x\psi_{yy}) + \tag{10.183}$$
$$\rho(\psi_{yyt} + \psi_y\psi_{yyx} - \psi_x\psi_{yyy}) - \rho_x(-\psi_{xt} - \psi_y\psi_{xx} + \psi_x\psi_{xy}) -$$
$$\rho(-\psi_{xxt} - \psi_y\psi_{xxx} + \psi_x\psi_{xxy}) = -J\{\phi, \rho\} + J\{\frac{1}{2}\omega^2 r^2, \rho\}$$

where $r^2 = x^2 + y^2$. The sum of the second and fourth terms in this equation can be rewritten as

$$\rho(\nabla^2\psi)_t + \rho J\{\nabla^2\psi, \psi\}. \tag{10.184}$$

To reduce the first and third terms in Equation (10.183) we use Equation (10.179). It follows that

$$\rho_y(\psi_{yt} + \psi_y\psi_{yx} - \psi_x\psi_{yy}) - \tag{10.185}$$
$$\rho_x(-\psi_{xt} - \psi_y\psi_{xx} + \psi_x\psi_{xy}) =$$
$$\rho_y\psi_{yt} + \rho_y\psi_y\psi_{yx} - (\rho_t + \rho_x\psi_y)\psi_{yy} + \rho_x\psi_{xt} + (\psi_x\rho_y - \rho_t)\psi_{xx} -$$
$$\rho_x\psi_x\psi_{xy} = \rho_y\psi_{yt} + \rho_x\psi_{xt} - \rho_t\nabla^2\psi + \frac{1}{2}J\{(\psi_x)^2 + (\psi_y)^2, \rho\}.$$

Combining the results of Equations (10.184), (10.185), Equation (10.183) becomes

$$\rho_y\psi_{yt} + \rho_x\psi_{xt} - \rho_t\nabla^2\psi + \rho(\nabla^2\psi)_t + \tag{10.186}$$
$$\rho J\{\nabla^2\psi, \psi\} + \frac{1}{2}J\{(\psi_x)^2 + (\psi_y)^2, \rho\}$$
$$= -J\{\phi, \rho\} + J\{\frac{1}{2}\omega^2 r^2, \rho\}.$$

Thus we have reduced the original five Equations (10.172)-(10.176) to three Equations (10.176), (10.179), and (10.186). Although Equation (10.186) is rather cumbersome in general, it can be simplified further when we consider only the steady state of the gas (a simplification for the time dependent flow is also possible under some constraints but will not be presented here).

10.11.2 Steady State Model Equations

When we consider only steady states of the flow Equation (10.179) implies that $\psi = \psi(\rho)$ and Equation (10.186) can be rewritten as

$$\rho J\{\nabla^2\psi, \psi\} + J\{\frac{1}{2}(\psi_x^2 + \psi_y^2), \rho\} = -J\{\phi, \rho\} + J\{\frac{1}{2}\omega^2 r^2, \rho\}. \qquad (10.187)$$

However, in view of eq. (10.179), $\psi = \psi(\rho)$, and this fact can be used to eliminate ψ from Equation (10.187). To this end we observe that

$$\psi_x = \psi_\rho \rho_x, \quad \psi_y = \psi_\rho \rho_y, \quad \nabla^2\psi = \psi_{\rho\rho}[\rho_x^2 + \rho_y^2] + \psi_\rho \nabla^2\rho. \qquad (10.188)$$

Note also that for any function of $F(\rho)$ we have $J\{F(\rho), \rho\} = 0$. This leads after some algebra to the following relation

$$J\left\{(\rho\psi_\rho^2)\nabla^2\rho + \frac{1}{2}(\rho\psi_\rho^2)_\rho[\rho_x^2 + \rho_y^2] + \phi - \frac{1}{2}\omega^2 r^2, \rho\right\} = 0. \qquad (10.189)$$

Hence we infer that

$$H(\rho)\nabla^2\rho + \frac{1}{2}H'(\rho)[\rho_x^2 + \rho_y^2] + \phi - \frac{1}{2}\omega^2 r^2 = S(\rho), \quad H' = \frac{dH(\rho)}{d\rho}. \qquad (10.190)$$

where

$$H(\rho) = \rho\psi_\rho^2 \qquad (10.191)$$

and $S(\rho)$ is some function of ρ. Equation (10.190) can be rewritten as

$$H(\rho)^{1/2}\nabla\cdot(H(\rho)^{1/2}\nabla\rho) + \phi - \frac{1}{2}\omega^2 r^2 = S(\rho). \qquad (10.192)$$

Thus the equations governing the steady state are Equations (10.192), (10.176). $H(\rho)$ and $S(\rho)$ are "parameter functions" which determine the nature of the steady state.

10.11.3 Physical Meaning of the Functions $H(\rho)$, $S(\rho)$

The function $H(\rho)$ is a parameter function which is determined by the momentum (and angular momentum) distribution in the fluid. From a practical point of view the choice of this function determines the structure of the steady state density distribution. The corresponding flow field can be computed then aposteriori (that is, after solving for ρ) from the following relations.

$$u = \sqrt{\frac{H(\rho)}{\rho}}\frac{\partial\rho}{\partial y}, \quad v = -\sqrt{\frac{H(\rho)}{\rho}}\frac{\partial\rho}{\partial x}. \qquad (10.193)$$

The function $S(\rho)$ that appears in Equation (10.192) can be determined from the asymptotic values of ρ and ϕ on the boundaries of the domain on which Equations (10.176),(10.192) are solved. When these asymptotic values are imposed or known, one can evaluate the left hand side of Equation (10.192) on the domain boundaries and re-express it in terms of ρ only to determine $S(\rho)$ on the boundary of the domain. However, the resulting functional relationship of S on ρ must then hold also within the domain itself since S does not depend on x, y directly.

For example, if we assume that on an infinite domain $h(\rho) = 1$, $\omega = 0$, and the asymptotic behaviour of ρ and ϕ is given by

$$\lim_{r\to\infty} \rho(r) = e^{-ar^2}, \quad \lim_{r\to\infty} \phi(r) = 4a^2r^2e^{-ar^2} \qquad (10.194)$$

then (asymptotically), Equation (10.192) evaluates to

$$S(\rho) = -4\alpha e^{-ar^2} = -4\alpha\rho. \qquad (10.195)$$

10.11.4   Radial Solutions for the Steady State Model

It is natural to consider this special case using polar coordinates. Then $\rho = \rho(r)$ and $\phi = \phi(r)$. The system consisting of Equations (10.192)-(10.176) with $H(\rho) = 1$ reduces to

$$\rho'' = -\frac{\rho'}{r} + S(\rho) - \phi + \frac{1}{2}\omega^2r^2 \qquad (10.196)$$

$$\phi'' = -\frac{1}{r}\phi' + c\rho, \quad c = 4\pi G. \qquad (10.197)$$

To solve this system of equations we let $S(\rho) = \alpha\rho$, solve Equation (10.196) for ϕ, and substitute the result in Equation (10.197). This leads to the following fourth order equation for ρ

$$\rho'''' + \frac{2}{r}\rho''' - \left(\alpha + \frac{1}{r^2}\right)\rho'' + \left(\frac{1}{r^3} - \frac{1}{r}\right)\rho' + c\rho = 2\omega^2. \qquad (10.198)$$

The general solution of this equation is

$$\rho = \frac{2\omega^2}{c} + C_1J_0(a_1r) + C_2J_0(b_1r) + C_3Y_0(a_1r) + C_4Y_0(b_1r) \qquad (10.199)$$

where J_0 and Y_0 are Bessel functions of the first and second kind of order 0 and

$$a_1 = \frac{1}{2}\sqrt{-2\alpha + 2\sqrt{\alpha^2 - 4c + \alpha^2}}, \quad b_1 = \frac{1}{2}\sqrt{-2\alpha - 2\sqrt{\alpha^2 - 4c + \alpha^2}}.$$

Assuming no singularity at the origin we set $C_3 = C_4 = 0$. To assess the

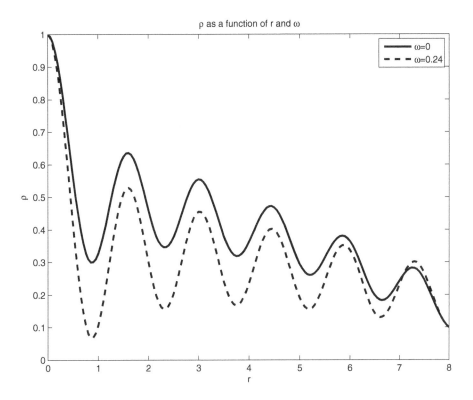

Figure 10.4 Steady state of the interstellar gas with $\alpha = -19.4$, $c = 1$, and boundary conditions $\rho(0) = 1$, $\rho(8) = 0.1$ with different values of ω

impact of the rotation term on the steady state we solved this system for C_1, C_2 on a circular disk using the boundary conditions $\rho(0) = 1$ and $\rho(8) = 0$ with $c = 1$, $\alpha = -19.4$. The results of these computations for different values of ω are plotted in Fig. 10.4. In this figure we see that the separation between the density peaks becomes more pronounced as ω increases. This might be interpreted as leading to the creation of protoplanets around the central core.

A strong dependence on ω is shown in Fig. 10.5 which has the same parameters as Fig. 10.4 except that the boundary conditions on ρ are: $\rho(0) = 0.35$ and $\rho(8) = 0.25$. This figure clearly illustrates the effect that rotation can have

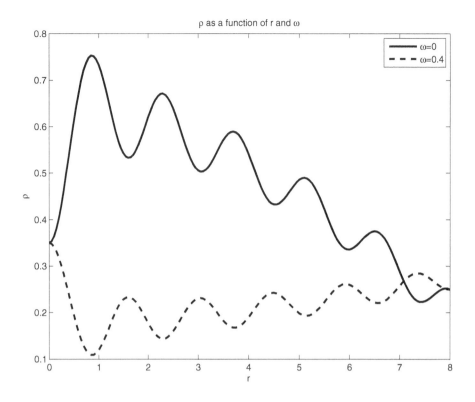

Figure 10.5 Steady state of the interstellar gas with $\alpha = -19.4$, $c = 1$ and boundary conditions $\rho(0) = 0.35$, $\rho(8) = 0.25$ with different values of ω

on the pattern of density fluctuations within the cloud. Furthermore, in this figure the magnitude of the density fluctuations reverses itself as ω becomes larger viz. the higher density peaks are placed at larger values of r (which is reminiscent of the situation in the solar system).

10.12   APPENDIX A - GAUSS THEOREM AND ITS VARIANTS

Notation: Let

$$
\begin{aligned}
&\mathbf{x} = (x_1, x_2, x_3) = (x, y, z) \\
&V = \text{a three dimensional domain} \\
&\partial V = \text{the boundary of } V \\
&\mathbf{n} = (n_1, n_2, n_3) \text{ the outward normal to } V.
\end{aligned}
$$

1. Basic Gauss Theorem: If $\phi(\mathbf{x})$ is a scalar field then

$$
\int_{\partial V} \phi n_i dS = \int_V \frac{\partial \phi}{\partial x_i} dV \tag{A. 1}
$$

(this is basically the fundamental theorem of the calculus in 3-D).

2. Gradient form: Use the basic theorem for $i = 1, 2, 3$

$$
\int_{\partial V} \phi n_1 dS = \int_V \frac{\partial \phi}{\partial x_1} dV
$$

$$
\int_{\partial V} \phi n_2 dS = \int \frac{\partial \phi}{\partial x_2} dV
$$

$$
\int_{\partial V} \phi n_3 dS = \int \frac{\partial \phi}{\partial x_3} dV.
$$

Hence

$$
\left(\int_{\partial V} \phi n_1 dS \right) \mathbf{i} \;+\; \left(\int_{\partial V} \phi n_2 dS \right) \mathbf{j} + \left(\int_{\partial V} \phi n_3 dS \right) \mathbf{k}
$$

$$
= \left(\int_V \frac{\partial \phi}{\partial x_1} \mathbf{i} dV \right) + \left(\int_V \frac{\partial \phi}{\partial x_2} \mathbf{j} dV \right) + \left(\int_V \frac{\partial \phi}{\partial x_3} \mathbf{k} dV \right)
$$

or in vector notation

$$
\int_{\partial V} \phi \mathbf{n} dS = \int_V \operatorname{grad} \phi \, dV. \tag{A. 2}
$$

3. Divergence form:

Let $\mathbf{F} = (f_1, f_2, f_3)$ then from the basic theorem we have

$$
\int_{\partial V} f_1 n_1 dS = \int_V \frac{\partial f_1}{\partial x_1} dV
$$

$$\int_{\partial V} f_2 n_2 dS = \int_V \frac{\partial f_2}{\partial x_2} dV$$

$$\int_{\partial V} f_3 n_3 dS = \int_V \frac{\partial f_3}{\partial x_3} dV.$$

Summing these three equations we have

$$\int_{\partial V} \mathbf{F} \cdot \mathbf{n} dS = \int_V div \mathbf{F} dV. \tag{A. 3}$$

4. Tensor form:

Let T be a second rank tensor with vector components $(\boldsymbol{\tau}_1, \boldsymbol{\tau}_2, \boldsymbol{\tau}_3)$ then from (A.3) we have

$$\int_{\partial V} \boldsymbol{\tau}_i \cdot \mathbf{n} dS = \int_V \text{div} \boldsymbol{\tau}_i dV =$$

or in component form

$$\int_{\partial V} \boldsymbol{\tau}_i \cdot \mathbf{n} dS = \int_V \sum_j \frac{\partial \tau_{ij}}{\partial x_j} dV$$

or in tensor form

$$\int_{\partial V} \mathbf{T} \cdot \mathbf{n} dS = \int_V \text{div} \mathbf{T} dV. \tag{A. 4}$$

5. Curl form:

$$\int_{\partial V} f_1 n_2 dS = \int_V \frac{\partial f_1}{\partial x_2} dV$$

$$\int_{\partial V} f_2 n_1 dS = \int_V \frac{\partial f_2}{\partial x_1} dV.$$

Hence

$$\int_{\partial V} (n_1 f_2 - n_2 f_1) dS = \int_V \left(\frac{\partial f_2}{\partial x_1} - \frac{\partial f_1}{\partial x_2} \right) dV.$$

This is the **k** component of

$$\int_{\partial V} \mathbf{n} \times \mathbf{F} dS = \int_V \text{curl} \mathbf{F} dV. \tag{A. 5}$$

10.13   APPENDIX B - POINCARE INEQUALITY AND BURGER'S EQUATION

Theorem (Poincare): Let $u(x)$ be a bounded differentiable function on $[0,1]$ with $u(1) = 0$. Then

$$\int_0^1 [u'(x)]^2 dx \geq \frac{\pi^2}{4} \int_0^1 u^2(x) dx. \tag{B.1}$$

Proof: To prove this inequality we introduce an "auxiliary function"

$$h(x) = \frac{\pi}{2} \tan\left(\frac{\pi x}{2}\right). \tag{B.2}$$

This function satisfies $h(0) = 0$ and the differential equation

$$h' - h^2 = \frac{\pi^2}{4}. \tag{B.3}$$

We now consider the integral

$$\int_0^1 [uh + u']^2 dx \geq 0 \tag{B.4}$$

$$\int_0^1 [uh + u']^2 dx = \int_0^1 u^2 h^2 dx + \int_0^1 (u')^2 dx + 2\int_0^1 uhu' dx \geq 0 \tag{B.5}$$

but

$$\begin{aligned}
\int_0^1 uu'h dx &= \left.\frac{u^2 h}{2}\right|_0^1 - \frac{1}{2}\int_0^1 u^2 h' dx \\
&= \frac{1}{2}[u^2(1)h(1) - u^2(0)h(0)] - \frac{1}{2}\int_0^1 u^2 h' dx \\
&= -\frac{1}{2}\int_0^1 u^2 h' dx.
\end{aligned} \tag{B.6}$$

Hence from (B.5)

$$\int_0^1 (u')^2 dx \geq \int_0^1 u^2 h' dx - \int_0^1 u^2 h^2 dx = \int_0^1 u^2(h' - h^2)dx = \frac{\pi^2}{4}\int_0^1 u^2 dx. \tag{B.7}$$

Remark: To obtain the same result on $[0,a]$ we let $z = ax$

$$\int_0^1 [u'(x)]^2 dx = \int_0^a a^2 \left(\frac{du}{dz}\right)^2 \frac{1}{a} dz = a\int_0^a (u'(z))^2 dz$$

$$\int_0^1 u^2(x)dx = \int_0^a u^2(z)\frac{1}{a}dz.$$

Hence

$$a\int_0^a (u'(z))^2 dz \geq \frac{\pi^2}{4}\frac{1}{a}\int_0^a u^2 dz,$$

i.e.

$$\int_0^a (u'(z))^2 dz \geq \frac{\pi^2}{4a^2}\int_0^a u^2(z)dz.$$

Burger's Equation.

Burger's equation is:

$$u_t + uu_x = \nu u_{xx}. \tag{B.8}$$

Observe that this is a nonlinear partial differential equation which contains a convective term uu_x. It was derived originally as a prototype, in one dimension, that can provide analytic insight about the nature of turbulence and its modeling. Such equations with convective terms appear in applied mathematics and theoretical physics (e.g. gas dynamics and traffic flow).

Lemma Let $u = u(x,t)$ be a solution of Burger's equation on the interval $[0, a]$, for $t > 0$. Show that if $u(0, t) = u(a, t) = 0$ then

$$\int_0^a u^2 dx \leq Ce^{-\frac{\nu\pi^2}{2a^2}t}, \quad C \text{ is a constant}. \tag{B.9}$$

Solution: Multiply Equation (B.8) by u and integrate over $[0, a]$

$$\int_0^a uu_t dx + \int_0^a u^2 u_x dx = \nu \int_0^a uu_{xx} dx.$$

But

$$\int_0^a u_t u dx = \frac{1}{2}\frac{d}{dt}\int_0^a u^2 dx$$

$$\int_0^a u^2 u_x dx = \frac{u^3}{3}\Big|_0^a$$

$$\int_0^a uu_{xx} dx = uu_x\Big|_0^a - \int_0^a u_x^2 dx.$$

Hence

$$\frac{1}{2}\frac{d}{dt}\int_0^2 u^2 dx + \frac{u^3}{3}|_0^a = \nu uu_x|_0^a - \nu\int_0^a u_x^2 dx$$

$$\frac{1}{2}\frac{d}{dt}\int_0^a u^2 dx + \nu\int_0^a u_x^2 dx = u\left(\nu u_x - \frac{u^2}{3}\right)\Big|_0^a \tag{B.10}$$

with the boundary conditions $u(0, t) = u(a, t) = 0$ this leads to

$$-\frac{1}{2}\frac{d}{dt}\int_0^a u^2 dx = \nu \int_0^a u_x^2 dx \geq \frac{\nu \pi^2}{4a^2}\int_0^a u^2 dx.$$

If we let

$$F(t) = \int_0^a u^2(x, t) dx \geq 0$$

then

$$-\frac{d}{dt}F(t) \geq \frac{\nu \pi^2}{2a^2}F(t)$$

(multiply by -1)

$$\frac{F'(t)}{F(t)} \leq -\frac{\nu \pi^2}{2a^2}.$$

Integrate with respect to t

$$\ln F(t)|_0^t \leq -\frac{\nu \pi^2}{2a^2}t$$

$$\ln \frac{F(t)}{F(0)} \leq -\frac{\nu \pi^2}{2a^2}t$$

$$F(t) \leq F(0)exp(-\frac{\nu \pi^2}{2a^2}t).$$

Observe that if $u(0, t)$ or $u(a, t)$ are not equal to 0, one can still use Poincare inequality to obtain a similar result.

10.14   APPENDIX C - GRONWELL INEQUALITY

Theorem (Gronwell): Let $u : [0, a] \to R$ be a continuous non-negative function and let K, C be non-negative constants so that

$$u(t) \leq C + \int_0^t Ku(s)ds \qquad (C.\ 1)$$

for all $t \in [0, a]$. Then

$$u(t) \leq Ce^{Kt}. \qquad (C.\ 2)$$

Proof: Define

$$U(t) = C + \int_0^t Ku(s)ds; \qquad (C.\ 3)$$

then it is obvious that $u(t) \leq U(t)$ for all t. However

$$\frac{dU(t)}{dt} = U'(t) = Ku(t). \tag{C.4}$$

Hence

$$\frac{U'(t)}{U(t)} = \frac{Ku(t)}{U(t)} \leq K. \tag{C.5}$$

Integrating both sides of this inequality with respect to t we obtain

$$\ln U(t) \leq \ln U(0) + Kt = \ln C + Kt. \tag{C.6}$$

Exponentiating this inequality we finally obtain

$$u(t) \leq U(t) \leq Ce^{Kt}. \tag{C.7}$$

As an application of this theorem we now prove the following
Theorem: Let $f(x)$ be Lipschitz, i.e.

$$|f(x) - f(y)| \leq K|x - y|, \quad for\ all\ x, y. \tag{C.8}$$

Then for any two solutions v(t),w(t) of

$$\frac{du(t)}{dt} = f(u) \tag{C.9}$$

with initial conditions $v(t_0) = v_0$, $w(t_0) = w_0$ we have

$$|v(t) - w(t)| \leq |v_0 - w_0|e^{K(t-t_0)}. \tag{C.10}$$

The solutions $v(t), w(t)$ can be expressed as

$$v(t) = v_0 + \int_{t_0}^{t} f(v(s))ds, \quad w(t) = w_0 + \int_{t_0}^{t} f(w(s))ds. \tag{C.11}$$

Therefore

$$v(t) - w(t) = v_0 - w_0 + \int_{t_0}^{t} [f(v(s)) - f(w(s)]ds. \tag{C.12}$$

Taking the absolute values on both sides of this equation we infer that

$$|v(t) - w(t)| \leq |v_0 - w_0| + \int_{t_0}^{t} |[f(v(s)) - f(w(s)]|ds. \tag{C.13}$$

Using the fact that $f(x)$ is Lipschitz, it follows that the function $g(t) = |v(t) - w(t)|$ satisfies the inequality

$$g(t) \leq C + \int_{t_0}^{t} K g(s) ds \tag{C. 14}$$

where $C = |v_0 - w_0|$. All the conditions of Gronwell lemma are satisfied and therefore

$$g(t) \leq C e^{K(t-t_0)},$$

i.e.

$$|v(t) - w(t)| \leq |v_0 - w_0| e^{K(t-t_0)} \tag{C. 15}$$

which is the desired result.

10.15 APPENDIX D - THE SPECTRUM

Given a physical quantity, such as Energy, we can always rewrite it as a Fourier integral

$$E(\tau) = \frac{1}{2} \int_{-\infty}^{\infty} \Phi(\omega) e^{-i\omega\tau} d\omega \tag{D. 1}$$

where $\Phi(\omega)$ is the Fourier transform of $E(\tau)$ which is given explicitly by the "inverse transform"

$$\Phi(w) = \frac{1}{\pi} \int_{-\infty}^{\infty} E(\tau) e^{i\omega\tau} d\tau. \tag{D. 2}$$

From (D. 1) we see that $\Phi(\omega) d\omega$ is the contribution to the energy from harmonic oscillations in the frequency interval $[\omega, \omega + d\omega]$. By analogy with the spectra of light we therefore call $\Phi(\omega)$ the "spectral density function."

Bibliography

1 A. J. Chorin, J.E. Marsden, (1993) A Mathematical Introduction to Fluid Mechnics Springer Verlag NY

2 M. Humi, (2006) Steady States of self gravitating incompressible fluid. J. Math. Phys. 47, 093101 (10 pages).

3 H. Lamb, (1945) Hydrodynamics, Dover Publications NY

4 L.D. Landau and E.M. Lifshitz, (1987) Fluid Mechanics Pergamon Press, NY

5 Lissauer J.J, (1993) Planet formation, Ann. Rev. Astron. 31, pp. 129-174

6 M. Ya Marov and A.V. Kolesnichenko (2013) Turbulence and Self-Organization, Modeling Astrophysical Objects, Springer, NY.

7 L.M. Milne-Thompson, (1996) Theoretical Hydrodynamics, Dover Publications,NY

8 C. Pozrikidis (1997) Introduction to Theoretical and Computational Fluid Dynamics, Oxford Univ. Press

CHAPTER 11

Modeling Geophysical Phenomena

CONTENTS

11.1   ATMOSPHERIC STRUCTURE

On the large scale (in height), the atmosphere is divided into three sections. These are

1. Homosphere (up to a height of 100km),

2. Heterosphere (100 to 500km),

3. Exosphere (above 500km).

In the exosphere, the air density is very low (and the mean free path is large). As a result, molecules in this region have a "fair chance" to escape into space. In the heterosphere, the strong ultraviolet radiation from the sun dissociates the H_2O and O_2 molecules. By this process, part of this harmful radiation is filtered out and does not reach the lower levels.

In the homosphere, the molecular mean free path is small. As a result, bulk transport by turbulent air motion dominates the diffusive processes. This turbulent mixing "homogenizes" the passive constituents of the atmosphere; i.e., their densities decrease exponentially with altitude at the same rate which gives air a homogeneous composition of 78% N_2 and 21% O_2.

Due to its importance, the homosphere is further divided into

1. Troposphere (between 0-10km in height),

2. Stratosphere (between 10-50km),

3. Mesosphere (above 50km).

Remark: The boundary layers between these are referred to as tropopause and stratopause.

In the troposphere (which is also referred to as the biosphere), the temperature decreases at a rate of 6.5° Kelvin/km. In the stratosphere, on the other hand, the temperature increases with height due to ozone heating (as a result, the stratosphere is "stably stratified; " i.e. "lighter air" is on top of the "denser" air). In the mesosphere, the temperature again decreases with height.

11.2 THERMODYNAMICS AND COMPRESSIBILITY

In the previous section, we gave a qualitative overview of the atmospheric structure. In this section, we concentrate on the troposphere and its quantitative modeling.

11.2.1 Thermodynamic Modeling

The "ideal gas law" states that for a parcel of gas

$$PV = \frac{m}{M}RT \tag{11.1}$$

where m is the mass of the parcel, M is the molecular weight of the gas in the parcel, P is the pressure, V is the volume, T is the temperature in Kelvins (K), and R is the (universal) "gas constant"

$$R = 8.314\frac{Joule}{molK}.$$

The gas density in the parcel is

$$\rho = \frac{m}{V} = \frac{MP}{RT}. \tag{11.2}$$

Let U be the (internal) energy of a mole ideal gas. We define the heat capacity of a gas held at constant volume as

$$C_V = \frac{dU}{dT}.$$

In the following, we assume that C_V is constant.

For a mole of an adiabatic parcel of gas (where no heat (Q) is exchanged between the parcel and its surroundings), the first law of thermodynamics implies that

$$dQ = C_V dT + P dV = 0; \tag{11.3}$$

i.e. the sum of the change in the internal energy and the work done by the pressure is zero due to the fact that there is no heat exchange with the environment. Using Equation (12.1) leads to

$$C_V dT + \frac{RT}{V}dV = 0. \tag{11.4}$$

Hence,

$$\frac{dT}{T} = -\frac{R}{C_V V} dV. \tag{11.5}$$

Integrating this equation, we obtain

$$TV^{k-1} = constant \tag{11.6}$$

where $k = 1 + \frac{R}{C_V}$. Using Equation (12.1) we can rewrite Equation (12.8) as

$$TP^{-\frac{k-1}{k}} = constant. \tag{11.7}$$

Taking the logarithmic derivative of Equation (12.9) yields

$$\frac{dT}{T} = \frac{k-1}{k} \frac{dP}{P}. \tag{11.8}$$

We assume now that the troposphere is composed of a gas which obeys the "ideal gas law." For an adiabatic parcel of this gas between heights h and $h + dh$, the pressure difference between the top and bottom is

$$dP = -\rho g dh \tag{11.9}$$

where g is the acceleration due to gravity $g = 9.8 m/sec^2$. Substituting for ρ from Equation (12.2) yields

$$dP = -\frac{gMP}{RT} dh. \tag{11.10}$$

Using Equation (12.10) to substitute for $\frac{dP}{P}$ in Equation (12.4) leads to

$$\frac{dT}{dh} = -\frac{k-1}{k} \frac{gM}{R}. \tag{11.11}$$

In the troposphere where the air is composed mostly of diatomic molecules $k = 1.4$, $M = 28.88$ which yield

$$\frac{dT}{dh} \approx -9.8 \ K/km.$$

This value is greater (in absolute value) than the observed value quoted in the previous section of $-6.5 \ K/km$. The difference is due to the fact that this "simplistic" model neglects air moisture.

Observe that by integrating Equation (12.11), one can compute P as a function of height using Equation (12.4).

11.2.2 Compressibility

In the previous chapter, we considered several aspects of Navier-Stokes equations (NSE) under the assumption of incompressibility. However, in many applications (e.g., gas dynamics or atmospheric applications), compressibility has to be taken into account. Boussinesq Approximation, in this context, enables us to take into account some of these compressibility effects while retaining the "flavor" of the incompressible equations (thereby reducing the nonlinearities in the equations). To this end, we neglect the density variations in the continuity equation and the inertia term in the momentum equations. However, these density variations do give rise to buoyancy forces in the momentum equations.

The basic equations of the flow $\mathbf{u} = (u, v, w)$ are

$$\frac{\partial \rho}{\partial t} + \nabla \cdot (\rho \mathbf{u}) = 0$$

$$\frac{\partial \mathbf{u}}{\partial t} + (\mathbf{u} \cdot \nabla)\mathbf{u} = -\frac{1}{\rho}\nabla p - g\mathbf{k}$$

where p is the pressure, ρ is the density, $\mathbf{k}$ is a unit vector in the z direction, and g is the acceleration due to gravity.

As a first step in this approximation, the mass continuity equation

$$\frac{\partial \rho}{\partial t} + \nabla \cdot (\rho \mathbf{u}) = 0$$

is split into two equations

$$\nabla \cdot \mathbf{u} = 0, \quad \frac{\partial \rho}{\partial t} + (\mathbf{u} \cdot \nabla)\rho = 0.$$

That is we consider the flow to be incompressible and ρ as a scalar that is carried over by the flow.

Furthermore, we assume that the density and pressure of the atmospheric fluid can be written as

$$\rho(\mathbf{x}, t) = \rho_0(z) + \rho'(\mathbf{x}, t), \quad p(\mathbf{x}, t) = p_0(z) + p'(\mathbf{x}, t)$$

and

$$\frac{dp_0(z)}{dz} = -\rho_0 g$$

where $\rho'(\mathbf{x}, t)$ and $p'(\mathbf{x}, t)$ are small perturbations from $\rho_0(z)$ and $p_0(z)$ respectively. Introducing the following definitions

$$P = \frac{p}{\bar{\rho}}, \quad N = -g\frac{\rho - \rho_0}{\bar{\rho}}, \quad N_0^2 = -\frac{g}{\bar{\rho}}\frac{d\rho_0(z)}{dz}$$

where $\bar{\rho}$ is a *constant* "reference density", this leads to the following approximate system of equations for the flow:

$$\nabla \cdot \mathbf{u} = 0$$

$$\frac{\partial \mathbf{u}}{\partial t} + (\mathbf{u} \cdot \nabla)\mathbf{u} = -\nabla P + N\mathbf{k}$$

$$\frac{\partial N}{\partial t} + \mathbf{u} \cdot \nabla N + N_0^2 w = 0$$

where N_0^2 is called the Brunt-Vaisala frequency. This system of equations yields a reasonable approximation to the exact equations when the fluid is "almost incompressible."

11.3 GENERAL CIRCULATION

The gross features of the atmospheric circulation are driven by convective currents due to differential heating, Earth rotation, and the asymmetric distribution of land and sea. Due to these factors, the atmosphere is divided into meridional and longitudinal convection cells.

To see how the meridional cells are formed, we observe that the equator receives much more heat than the poles. Accordingly, we expect hot air to rise at the equator and travel in the upper troposphere towards the poles, sink there, and then return along the surface to the equator. This global picture was formulated by G. Hadley in the early 18th century and is named after him as "Hadley circulation" (as a matter of fact such one-cell atmosphere exists on Venus).

On earth, the hot air rising at the equator sinks at about 30° (north and south) and thus forms the tropical Hadley cell. At the descending branch of the Hadley cell, the air is dry. As a result, it creates the great subtropical deserts on earth such as the Sahara and Gobi deserts.

A second meridional cell forms in the subtropics (between 30° and 60°). It is called the Ferrel cell. Finally, there exists the polar cell in the arctic region.

The junction between the polar and Ferrel cells is called the polar front. Along this front in the upper troposphere there is a strong band of westerly winds (i.e., winds blowing from the west) called the jet stream. When the jet stream "meanders" south over the northern US, that region experiences very cold weather.

We should note that due to the earth rotation, the winds in the upper troposphere of the Hadley cell have a westerly direction. However, the returning surface wind is easterly (i.e., blowing from the east). To understand this, observe that the tangential velocity of the earth is maximum at the equator. For the Ferrel cell, this situation is reversed, and therefore we have "surface westerlies" in the mid-latitudes. These are referred to as the "trade winds."

Nonuniform heating due to the uneven distribution of land and sea drives the "zone overturning" or Walker Circulation. In these cells, air rises at longitudes of heating (e.g., Indonesia) and sinks at longitudes of cooling (west of South America). This circulation in normal years reinforces the easterly trade wind across the equator. However, when this circulation reverses itself, it causes the "El-Nino" current in the Pacific Ocean.

11.4 CLIMATE

There exist various models for climate predictions in general and the computation of the mean temperature of the earth. These models are usually classified by their degree of sophistication as 0-dim, 1-dim, etc. The most sophisticated current models are the "Community Atmospheric Models" (CAM) which were written at the National Center for Atmospheric Research (NCAR) [1].

In the 0-dimensional models, only the time dependence of the mean temperature is modeled and an average is taken over the spatial dependence. For the one dimensional models, the dependence of the temperature on the latitude and time is taken into consideration and so on.

In the following we provide a short narrative for Earth "climate modeling"

A key parameter in all these models is the albedo, which is the fraction of solar energy in the short wave band which is reflected from Earth back into space. The value of the albedo depends on the nature of the surface (ocean and different types of land, e.g., forests, deserts, etc.) and time (extent of ice and snow cover). In the past, this "parameter" was modeled by various means, however, due to recent advances in satellite imagery, it is now possible to compute the albedo accurately for each location of the Earth.

In the following, we consider models with zero and one dimensions.

For the zero dimensional model, we take into consideration only the balance between the total incoming and outgoing radiations which we denote by R_i and R_o respectively.

$$C\frac{dT_m}{dt} = R_i - R_o \tag{11.12}$$

where T_m is the global mean temperature, t is the time, and C is the heat capacity of the Earth system (more precisely land, air, and oceans). To model the incoming and outgoing radiations, we let

$$R_i = Q\{1 - A(T_m)\} \tag{11.13}$$

$$R_o = \sigma g(T_m)T_m{}^4. \tag{11.14}$$

Here, Q is the flux of the solar radiation, C is the heat capacity of the Earth system (more precisely land, air, and oceans). $A(T_m)$ is the mean albedo and $g(T_m)$ is a "grayness-factor," which measures the deviation of Earth emissions from black body radiation due to the greenhouse effect.

Neglecting the spatial heat distribution on Earth, we can formulate a prototype model for the global mean temperature ($\cong$ climate). This can be done using the basic facts about the radiative balance of Earth which were described in the previous sections. Thus, from Equations. (12.39)-(12.40) we infer that

$$C\frac{dT_m}{dt} = Q[1 - A(T_m)] - \sigma.g(T_m)T_m{}^4 \tag{11.15}$$

The factor $g(T_m)$ was modeled by Sellers [2] by the following formula

$$g(T_m) = 1 - k\tanh[(T_m/T_0)^6], \; T_0^6 = 0.53 \cdot 10^{15}K^6 \tag{11.16}$$

where k is the portion of Earth covered by clouds (under present conditions $k \sim 0.5$). In this model, $g(T_m)$ decreases as T_m increases as the greenhouse effect becomes more pronounced.

As to the albedo, the following linear interpolation function was formulated by Sellers

$$A(T) = \begin{cases} \alpha_M, & T < T_1 \\ \alpha_M - \frac{T-T_1}{T_2-T_1}(\alpha_M - \alpha_m), & T_1 < T < T_2 \\ \alpha_m & T_2 < T \end{cases} \tag{11.17}$$

where α_M and α_m are the albedo values assigned to ice-covered and ice-free surfaces respectively ($\alpha_M \cong 0.85$, $\alpha_m \cong 0.25$, $T_1 \cong 210°K$, and $T_2 = 275°K$).

In a steady state, $\frac{dT_m}{dt} = 0$ and Equations (11.15) and (11.17) reduce to an algebraic equation which can be solved for T_m. We find then that this model has three equilibrium points, two of which are stable while the third (in between) is unstable. The two stable equilibria correspond to "glaciation period" and "present day" conditions.

Refinements to the model given by Equations (11.15), (11.17) are obtained when we let Q depend on a parameter $\lambda(t)$

$$Q = \lambda(t)Q_0 \tag{11.18}$$

to take into account possible variations in the Sun radiative output with time.

We see that even this "prototype" model for Earth climate depends on many parameters whose exact value is not known (and subject to change). This renders the predictions of this and more sophisticated models somewhat unreliable with a large margin of error.

Many more elaborate models for the albedo have appeared in the literature. One of the major sticking points in these models is the clouds that cover the Earth and their actual impact on the albedo. For example, Bhattacharaya, et al. suggested a model for $A(T_m)$ which is described by Fig. 11.1. In this figure, the peak in the albedo near $T = 220°K$ is attributed to the increased cloudiness near the "ice-margin." The use of this albedo model and Equation (11.15) yield five steady state points, three of which are stable.

A more detailed model for the albedo takes into account the different albedos of ocean, land, and ice and the meridional extent of the ice cover of the earth. Thus,

$$A = a_L\beta + (1 - \beta)a_{oc}, \quad a_L = a_1 + a_2M \tag{11.19}$$

where β is the land and ice percentage of the earth surface and a_{oc} is the albedo of the ocean. The albedo of the land (a_L) is composed of two parts: a_1 is the albedo of the ice free land for which $M = 0$, and the albedo of the ice sheet $a_1 + a_2M$ where M is the meridional extent of the ice sheet.

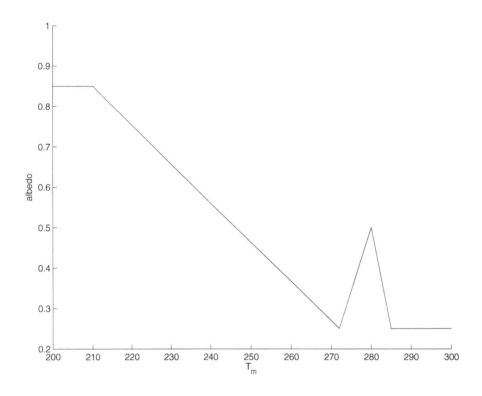

Figure 11.1   Bhattacharaya, et al. model for the albedo

In a more refined model for the albedo of the land, M is a function of time and is governed by the following differential equation

$$\frac{dM}{dt} = \lambda M^{-1/2}[(1 + \epsilon(T))M_T - M].\qquad(11.20)$$

Here, M_T is the meridional extent of the ice accumulation zone and $\epsilon(T)$ is a ramp function. Thus, in this formulation the earth climate is governed by two differential Equations, (11.20) and (11.20), which depend on several parameters. As these parameters can vary, the climate can pass through various bifurcations.

To introduce spatial dependence in these models, we have to change dT/dt into

$$DT/Dt = \frac{\partial T}{\partial t} + (\mathbf{u} \cdot \nabla)T\qquad(11.21)$$

where $\mathbf{u}$ is the wind speed. However, as the climate time-scale is long ($\approx O(10^4 yrs)$), it is usual in this context to eliminate $\mathbf{u}$ by applying the "eddy diffusivity approximation"

$$-(\mathbf{u} \cdot \nabla)T \cong \nabla \cdot (\nu_e \nabla T) \tag{11.22}$$

where ν_e is the "eddy diffusivity coefficient." With this approximation, Equation (11.15) takes the form

$$C(\mathbf{x})\frac{\partial T}{\partial t} = QS(\mathbf{x})\{1 - A(\mathbf{x}, T)\} - \tag{11.23}$$
$$\sigma g(x, T)T^4 + \nabla \cdot (\nu_e \nabla T)$$

where $S(\mathbf{x})$ is the distribution of the solar flux on earth (if the earth axis had no tilt then $S(x) = \frac{1}{2}\cos\phi$, where ϕ is the latitude). To simplify, to some extent, one can assume that all quantities in Equation (11.23) depend only on time and latitude. Equation (11.23) then takes the following form:

$$C(\phi)\frac{\partial T}{\partial t} = QS(\phi)\{1 - A(\phi, T)\} - \tag{11.24}$$
$$\sigma g(\phi, T)T^4 + \frac{1}{\cos\phi}\frac{\partial}{\partial\phi}\left\{\nu_e(\phi)\cos\phi\frac{\partial T}{\partial\phi}\right\}.$$

We see that due to the extreme complexity of the climate system, there remains a lot of uncertainty about our ability to predict the future climate of the earth. In particular, it is questionable whether current climate models can reliably predict the impact of man-made inputs to this system.

Bibliography

[1] CCSM3.0 Community Atmosphere Model (CAM), http://www.ccsm. ucar.edu/models/atm-cam

[2] K. Bhattacharaya, et al.(1982) J. Atmos. Sci. **39** p. 1747-1773.

[3] J. Pedlosky - Geophysical Fluid Dynamics 2nd edition. Springer,NY.

[4] W.D. Sellers (1969) J. App. Met. **8** p. 392-400.

[5] J. Pedlosky - Geophysical Fluid Dynamics 2nd edition. Springer NY.

CHAPTER 12

Stochastic Modeling

CONTENTS

12.1 INTRODUCTION

In previous chapters, we considered the modeling of deterministic systems. For these systems, information about the state of the system at time t determines with certainty its state at any later time. For stochastic systems, on the other hand, no such certainty can be achieved, viz. the knowledge of the state of the system at time t, enables us to predict only the probability that the system be in any of several possible states in the future.

Our objective in this chapter is, therefore, to describe the basic logical steps that lead to such stochastic models. To achieve this objective, we describe several stochastic models for various growth and decay processes and compare to some extent their predictions with the corresponding deterministic ones.

12.2 PURE BIRTH PROCESS

Problem: A biologist is experimenting with a new hybrid tree of which there are very few plants. We are asked to build a model which describes the population size of the trees in the next few years.

Approximations and Simplifications:

1. Since the size of the population $N(t)$ at time t is expected to remain small, we must treat $N(t)$ as a discrete variable.

2. In order to build a prototype model for the tree population, we shall ignore the death process. This is justified since trees are expected to live more than just a "few years."

3. The basic assumption in the deterministic treatment of such population models (that is when $N(t)$ is large and can be treated as a continuous variable) was that the change in the population size on the time interval $[t, t + \Delta t]$ is proportional to the size of the population and the time interval Δt, i.e.,

$$N(t + \Delta t) - N(t) = aN(t)\Delta t \qquad (12.1)$$

Assumptions and Abstractions:

1. Reproduction is an individual process, viz. reproduction of one tree does not affect the reproduction of another tree in the population.

2. The probability $P(t)$ that a tree reproduces itself in the time interval $[t, t + \Delta t]$ is given by

$$P(t) = k\Delta t + p_1(\Delta t) \tag{12.2}$$

where k is a constant and

$$\lim_{\Delta t \to 0} \frac{p_1(\Delta t)}{\Delta t} = 0 \ \ i.e. \ \ p_1(\Delta t) = O((\Delta t)^2). \tag{12.3}$$

3. The probability $p_2(t)$ that a tree reproduces more than once on $[t, t+\Delta t]$ becomes negligible, as $\Delta t \to 0$, i.e.,

$$\lim_{\Delta t \to 0} \frac{p_2(\Delta t)}{\Delta t} = 0 \ \ or \ \ p_2(t) = O((\Delta t)^2). \tag{12.4}$$

4. The probabilities of reproduction of a tree on two disjoint time intervals are independent of each other.

Remark: A process that satisfies assumptions $1 - 4$ or their equivalents is referred to as a "**Poisson process.**"

Mathematical Model:

Our basic objective here is to derive differential equations for the probability $P_N(t)$ that the tree population at time t is equal to N.

To begin with, we compute the probability that the tree population will increase from N to $N+1$ on the interval $[t, t + \Delta t]$. To do so, we observe that the population will increase by one on $[t, t + \Delta t]$ if one tree reproduces itself once and all the others do not reproduce themselves on this time interval. Hence, since there are N possibilities to choose the tree that reproduces itself, this probability is given by

$$P(N \to N+1) = \tag{12.5}$$
$$N(k\Delta t + p_1(t))[1 - (k\Delta t + p_1(t)) - p_2(t)]^{n-1} \approx Nk\Delta t + p(\Delta t)$$

where $p(\Delta t) = O((\Delta t)^2)$. From this result, we infer that if the tree population at time t is N, then the probability that the population remains unchanged at a later time is

$$P_N(t + \Delta t) = P_N(t) \cdot (1 - k\Delta t N) + O((\Delta t)^2). \tag{12.6}$$

Similarly, $P_N(t+\Delta t)$, viz. the probability that the tree population at $t+\Delta t$ is (exactly) $N(\geq N_0)$ is the sum of:

1. The probability that at t the tree population is N and there was no reproduction on $[t, t + \Delta t]$.

2. The probability that at t the tree population is $N - 1$ and there was exactly one reproduction on $[t, t + \Delta t]$ (remember that the probability of more than one reproduction in Δt is $O((\Delta t)^2)$.

Hence,

$$P_N(t + \Delta t) = P_N(t)(1 - kN\Delta t) + P_{N-1}(t)k(N - 1)\Delta t + O((\Delta t)^2). \quad (12.7)$$

If the tree population at time $t = 0$ is N_0, Equation (12.6) implies that it remains unchanged at time Δt if

$$P_{N_0}(\Delta t) - P_{N_0}(0) = -P_{N_0}(0)k\Delta t N_0.$$

Dividing this equation by Δt and letting $\Delta t \to 0$, we obtain the following differential equation for the tree population to be the same at time t

$$\frac{dP_{N_0}(t)}{dt} = -kN_0, \quad P_{N_0}(0) = 1 \quad (12.8)$$

where we used the fact that $P_{N_0}(0) = 1$.

Similarly, dividing Equation (12.7) by Δt and taking the limit as $\Delta t \to 0$, we obtain the following differential equation

$$\frac{dP_N(t)}{dt} = k[(N - 1)P_{N-1}(t) - NP_N(t)], \quad P_N(0) = 0, \quad N > N_0. \quad (12.9)$$

Equations (12.8) and (12.9) describe the stochastic process under consideration.

Analysis of the Model:

Since N is an integer, the system Equations (12.8) and (12.9) can be solved recursively for $N = N_0, N_0 + 1, \ldots$ Thus, from Equation (12.8) we conclude that

$$P_{N_0}(t) = e^{-kN_0 t}. \quad (12.10)$$

Substituting this result in the differential equation for $N_0 + 1$, we obtain

$$\frac{dP_{N_0+1}}{dt} + k(N_0 + 1)P_{N_0+1} = kN_o e^{-kN_0 t} \quad (12.11)$$

whose solution is

$$P_{N_0+1} = N_0 e^{-kN_0 t}(1 - e^{-kt}). \qquad (12.12)$$

In general, however, it is possible (after a long algebra) to show that

$$P_N(t) = \frac{(N-1)!}{(N-N_0)!(N_0-1)!}e^{-kN_0 t}(1 - e^{-kt})^{N-N_0}. \qquad (12.13)$$

Other related quantities which are important in the analysis of such models are the expected value of the population size at time t

$$\mu(t) = \sum_{N=N_0}^{\infty} N P_N(t) \qquad (12.14)$$

and the variance

$$\sigma^2(t) = \sum_{N=N_0}^{\infty} (N - \mu(t))^2 P_N(t). \qquad (12.15)$$

To evaluate $\mu(t)$, we differentiate Equation (12.14) with respect to t and use Equations (12.8), (12.9) to obtain

$$\frac{d\mu}{dt} = \sum_{N=N_0}^{\infty} N \frac{dP_N}{dt} = \qquad (12.16)$$

$$= -kN_0^2 P_{N_0} - k \sum_{N>N_0} N^2 P_N + k \sum_{N>N_0} N(N-1)P_{N-1}$$

but

$$\sum_{N>N_0} N(N-1)P_{N-1} = \sum_{N>N_0} (N-1)^2 P_{N-1} + \qquad (12.17)$$

$$\sum_{N>N_0} (N-1)P_{N-1} = \sum_{N=N_0}^{\infty} N^2 P_N + \sum_{N=N_0}^{\infty} N P_N.$$

Hence,

$$\frac{d\mu}{dt} = k\mu, \quad \mu(0) = N_0 \qquad (12.18)$$

which leads to

$$\mu = N_0 e^{kt}. \qquad (12.19)$$

Similarly, for σ^2 we obtain the differential equation

$$\frac{d\sigma^2}{dt} - 2k\sigma^2 = k\mu, \quad \sigma^2(0) = 0 \qquad (12.20)$$

whose solution is

$$\sigma^2(t) = \frac{N_0}{2}e^{2kt}(1 - e^{-kt}). \qquad (12.21)$$

From Equation (12.18) we see that the deterministic version of this stochastic model deals only with the expected value of the population size. This can be justified for large populations since then as t becomes large

$$\frac{\sigma(t)}{\mu(t)} \approx \frac{1}{\sqrt{2N_0}} \approx 0 \qquad (12.22)$$

viz. for large N_0 the probability distribution is sharply centered around $\mu(t)$.

Exercises

1. Build and solve a stochastic model for a pure death process (e.g., radioactive decay).

2. Build a model for a population in which both birth and death occur.

3. In our discussion of the pure birth process we assumed that k is constant. Discuss what happens if $k = k(N)$ or $k = k(t)$.

4. Derive and solve Equation (12.20).

12.3   KERMACK AND MCKENDRICK MODEL.

In 1927, Kermack and McKendrick constructed a general mathematical model for the spread of epidemics, rumors, etc. within a given population of $M + 1$ individuals. We shall describe here a version of this model for the spread of rumors.

The basic starting point of this model is that at any time t the population can be divided into three groups.

1. $m(t)$ consists of persons who have not heard the rumor.

2. $n(t)$ people who heard the rumor and are actively spreading it.

3. $s(t)$ people who heard the rumor but stopped spreading it.

Remarks:

1. The corresponding version of this model for the spread of epidemics will have $m(t)$ as those who are susceptible to the disease but are not infected as yet, $n(t)$ are people who are infected and are spreading the disease, $s(t)$ are people who are not susceptible to the disease (due to inoculation, etc.). However, sometimes to make the model more realistic a fourth group is added as those people who are infected but are not capable of spreading the disease (hospitalized).

2. Since the total population is $M + 1$ and

$$m + n + s = M + 1 = constant,$$

it is obvious that a knowledge of two of these three quantities will automatically give the third. However, to simplify our treatment further, we shall assume in the following that $s(t) = 0$.

Assumptions and Abstractions:

1. At time $t = 0$, $\ m(0) = M$, $\ and \ n(0) = 1$.

2. An individual in the sub-population m hears the rumor upon contact with a member of the sub-population n.

3. Once an individual is in the sub-population n, he remains there for all future times.

4. If I_1 and I_2 are disjoint time intervals, then the number of contacts between the sub-populations n and m on the interval I_1 does not affect the number of contacts on the interval I_2.

5. The probability of exactly one contact between the sub-populations n and m on $[t, t + \Delta t]$ is given by $kn(t)m(t)\Delta t + O((\Delta t)^2)$ where k is constant.

6. The probability of more than one contact between the sub-populations n and m on $[t, t + \Delta t]$ is $O((\Delta t)^2)$.

Mathematical Model:

To begin our derivation, we observe that if $m(t) = R$ (and therefore $n(t) = M + 1 - R$) then the probability of no contacts between the sub-populations n and m on the time interval $[t, t + \Delta t]$ is given by (using assumption 5. above)

$$1 - kR(M + 1 - R)\Delta t + O((\Delta t)^2). \tag{12.23}$$

Hence, if we denote by $P_M(t)$ the probability that $m(t) = M$ at t, then the probability that $m(t + \Delta t) = M$ is

$$P_M(t + \Delta t) = P_M(t)[1 - kM]\Delta t] + O((\Delta t)^2). \tag{12.24}$$

Similarly, if $m(t + \Delta t) = R, \ R \neq M$ we have three possibilities:

1. $m(t) = R$ and there was no contact between the sub-populations n and m on $[t, t + \Delta t]$.

2. $m(t) = R + 1$ and there was only one contact between the sub-populations n and m on $[t, t + \Delta t]$.

3. $m(t) = R+2$ and there were two contacts between n and m on $[t, t+\Delta t]$, etc.. However, the probability of this to happen is $O((\Delta t)^2)$ in view of assumption 6.

Thus, we infer that:

$$P_R(t + \Delta t) = P_R(t)[1 - kR(M + 1 - R)\Delta t] + \tag{12.25}$$
$$P_{R+1}(t)k(R + 1)(M - R)\Delta t + O((\Delta t)^2).$$

Dividing Equation (12.24) and Equation (12.25) by Δt and letting $\Delta t \to 0$, we obtain the following differential equations for $P_M(t)$ (viz. m(t) equals its initial size M at time t) and $P_R(t), R \neq M$

$$\frac{dP_M}{dt} = -kM, \ P_M(0) = 1 \tag{12.26}$$

$$\frac{dP_R}{dt} = k\{(R + 1)(M - R)P_{R+1} - R(M + 1 - R)P_R\}, P_R(0) = 0, \ R \neq M \tag{12.27}$$

Equations (12.26) and (12.27) can be solved recursively as in the previous section, viz.

$$P_M(t) = e^{-kMt} \tag{12.28}$$

$$P_{M-1}(t) = \frac{M}{M-2} e^{-kMt} \left[1 - e^{-k(M-2)t} \right].$$

etc.

Exercises

1. The expected size of the population m at t is defined as

$$\mu(t) = \sum_{k=0}^{M} k P_k(t). \tag{12.29}$$

   Evaluate $\mu(t)$.

2. Evaluate $P_{M-2}(t)$. Hint: use a computer algebra package.

3. A population of certain species consists of males and females. In a small colony, any male is likely to mate with any female in any time interval of length Δt with probability $k\Delta t + O((\Delta t)^2)$. Each such mating produces immediately one offspring which is equally likely to be male or female. If $M(t)$ and $F(t)$ denote the number of males and females in the population at time t, derive differential equations for $P_{M,F}(t)$.

12.4   QUEUING MODELS

Problem: A small bank has one teller. Recently, however, the manager received several complaints about the time that customers have to wait in line for service. To determine if a second teller is needed (or whether the current teller is slow), we are asked to formulate a model for the queue length and waiting time in the bank.

Assumptions and Abstractions:

Let $\ell(t)$ be the queue length at time t, viz. the number of customers in line including the customer that is being served. Thus, $\ell(t) = 0$ if nobody is in line or being served at time t. Let $P_\ell(t)$ be the probability that the queue length at t is ℓ. We assume the following regarding the arrival rate of new customers:

A1. The probability that one customer arrives at the queue in $[t, t + \Delta t]$ is $k\Delta t + O((\Delta t)^2)$ where k is a constant which is referred to as the mean arrival rate.

A2. The probability that more than one customer arrives to the queue in $[t, t + \Delta t]$ is $O((\Delta t)^2)$.

A3. If I_1 and I_2 are two disjoint time intervals, then the number of customers which arrive in I_1 does not affect the number of arrivals in I_2.

Similarly, regarding the service time, we make the following assumptions:

S1. If a customer is being serviced at time t, then the probability that the service is completed in $[t, t + \Delta t]$ is $s\Delta t + O((\Delta t)^2)$ where s is a constant representing the mean service time.

S2. The probability that service to more than one customer is completed in $[t, t + \Delta t]$ is $O((\Delta t)^2)$.

S3. If I_1 and I_2 are two disjoint time intervals, then the number of customers whose service is completed in I_1 does not affect the number of customers whose service is completed in I_2.

Remark: We note that by our assumption the probability of one arrival and one completion in $[t, t + \Delta t]$ is

$$[k\Delta t + O((\Delta t)^2)][s\Delta t + O((\Delta t)^2)] = O((\Delta t)^2). \tag{12.30}$$

Mathematical Model: To derive differential equations for $P_\ell(t)$, we consider now the conditions under which the queue length at $t + \Delta t$ is $\ell > 0$. These are:

1. The queue length at t is ℓ and there were no arrivals or departures during $[t, t + \Delta t]$.

2. The queue length at t is $\ell - 1$ and there was one arrival and no departures during $[t, t + \Delta t]$.

3. The queue length at t is $\ell + 1$ and there was one departure on $[t, t + \Delta t]$.

Other possible events are $O((\Delta t)^2)$ by the assumptions and as noted in Equation (12.30) Hence,

$$P_\ell(t + \Delta t) = P_\ell(t)(1 - k\Delta t)(1 - s\Delta t) + \tag{12.31}$$
$$P_{\ell-1}(t)(k\Delta t)(1 - s\Delta t) + P_{\ell+1}(t)(s\Delta t)(1 - k\Delta t) + O((\Delta t)^2).$$

Dividing by Δt and letting $\Delta t \to 0$ we obtain

$$\frac{dP_\ell}{dt} = kP_{\ell-1} + sP_{\ell+1} - (k+s)P_\ell, \quad \ell > 0. \tag{12.32}$$

Similarly, if $\ell = 0$ then

$$P_0(t + \Delta t) = P_0(t)(1 - k\Delta t) + P_1(t)(s\Delta t)(1 - k\Delta t) + O((\Delta t)^2). \tag{12.33}$$

Hence,

$$\frac{dP_0}{dt} = -kP_0 + sP_1. \tag{12.34}$$

Although this system of differential equations given by Equations (12.32) and (12.34) can be solved "in principle," it is not possible to solve it recursively as we did in the previous sections. Hence, we consider only the steady state solution for the queue, viz. the solution for $P_\ell(t)$ when $dP_\ell/dt = 0$. Under these conditions, Equations (12.32) and (12.34) reduce to an algebraic system of equations:

$$-kP_0 + sP_1 = 0 \tag{12.35}$$

$$kP_{\ell-1} + sP_{\ell+1} - (k+s)P_\ell = 0. \tag{12.36}$$

The solution of these equations is

$$P_\ell = \left(\frac{k}{s}\right)^\ell P_0. \tag{12.37}$$

The ratio $q = \frac{k}{s}$ is called the **"traffic intensity"** or the **"utilization factor"** of the queue.

Remarks: 1. If $q > 1$, then obviously the solution of Equation (12.37) is meaningless since $\sum P_\ell = \infty$ unless $P_0 = 0$.

2. If $q < 1$, then we must set $\sum P_\ell = 1$ which yields $P_0 = 1 - q$ and hence

$$P_\ell = q^\ell(1 - q).$$

For the expected queue length in the steady state, we obtain

$$E(\ell) = \sum_{\ell=0}^{\infty} \ell P_\ell = \sum \ell q^\ell(1 - q) = \tag{12.38}$$

$$(1 - q)q \sum \ell q^{\ell-1} = (1 - q)q \frac{d}{dq}\left(\sum q^\ell\right) =$$

$$(1 - q)\frac{q}{(1 - q)^2} = \frac{q}{1 - q}.$$

Exercises

1. Build a model for N server queue viz. a bank with one line and N tellers.

   Hint: Only S1 has to be modified as follows:

   S1': If customers are being served at time t, then the probability that service to at least one of them is completed in $[t, t + \Delta t]$ is

   a. $\ell s \Delta t + O((\Delta t)^2)$ if $\ell < N$

   b. $2s \Delta t + O((\Delta t)^2)$ if $\ell > N$.

2. Compute the steady state solution and the expected queue length for the queue in exercise 1.

3. What happens if $N = \infty$ in exercise 1 (infinite server queue)?

12.5   MARKOV CHAINS

We start with some general definitions and background.

Consider a system which can be in any of a finite or countable number of states, and let S denote this set of states (S is called the state space of the system). Obviously, we can identify S with

$$S = \{1, 2, \ldots, n\}$$

or

$$S = \{1, 2, \ldots, \infty\}.$$

Furthermore, assume that the system changes its state at discrete times t_i, $i = 1, 2, \ldots$ due to some process occurring in the system. For brevity, however, we refer to these times by $0, 1, 2, \ldots$. **Definition**: If the probability that the

system is in the state y at time $k + 1$ depends only on its state at time k, we shall then say that the system has the "**Markov property**" and refer to the system (or the process which governs the system) as a "**Markov chain**". **Definition**: Let a system S be a Markov chain. The conditional probabilities

$P(x, y)$ that the system in a state x at time k will transition to the state y at time $k + 1$ are called the **transition probabilities** of the system.

Example 1: Birth-Death Process.

Consider a Markov chain with state space $S = \{1, \ldots, n\}$ so that if the state of the system at time k is x, then its state at $k+1$ will be at $x+1, x-1$, and x. The transition function for this chain is given by

$$P(x, y) = \begin{cases} p_x & y = x+1 \\ q_x & y = x-1 \\ r_x & y = x \\ 0 & otherwise \end{cases}$$

where p_x, q_x and r_x are the probabilities of birth, death, or neither birth nor death during $[t_k, t_{k+1}]$. Also, note that $p_x + q_x + r_x = 1$. With these transition probabilities we define

$$w(x) = P_x(t_a < t_b) \tag{12.39}$$

viz. the probability that a system which starts in a state x ($a \leq x \leq b$) will arrive to the state a before arriving to the state b. (The states a, b are fixed.)

We observe that $w(a) = 1$ and $w(b) = 0$. Moreover, since a system in the state y can transit to the states $y+1$, $y-1$, and y with probabilities p_y, q_y, and r_y, we infer that

$$w(y) = q_y w(y-1) + p_y w(y+1) + r_y w(y). \tag{12.40}$$

Example 2: A gambler starts with m dollars and makes a series of one dollar bets against the House until either he has n dollars ($n > m$) or his money runs out. Obviously, this is a Markov chain with state space $\{0, \ldots, n\}$. Furthermore, if p and q are the probabilities of winning and losing at each bet ($p + q = 1$), then the transition probabilities of the chain are

$$P(x, y) = \begin{cases} p & y = x+1 \\ q & y = x-1 \\ 0 & otherwise \end{cases}$$

(note that $r_x = 0$).

Hence, Equation (12.40) implies that

$$w(y+1) - w(y) = \frac{q_y}{p_y}[w(y) - w(y-1)]. \tag{12.41}$$

Introducing

$$\alpha_y = \frac{q_1 \cdots q_y}{p_1 \cdots p_y}, \tag{12.42}$$

we can rewrite Equation (12.41) as

$$w(y+1) - w(y) = \frac{\alpha_y}{\alpha_{y-1}}[w(y) - w(y-1)] \tag{12.43}$$

and by recursion it follows that

$$w(y+1) - w(y) = \frac{\alpha_y}{\alpha_{y-1}}[w(y) - w(y-1)] = \tag{12.44}$$

$$\frac{\alpha_y}{\alpha_{y-2}}[w(y-1) - w(y-2)] = \ldots = \frac{\alpha_y}{\alpha_a}[w(a+1) - w(a)].$$

Summing Equation(12.44) over $y = a, \ldots, b-1$ we conclude that

$$w(a) - w(a+1) = \frac{\alpha_a}{A} \tag{12.45}$$

where $A = \sum_{y=a}^{b-1} \alpha_y$. Substituting Equation (12.45) in Equation (12.44) yields

$$w(y) - w(y+1) = \frac{\alpha_y}{A}. \tag{12.46}$$

Summing Equation (12.46) for $y = x, \ldots, b-1$ and using $w(b) = 0$ finally leads to

$$w(x) = P_x(t_a < t_b) = \frac{1}{A}\sum_{y=x}^{b-1} \alpha_y. \tag{12.47}$$

Example 2 (continued): Assume that $m = 100$, $n = 200$, $p = 0.4$, and $q = 0.6$. The probability that the gambler is ruined before having \$200 is

$$w(0) = P_{100}(t_0 < t_{200}) = \frac{\sum\limits_{y=100}^{199}(6/4)^y}{\sum\limits_{y=0}^{199}(6/4)^y} = \tag{12.48}$$

$$\frac{(3/2)^{200} - (3/2)^{100}}{(3/2)^{200} - 1} \approx 1;$$

i.e., the gambler has very little chance of winning.

Exercises

1. Consider a Markov chain with state space $S = \{0, 1, 2\}$ and transition matrix (i.e., a matrix whose entries are the probabilities $P(x, y)$) in the form

$$
P(x, y) = \begin{pmatrix} 0 & 1 & 0 \\ p & 0 & 1-p \\ 0 & 1 & 0 \end{pmatrix}
$$

(thus $P(0, 1) = 1$, $P(1, 0) = p$ $P(1, 1) = 0$, $P(1, 2) = 1-p$, etc.) Compute P^n for all n and interpret your results.

Hint: Show that $P^4 = P^2$

2. What strategy should the gambler in Example 2 use to maximize the chances of winning?

CHAPTER 13

Answers to Problems

CONTENTS

ANSWERS TO CHAPTER 1 EXERCISES

Sec. 2

All the problems in this section do not have a unique answer and in some cases they are "ill stated." We present some pointers.

Ex. 1. The problem statement does not specify the initial location of the "tour around the world." Thus if the this location is at one of the Earth's poles, It will take zero time to "go around the world." On the other hand if the tour is around the Earth equator, the answer will depend on some assumptions. Thus if we assume that one can walk $5km$ an hour for 8 hours a day then since the Earth radius is $6400km$ then

$$Time\,in\,days = \frac{(2\pi \cdot 6400)}{(5 \cdot 8)} \approx 1005\,days$$

Assuming a variation of 10% in the daily covered distance (i.e. the person covers $40 \pm 4kms$ a day) then the answer becomes 1005 ± 111 days.

Ex. 2. Use the Internet to find the size of the opening of the Mississippi to the ocean and the average speed of the water flow in the river.

Ex. 3. For this project assume that one is required to remove only the mountain "cone" from it surroundings (not sea level). Then find the average size of a dump truck. Also take into account that the loading of a dump truck is not precise.

Ex. 4. One has to estimate the average volume of the human body to find the dimension of the box. However, there are different levels of sophistication for the answer (e.g. taking into account gender size, size of people from different cultures, etc.)

Ex. 5. The distance of Earth from the Sun is approximately $150 \cdot 10^6$km (however the Earth orbit is somewhat elliptical. The angular velocity of Earth is (approximately) $2\pi/365$ radians per day.

Ex. 6. In this problem the concept of a "drop" has no exact value. Assuming that the volume of a drop is $1cm^3$, find the average depth of the Pacific ocean and it approximate size (as a portion of the Earth surface) to estimate the answer.

Ex. 7. Assume that the average size of a bookstore is $30'x30'x12'$ and the walls are covered completely with books whose average thickness is 1.5 inches and height is 15 inches. (This includes spaces between the shelves)

Ex. 8. The radius of an atom is about $10^{-8}\,cm$ assuming that an atom is a sphere (simplification) and a volume of a cell is $10^{-3}\,cm^3$ we can estimate the number of atoms in a cell.

Ex. 9. Use the data from ex. 4 and 8 to carry out this estimate. A more sophisticated model will use different cell sizes for different organs in the body.

Ex. 10. Estimate that each human on average uses two light bulbs and each bulb has a life time of 1000 ± 100 hours. Each day these light bulbs is used on average $8\,hours$ daily.

Ex. 11. Use the fact that the 'radius' of electron protons, and nuclei is about 10^{-13}cm. Assume that the number of atoms in this volume is around 10^{25} to carry out this estimate.

Ex. 12. Make sure to list the different uses of the car (e.g. driving to work, leisure, etc.) and the features that are important to you, (e.g. low maintenance, miles per gallon, etc.). Then follow each of the model building steps that were discussed in this section.

Ex. 13. Follow essentially the same steps as in Ex. 12.

Sec. 3

Ex. 1(a). In this problem $p_1 = SW$ and $p_2 = WW$. Using the same notation as in the text we have

$$R_1 = P(S \mid p_1)P(S \mid p_2) = 0,$$

$$R_2 = P(S \mid p_1)P(W \mid p_2) + P(W \mid p_1)P(S \mid p_2) = \frac{1}{2},$$

$$R_3 = P(W \mid p_1)P(W \mid p_2) = \frac{1}{2}.$$

Ex. 1(b). In this problem $p_1 = SW$ and $p_2 = SS$, therefore,

$$R_1 = \frac{1}{2}, \quad R_2 = \frac{1}{2}, \quad R_3 = 0.$$

Ex. 2. In this example there are nine possible geno types of beans:

$$(SS, CC), (SW, CC), (WW, CC), (SS, Cc), (SW, Cc), (WW, Cc),$$

$$(SS, cc), (SW, cc), (WW, cc).$$

Therefore the reproduction function $R(p_1, p_2)$ is a vector with nine components

$$R(p_1, p_2) = (R_1, \dots, R_9)$$

where

$$R_1 = P(S \mid p_1)P(S \mid p_2)P(C \mid p_1)P(C \mid p_2)$$

$$R_2 = P(S \mid p_1)P(W \mid p_2)P(C \mid p_1)P(C \mid p_2)+$$

$$P(W \mid p_1)P(S \mid p_2)P(C \mid p_1)P(C \mid p_2)$$

$$R_3 = P(W \mid p_1)P(W \mid p_2)P(C \mid p_1)P(C \mid p_2)$$

$$R_4 = P(S \mid p_1)P(S \mid p_2)P(C \mid p_1)P(c \mid p_2)+$$

$$P(S \mid p_1)P(S \mid p_2)P(c \mid p_1)P(C \mid p_2)$$

$$R_5 = P(S \mid p_1)P(W \mid p_2)P(C \mid p_1)P(c \mid p_2)+$$

$$P(W \mid p_1)P(S \mid p_2)P(C \mid p_1)P(c \mid p_2)+$$

$$P(S \mid p_1)P(W \mid p_2)P(c \mid p_1)P(C \mid p_2)+$$

$$P(W \mid p_1)P(S \mid p_2)P(c \mid p_1)P(C \mid p_2)$$

$$R_6 = P(W \mid p_1)P(W \mid p_2)P(C \mid p_1)P(c \mid p_2)+$$

$$P(W \mid p_1)P(W \mid p_2)P(c \mid p_1)P(C \mid p_2)$$

$$R_7 = P(S \mid p_1)P(S \mid p_2)P(c \mid p_1)P(c \mid p_2)$$

$$R_8 = P(S \mid p_1)P(W \mid p_2)P(c \mid p_1)P(c \mid p_2)+$$

$$P(W \mid p_1)P(S \mid p_2)P(c \mid p_1)P(c \mid p_2)$$

$$R_9 = P(W \mid p_1)P(W \mid p_2)P(c \mid p_1)P(c \mid p_2)$$

Ex. 3. Substitute $p_1 = (SS, CC)$ and $p_2 = (WW, cc)$ in the formulas of the previous exercise.

Ex. 4. The blood has six genotypes. These are AA, AO, BB, BO, AB and OO.

Therefore the reproduction function $R(p_1, p_2)$ is a vector with six components,

$$R(p_1, p_2) = (R_1, \ldots, R_6)$$

where

$$R_1 = P(A \mid p_1)P(A \mid p_2)$$

$$R_2 = P(A \mid p_1)P(O \mid p_2) + P(O \mid p_1)P(A \mid p_2)$$

$$R_3 = P(B \mid p_1)P(B \mid p_2)$$

$$R_4 = P(B \mid p_1)P(O \mid p_2) + P(O \mid p_1)P(B \mid p_2)$$

$$R_5 = P(A \mid p_1)P(B \mid p_2) + P(B \mid p_1)P(A \mid p_2)$$

$$R_6 = P(O \mid p_1)P(O \mid p_2).$$

ANSWERS TO CHAPTER 2 EXERCISES

Sec. 2.1

Ex. 1. Assuming that the projectile range is not large we can treat approximately the Earth as "flat". The gravitational force is then in the y-direction, and we have to modify only the second equation in (2.1.2) which is replaced by

$$\ddot{y} = -g(y), \quad g(y) = GM/(R+y)^2$$

where G is the gravitational constant, M is the Earth mass, and R is the Earth radius.

Ex. 2. At the maximum height $v_y = 0$. From Equation (2.1.7) we have

$$v_y = \dot{y} = -gt + v_0 \sin\theta$$

but $t = \frac{x}{v_0 \cos \theta}$ hence $v_y = 0$ when

$$\frac{gx}{v_0 \cos \theta} = v_0 \sin \theta,$$

i.e $x = \frac{v_0^2 \sin 2\theta}{2g}$.

Ex. 3. Assuming $x(0) = 0$ then, Equation (2.1.16) implies that $c_1 = m\dot{x}$. The value of c_2 can be obtained then by substituting $t = 0$ in Equation (2.1.17). Similarly from Equation (2.1.18) we have $c_3 = \dot{y}(0) + \frac{g}{b}$. c_4 is obtained by substituting $t = 0$ in Equation (2.1.19).

Ex. 4. The Taylor expansion of e^{-bt} around $bt = 0$ is

$$e^{-bt} = 1 - bt + \frac{(bt)^2}{2} - \cdots$$

Substituting the values of the constants c_i, $i = 1, 2, 3, 4$ in Equations (2.1.17) and (2.1.19) and approximating e^{-bt} by $1 - bt$ yields the desired results.

Ex. 5. The initial conditions in Equation (2.1.5) are modified as follows

$$\dot{x}(0) = v_0 \cos \theta + w_1, \quad \dot{y}(0) = v_0 \sin \theta + w_2.$$

Ex. 6. Two angles.

Ex. 7. $\frac{\pi}{4}$

Ex. 8. Since the pilot directs the plane always towards N and the position of the plane at time t is $(x(t), y(t))$, it follows that the angle of $\mathbf{u}$ with the x-direction satisfies $\tan \theta = -\frac{d-y(t)}{x(t)}$. Hence the components of $\mathbf{u}$ at this position are

$$u_x = |\mathbf{u}| \cos \theta, \quad u_y = |\mathbf{u}| \sin \theta.$$

The equations of motion of the plane are

$$\dot{x} = u_x - v, \quad \dot{y} = u_y.$$

Sec. 2

Ex. 1. $m\ddot{x} = -kx$ where $\frac{1}{k} = \frac{1}{k_1} + \frac{1}{k_2}$.

Ex. 2. If the displacement of the mass m from equilibrium is x. then $m\ddot{x} = -(k_1 + k_2)x$.

Ex. 3. The equivalent stiffness k of the springs in Ex. 1 satisfies

$$\frac{1}{k} = \frac{1}{k_1} + \frac{1}{k_2}$$

(parallel resistors). For Ex. 2, $k = k_1 + k2$ (resistors in series).

Ex. 8. From Equation (2.2.42) we have $x(0) = A\cos\phi$ and $\dot{x}(0) = -A\omega\sin\phi$

Ex. 9. Multiplying Equation (2.2.34) by $\dot{x}$ we obtain ($F_{ext} = 0$)

$$\frac{m}{2}\frac{d}{dt}(\dot{x}^2) + \frac{k}{2}\frac{d}{dt}(x^2) = -b\dot{x}^2$$

Hence

$$\frac{dE}{dt} = \frac{d}{dt}\left[\frac{m}{2}\dot{x}^2 + \frac{k}{2}x^2\right] = -b\dot{x}^2$$

This implies that the energy in the system decreases monotonically in time.

Ex. 15. Multiply the equation of motion by $\dot{x}$.

Sec. 3

Ex. 1. For a closed circuit in which R_1 and R_2 are in series the potential drop on each is respectively $V_1 = R_1 * i$ and $V_2 = R_2 * i$ where i is the current in the circuit. Hence

$$V_{ext} = V_1 + V_2 = (R_1 + R_2) * i.$$

It follows then that the equivalent resistance in the circuit is $R = R_1 + R_2$.

Similarly in a circuit where two resistors are connected in parallel to an external potential V_{ext} we have $V_{ext} = R_1 * i_1$ and $V_{ext} = R_2 * i_2$. However, the equivalent resistance in the circuit satisfies $V_{ext} = R * i$ and $i = i_1 + i_2$. Hence

$$\frac{1}{R} = \frac{1}{R_1} + \frac{1}{R_2}$$

Ex. 2.

$$e = L\frac{di_1}{dt} + R_3 * i_1 + R_2 * i_3, \quad \frac{Q}{C_1} + R_1 * i_2 + R_2 * i_3 = 0$$

$$i_1 = i_2 + i_3, \quad \frac{dQ}{dt} = i_2.$$

Ex. 4.

$$e = R_1 * i_1 + \frac{Q_1}{C_1} + L\frac{di_1}{dt} + R_2 * i_2 + \frac{Q_2}{C_2},$$

$$\frac{Q_2}{C_2} + R_2 * i_2 + L_2\frac{di_3}{dt} + \frac{Q_3}{C_3} = 0,$$

$$i_1 = i_2 + i_3, \quad \frac{Q_k}{dt} = i_k, \quad k = 1, 2, 3.$$

Sec. 4

Ex. 1. Assume that the rate of birth, death and self-cannibalization are proportional to the population size N(t) (with proportionality constants a, b, c). Also assume that there is no competition for plan food. Therefore

$$N(t + \Delta t) - N(t) = aN(t)\Delta t - bN(t)\Delta t - cN(t)\Delta t.$$

If competition for plant food exists, then additional term proportional to $N(t)^2$ has to be added to this equation.

Ex. 2.

$$\frac{dX}{dt} = \alpha_1 X - \beta_1 XY - \gamma_1 XZ, \quad \frac{dY}{dt} = \alpha_2 Y + \beta_2 XY - \gamma_2 YZ$$

$$\frac{dZ}{dt} = \alpha_3 Z + \beta_2 XZ + \gamma_3 YZ$$

Ex. 3.

$$\frac{dX}{dt} = \alpha_1 X - \beta_1 (X + Y)^2 + r, \quad \frac{dy}{dt} = \alpha_2 Y - \beta_2 (X + Y)^2$$

Ex. 4. The same number of people taken out of the S pool is added to the I pool.

Ex. 6. Substitute Equation (2.4.86) into Equation (2.4.85).

Ex. 7. Note that $\int \frac{dx}{x^2} dx = -\frac{1}{x} + C$.

Ex 10. Overproduction of Q depresses the production and price of fuel.

Sec. 5

Ex. 1. The following approximations were made: 1. Neglect the stretch in the rod, 2. Neglect friction in the joints (to the wall and the mass to the rod),3. Assume no air drag, 4."Small" angle of vibrations. 5. Neglect the mass of the rod and spring.

Ex. 3. Equation (2.5.106) has to be modified to include the force exerted by the spring

$$mL\dot{\theta}^2 = T - mg\cos\theta - k(L - L_0)$$

where L is a function of time and L_0 is the length of the rod and spring without the mass attached (T will not be present in this equation if we assume that the whole rod is replaced by the spring).

Ex. 5. When $f = -\frac{\mu}{r^2}$ Equation (2.5.129) becomes

$$\frac{d^2u}{d\theta^2} + u = \frac{\mu}{h^2}.$$

This is an inhomogeneous equation with constant coefficients which can be solved by standard methods.

Ex. 6. Substitute the approximations given in the text in Equations (2.5.137),(2.5.138) and keep only the linear terms in x_i, $i = 1, 2, 3, 4$ (Remember the x_i are assumed to be small).

Ex. 7. Use simple trigonometry to express the components of $e_{\mathbf{r}}$ and e_θ in terms of their Cartesian components.

Ex 8. For a one stage rocket with air friction Equation (2.5.131) is modified to,

$$\frac{d}{dt}(m(t)v(t)) = \frac{dm}{dt}(v(t) - u) - m(t)g(h(t)) - \alpha m(t)v(t).$$

If we neglect gravity this equation becomes

$$m(t)\frac{dv}{dt} = -u\frac{dm}{dt} - \alpha m(t)v(t)$$

changing the independent variable from time to the height of the rocket above ground and using the fact that $\frac{dh}{dt} = v$ yields

$$\frac{dv}{dh} = -u\frac{1}{m}\frac{dm}{dh} - \alpha.$$

Integrating this equation with respect to h leads to

$$v(h) = -u\ln(m(h)) - \alpha h + C$$

where C is a constant of integration which is determined by the initial conditions.

Ex. 9. Modify Equation (2.5.131) with u=u(t).

Ex. 11. The energy needed to overcome gravity can not be neglected for such a distance.

Ex. 13. Two different versions of Equation (2.5.131) modified to include frictional forces (see ex. 8) have to be used.

Ex. 14. $35,786$ km above sea level.

ANSWERS TO CHAPTER 3 EXERCISES

Sec. 1

Ex. 1. If $\lambda \neq 0$ is an eigenvalue of A then it is also an eigenvalue of A^T and $\frac{1}{\lambda}$ is an eigenvalue of A^{-1}.

Ex. 2. Observe that if $A\mathbf{v} = \lambda\mathbf{v}$ then

$$A^m\mathbf{v} = A^{m-1}(A\mathbf{v}) = \lambda A^{m-1}\mathbf{v} = \ldots = \lambda^m\mathbf{v}.$$

Ex. 3. (a) The eigenpairs are $(5, (-1/2, 1)^T)$ and $(-1, (1, 1)^T)$ (where T indicates a vector transpose).

   (b) $(0, (2, 1)^T$ and $(5, (-1/2, 1)^T)$

   (c) $(2, (1, 0, 1)^T) : (\sqrt{2}, (1, -\frac{2}{2+\sqrt{2}}, 1)^T): (-\sqrt{2}, (1, -\frac{2}{2-\sqrt{2}}, 1)^T)$

   (d) $(0, (-5, 4, 1)^T) : (2, (1, 0, 1)^T) : (2, (0, 0, 0)^T))$

Sec. 2

Ex. 1(a).

$$\frac{dx}{dt} = x_1, \quad \frac{dx_1}{dt} = -y_1 + 3xy + t$$

$$\frac{dy}{dt} = y_1, \quad \frac{dy_1}{dt} = -x_1 - 2x + 3y + e^{2t}$$

Initial conditions

$$x(1) = 1, \quad x_1(1) = 0, \quad y(1) = -2, \quad y_1 = 0.$$

Sec. 3

Ex. 1.

$$x = -\frac{3}{25}e^{5t} - e^{-t} - \frac{2t}{5} \mid \frac{3}{25}$$

$$y = -\frac{6}{25}e^{5t} + e^{-t} + \frac{t}{5} + \frac{6}{25}.$$

Ex. 2.

$$x = \frac{7}{30}e^{-4t} + \frac{3}{5}e^t + \frac{1}{6}e^{2t}.$$

Ex. 3.

$$x = -\frac{2}{5}e^t \cos t - \frac{4}{5}e^t \sin t - \frac{3}{5}e^{2t}$$

$$y = \left(\frac{2}{5}\cos t - \frac{6}{5}\sin t\right)e^t + \frac{3}{5}e^{-2t} - t.$$

Ex. 4.

$$x = \frac{3}{4}e^{2t} - \frac{1}{4}te^{2t} - \frac{3}{4} - \frac{t}{4}$$

$$y = \frac{5}{4}e^{2t} - \frac{1}{4}te^{2t} - \frac{5}{4} - \frac{t}{4}.$$

Sec. 4

Ex. 1(a). Exact solution y(1)=4.1548, Euler's solution y(1)=3.465.

Ex. 1(b). Exact solution y(1)=3.7844 Euler's solution y(1)=-15.32. (Solution has a singularity around x=0.7).

Ex. 1(c). Exact solution y(2)=-0.27634 Euler's solution y(2)=-0.1509.

Ex. 1(d). Exact solution y(1)=-0.89711 Euler's solution y(1)=-0.5700.

Ex. 1(e). Exact solution y(π) = 7.3891 Euler's solution y(π)=4.319.

Ex. 2.

(a) 3.781 (b) -108.6 (c)-0.2024 (d) -.7240 (e) 5.587

Ex. 3. (a)4.155 (7th order Taylor expansion) (b)-25.60 (order 5) (c) -.2763 (order 11) (d) -.8969 (order 16) (e) 7.389 (order 56).

Ex. 4. (a) 4.339 (b)= 3.622.

Ex. 6. (a) 8.620.

Ex. 7. (a)analyticaly u(1) = -3.389451301, v(1) = -2.492765599, Euler u(1)=-2.14016, v(1)=-1.56672,

(b) analytically u(1) = -1., v(1) = 0, Euler same as above.

Ex. 8. Extended Euler: 7.626026496, Euler 7.9863.

Sec. 3.5

Ex. 1. Use Taylor expansion around x_i

$$f_{i-1} = f_i - hf'(x_i) + O(h^2)$$

hence

$$f'(x_i) = \frac{f_i - f_{i-1}}{h} + O(h).$$

Ex. 2. Use Taylor expansion up to order 5 (i.e. with a tail of order $O(h^6)$) around x_i to express f_{i+2} etc. and then sum with the appropriate weights (as indicated by the formula in the text).

Ex. 3. Use the same strategy as indicated for Ex 2.

Ex. 4.

$$f_i'' = \frac{-f_{i-2} + 16f_{i-1} - 30f_i + 16f_{i+1} - f_{i+2}}{12h^2} + O(h^4).$$

Ex. 6.
$$f_i'' = \frac{f_{i+2} - 2f_{i+1} + 2f_{i-1} - f_{i-2}}{2h^3} + O(h^2).$$

Ex. 7.
$$f'(x) = \frac{h_2^2 f(x + h_1) + (h_1^2 - h_2^2)f(x) - h_1^2 f(x - h_2)}{h_2 h_1 (h_2 + h_1)}.$$

Sec. 3.6

Ex. 1. y(1)= 0.17141.

Ex. 2. y(1)= 0.78767.

Ex. 3. y(1)= -0.66741.

Ex. 4. y(2)=1.66677.

Ex. 5. y(1)=0.

Ex. 6. (1) 0.16612 (2) 0.78330 (3) -.6669 (4) 1.6644 (5) 0.

Ex. 7. (1) 0.1644 (2) 0.7819 (3) -.6667 (4) 1.664 (5) 0.

Ex. 8. u=1, v=0.

Ex 9. modified Euler:u=2.46345, v=.55591

Analytic : u=2.52800, v=.55314.

Ex 10. $u = 1.573312596 \cdot 10^{232}$, $v = 3.966500468 \cdot 10^{115}$

Sec 3.7

Ex. 2. u(0.2) = 1.945386550

u(0.5) = 2.707238530

u(0.8) = 2.978086205

ANSWERS TO CHAPTER 4 EXERCISES

Sec. 1

Ex. 1. (a)$N(t) = \frac{a}{b+Cae^{-at}}$, where C is an integration constant.

(c) $N(t) = \frac{a}{-b+Cae^{-at}}$, where C is an integration constant.

Ex 2. (a) The steady state $N = \frac{a}{b}$ is asymptotically stable. The steady state $N = 0$ is unstable.

(c) The steady state $N = -\frac{a}{b}$ is asymptotically stable. The steady state $N = 0$ is unstable.

Ex. 3.

(a) The steady states are $N = 0$ and

$$N = \frac{-\beta \pm \sqrt{\beta^2 - 4\alpha\gamma}}{2\gamma}$$

The steady state in between the other two states is stable.

Sec. 3

Ex. 1(a). A steady state of the system is $x = 2$, $y = 1$. To move this steady state to the origin we make the transformation

$$u = x - 2, \quad v = y - 1$$

the new system is

$$\frac{du}{dt} = (u + 2)u(v - 1), \quad \frac{dv}{dt} = (u + 4)v^2$$

Ex. 2(b). Trajectories

$$x = C_1 e^{-2t} - C_2 e^{-4t}$$

$$y = C_1 e^{-2t} + C_2 e^{-4t}$$

Integral curves

$$y(x) = \frac{1 - 2C_1 x \pm \sqrt{1 - 8C_1 x}}{C_1}$$

Ex. 2(c). To find the integral curves note that

$$x\frac{dx}{dt} + y\frac{dx}{dt} + z\frac{dx}{dt} = 0$$

and

$$m\frac{dx}{dt} + n\frac{dx}{dt} + k\frac{dx}{dt} = 0.$$

Hence along an integral curve

$$\frac{d}{dt}(x^2 + y^2 + z^2) = 0, \quad \frac{d}{dt}(mx + ny + kz) = 0.$$

This mean that on an integral curve of the system

$$x^2 + y^2 + z^2 = c_1, \quad mx + ny + kz = c_2.$$

This implies that the integral curve are the intersections of a sphere and a plane.

Ex. 3(c). The trajectories of this equation can be expressed in terms of elliptic functions. To find the integral curves we rewrite the Equation as

$$\frac{d\theta}{dt} = v, \quad \frac{dv}{dt} = -\nu^2 \sin \theta.$$

Hence

$$\frac{d\theta}{dv} = \frac{v}{-\nu^2 \sin \theta}$$

Therefore

$$2\nu^2 \cos \theta - v^2 = C$$

where C is a constant.

Sec. 4

Ex. 1. Observe that a solution of the equation is of the for $e^{\lambda t}$ where λ is a root of the polynomial $p(\lambda)$. Ex. 3. The point $(0,0)$ is a critical point of this system.

$$f_1 = y(1 + \cos x), \quad f_2 = x(1 + \sin y)$$

Hence the Jacobian at $(0,0)$ is

$$J(0, 0) = \begin{pmatrix} 0 & 1 \\ 1 & 0 \end{pmatrix}. \tag{13.1}$$

The determinant of the Jacobian is -1 and the system is linearizable around this point. Using Taylor expansions around $(0,0)$ we find that the linearized system is

$$\frac{dx}{dt} = 2y, \quad \frac{dy}{dt} = x$$

Exs. 4,5. Rewrite these equations as a system of two first order equation.

Sec. 5

Ex. 1. Inward spiral for $b > 0$. Outward spiral for $b < 0$.

Ex. 3. Unstable node.

Ex. 5. Saddle point (one direction is stable another is unstable).

Ex. 6. Saddle point.

Sec. 6.

Ex. 1. If $x > 0$ then the integral is positive and $F(x, y) > 0$. If $x < 0$ then

$$\int_0^x f(t)dt = -\int_x^0 f(t)dt$$

and since $f(t) < 0$ on [-b,0] the integral is positive.

Ex. 2. Follow similar steps as in example 4.6.4.

Ex. 3. From Ex. 1 we know that $F(x, \dot{x})$ is positive definite. Now apply Liapounov theorem.

Ex. 4. First rewrite the pendulum equation as a system

$$\dot{x} = y, \quad \dot{y} = -\omega^2 \sin x.$$

To determine the stability of (0,0) define on $[-\pi/2, \pi/2]$

$$F(x, y) = \frac{1}{2}y^2 + \int_0^x \sin t \, dt = \frac{1}{2}y^2 + (1 - cos(x))$$

and apply Liapounov theorem. This yields $\nabla F = gradF \cdot G = 0$.

Ex. 5. The results follow directly from Liapounov theorem.

Ex. 6. Use $F(x, y) = x^2 + y^2$.

Sec. 7

Ex. 1. When $|u| < 1$ the effective frictional term is positive and therefore u is decreasing. On the other hand if $|u| > 1$ the effective frictional term is negative and u is increasing. Therefore the limit cycle $u = 1$ is unstable.

Ex. 2. In polar coordinates, using Equations (4.7.87) and (4.7.88) we obtain,

$$\frac{dr}{dt} = rh(r) \quad \frac{d\theta}{dt} = 1$$

This implies that $\frac{dr}{dt} = 0$ whenever $h(r) = 0$.

Ex. 3.

(a) For the limit cycle $r = 1$: If $1 < r < 3$ then $\frac{dr}{dt} > 0$. If $r < 1$, $\frac{dr}{dt}$ is also positive. Hence this limit cycle is (one sided) unstable.

For the limit cycle $r = 3$: If $3 < r < 4$ then $\frac{dr}{dt} < 0$ on the other hand if $< 1r < 3$ then, $\frac{dr}{dt} > 0$. hence this limit cycle is asymptotically stable. (b) The limit cycle r=2 is unstable.

Ex. 4. In polar representation we have

$$\frac{dr}{dt} = -r(h(r) - 2).$$

Therefore limit cycles exist at the points where $h(r) = 2$.

Ex. 5. (a) Limit cycle exist only for n=2.

(b) Limit cycle exists for $r = 2k$ where k is an integer.

Ex. 6. When $b > 0$ the effective friction coefficient bx^2 is positive (for all values of $x(t)$, and therefore the steady state (0,0) is asymptotically stable. When $b < 0$ the effective friction coefficient is negative and (0,0) is unstable.

ANSWERS TO CHAPTER 5 EXERCISES

Sec. 2

Ex. 1. (a) There is competition for (plant) food within each species and each species cannibalize the other.

(b) One steady state is at (0,0) the others are a combination of the roots of the factors e.g. $x = 0$ and $y = a$.

The linearization of the system around (0,a) is

$$\dot{x} = xa(1 - \lambda), \quad \dot{z} = a(z - 2x)$$

where $z = y - a$. The bifurcation point is $\lambda = 1$.

Ex. 2. The system has two steady states $x = 0$ and $x = a + \lambda$. The linearization around $x = 0$ is

$$\dot{x} = x(\lambda + a)$$

and the bifurcation point is $\lambda = -a$. This is a transcritical bifurcation. The linearization around $x = a + \lambda$ is

$$\dot{z} = z(a + \lambda)$$

and the bifurcation point is $\lambda = -a$. This is a transcritical bifurcation.

Ex. 3. The linearized system around (0,0) is

$$\dot{x} = \lambda x + y, \quad \dot{y} = -x + \lambda y$$

The eigenvalues of this system are $\lambda = \pm i$.

Ex. 4. The steady state (0,0) is asymptotically stable for $\lambda < 0$ and unstable when $\lambda > 0$.

Ex. 6. The equation is equivalent to the system

$$\dot{x} = y, \quad \dot{y} = -(\lambda + x^2)y - 2x + x^3.$$

The linearization around (0,0) is

$$\dot{x} = y, \quad \dot{y} = -(1 + \lambda^2)y - 2x$$

The eigenvalues of this system are $\pm i$.

ANSWERS TO CHAPTER 6 EXERCISES

Sec. 3

Ex. 1. Writing the desired solution in the form

$$y = y_0 + \epsilon y_1 + \epsilon^2 y_2 + \dots$$

We have to the zeroth order of ϵ

$$\frac{dy_0}{dx} + 3y_0^2 = 0, y_0(0) = 1$$

and to order ϵ

$$\frac{dy_1}{dx} + 6y_0 y_1 = -y_0, \quad y_1(0) = 0.$$

Hence

$$y_0 = \frac{1}{3x + 1}, \quad y_1 = -\frac{1}{2} \frac{x(3x + 2)}{(3x + 1)^2}.$$

Ex. 2. Writing the desired solution in the form

$$u = u_0 + \epsilon u_1 + \epsilon^2 u_2 + \dots$$

We have to the zeroth order of ϵ

$$\frac{d^2 u_0}{dt^2} + u_0 = a$$

and to order ϵ

$$\frac{d^2 u_1}{d\theta^2} + u_1 = 2a u_0^2.$$

Hence

$$u_0 = a + C \cos(\theta + \phi),$$

where C, ϕ are constants. This solution has a period of 2π. The solution for u_1 is

$$u_1 = C_1 \cos(\theta + \phi_1) - \frac{a}{6}(C^2 \cos(2\theta + 2\phi) - \qquad (13.2)$$
$$3Ca \cos(\theta + \phi) - 3C^2 - 6a^2 - 6Ca\theta \sin(\theta + \phi)).$$

Assuming $a = 1$, $C = 1$, $C_1 = 0$(this term can be absorbed in u_0), $\phi = \phi_1 = 0$ and $\epsilon = 0.01$ plot $u_0 + \epsilon u_1$ on the interval $[0, 200\pi]$ to estimate the deviation of the period from 2π.

Remark: Observe that this solution contains a (secular)term $\theta \sin(\theta + \phi)$ which is unbounded as θ grows.

Ex. 3. The equations for y_0 and y_1 are

$$\frac{d^2 y_0}{dx^2} + 4y_0 = 0, \quad y_0(0) = 1, \frac{dy_0}{dx}(0) = 0,$$

$$\frac{d^2 y_1}{dx^2} + 4y_1 = y_0^2, \quad y_1(0) = 1, \frac{dy_1}{dx}(0) = 0.$$

To order ϵ the solution is

$$y_0 = \cos 2t, \quad y_1 = \frac{11}{12} \cos 2t - \frac{1}{24} \cos 4t + \frac{1}{8}.$$

Sec. 4

Ex. 1. Using the expansions in Equations (6.4.25) and (6.4.26) we have

$$\left(\frac{d^2 y_0(s)}{ds^2} + \epsilon \frac{d^2 y_1(s)}{ds^2} + \ldots \right) (1 - 2\epsilon a_1 + \ldots) + \quad (13.3)$$

$$\epsilon \left[\frac{dy}{ds} (1 - \epsilon a_1 + \ldots) \right]^2 + k^2 ((y_0(s) + \epsilon y_1(s) + \ldots)).$$

To order ϵ we obtain the following two equations

$$\frac{d^2 y_0}{ds^2} + k^2 y_0 = 0, \quad y_0(0) = 1, \frac{dy_0}{dt}(0) = 0$$

$$\frac{d^2 y_1}{ds^2} + k^2 y_1 = \left(\frac{dy_0}{ds} \right)^2 + 2a_1 \frac{d^2 y_0}{ds^2}, \quad y_1(0) = 0, \frac{dy_1}{dt}(0) = 0$$

where a_1 has to be determined so that the secular terms disappear. Hence $y_0 = \cos ks$ and

$$y_1 = C_1 \cos ks + C_2 \sin ks - a_1 \cos ks - a_1 ks \sin ks - \frac{1}{6} \cos 2ks - \frac{1}{2}.$$

We see that in order to get rid of the secular term we must have $a_1 = 0$. Applying the initial conditions we finally have

$$y_1 = \frac{2}{3} \cos ks - \frac{1}{6} \cos 2ks - \frac{1}{2}.$$

Sec. 5

Ex. 1. In the outer region one has to satisfy to order ϵ

$$(1 + ax) \frac{dy_0}{dx} + ky_0 = 0, \quad y(1) = 1$$

$$(1 + ax)\frac{dy_1}{dx} + ky_1 = -\frac{d^2y_0}{dz^2}, \quad y_1(0) = 0$$

whose solution is

$$y_0 = D_1(1 + ax)^{-k/a}, \quad y_1 = \left(D_2 + \frac{D_1k(a + k)}{(2a(1 + ax)^2} + \right)(1 + ax)^{-k/a}$$

where

$$D_1 = (1 + a)^{k/a}, \quad D_2 = -\frac{D_1k(a + k)}{2a}.$$

Performing the stretching transformation $z = \epsilon^b x$ in the inner region the equation becomes

$$\epsilon^{1+2b}\frac{d^2y}{dz^2} + \epsilon^b(1 + a\epsilon^{-b}z)\frac{dy}{dz} + ky = 0, \quad y(0) = 0.$$

Hence we must choose $b = -1$. The equation for the inner region becomes

$$\frac{d^2y}{dz^2} + (1 + a\epsilon z)\frac{dy}{dz} + k\epsilon y = 0, \quad y(0) = 0.$$

If we write $y = y_0 + \epsilon y_1$ we obtain to order ϵ the following system of equations

$$\frac{d^2y_0}{dz^2} + \frac{dy_0}{dz} = 0, \quad y_0(0) = 0$$

$$\frac{d^2y_1}{dz^2} + \frac{dy_1}{dz} + az\frac{dy_0}{dz} + ky_0 = 0, \quad y_1(0) = 0.$$

Hence

$$y_0 = C_1z, \quad y_1 = C_2(e^{-z} - 1) - C_1(a + k)z(z - 2)$$

To determine the constants C_1, C_2 apply Equation (6.5.52).

ANSWERS TO CHAPTER 7 EXERCISES

Sec. 1

Ex. 1.

$$\frac{1}{k}\frac{\partial u}{\partial t} = \frac{\partial^2 u}{\partial x^2} + r(x, t).$$

Ex. 4.

$$\frac{1}{k}\frac{\partial u}{\partial t} = \frac{\partial}{\partial x}\left(A(x)\frac{\partial u}{\partial x}\right).$$

Ex. 7.

$$\frac{1}{k}\frac{\partial u}{\partial t} = \nabla^2 u = \frac{\partial^2 u}{\partial x^2} + \frac{\partial^2 u}{\partial y^2} + \frac{\partial^2 u}{\partial z^2}.$$

Sec. 2

Ex. 1.
$$\frac{1}{c^2}\frac{\partial^2 u}{\partial t^2} = \frac{\partial^2 u}{\partial x^2} + g.$$

Ex. 2.
$$\frac{\partial^2 u}{\partial t^2} = \rho\frac{\partial^2 u}{\partial x^2} - ku - b\frac{\partial u}{\partial t}.$$

Ex. 5.
$$\frac{1}{c^2}\frac{\partial^2 u}{\partial t^2} = \nabla^2 u.$$

Ex. 7. Let $w = x - ct$, $z = x + ct$.

Sec. 3

Ex. 3.
$$h_{tt} = g(ax + b)h_{xx} + agh_x.$$

Sec. 4

Ex. 2.

Mass $\approx LC$, spring constant $\approx RG$, coefficient of friction $\approx LG + RC$

Sec. 5

Ex. 1. Zero inside the cavity; M/r outside the cavity.

Ex. 4. $\frac{1}{r}\frac{\partial}{\partial r}\left(r\frac{\partial u}{\partial r}\right) + \frac{1}{r^2}\frac{\partial^2 u}{\partial r^2} = 0.$

Ex. 7. Substitute these functions in Laplace equation.

Sec. 6

Ex. 1. $u \approx \frac{1}{\rho}.$

Ex. 3.
$$\frac{\partial u}{\partial t} + \frac{\partial(\rho u)}{\partial x} = -a.$$

Ex. 5.
$$\frac{\partial \rho_1}{\partial t} + \frac{\partial(\rho_1 u_1)}{\partial x} = -a_1(\rho_1) + a_2(\rho_2)$$
$$\frac{\partial \rho_2}{\partial t} + \frac{\partial(\rho_2 u_2)}{\partial x} = -a_2(\rho_2) + a_1(\rho_1).$$

ANSWERS TO CHAPTER 8 EXERCISES

Sec. 1.1.

Ex. 1.
$$\frac{400}{\pi} \sum_{n=1,3,5,\ldots}^{\infty} \frac{1}{n} exp\left(\frac{-kn^2\pi^2 t}{L^2}\right) \sin\frac{n\pi x}{L}$$

Ex. 3.
$$\pi \sin 30 \sum_{n=1}^{\infty} \frac{(-1)^n n}{900 - n^2\pi^2} exp\left(\frac{-n^2\pi^2 t}{9 \cdot 10^6 RC}\right) \sin\left(\frac{n\pi x}{3000}\right)$$

Ex. 5.
$$\frac{3200}{\pi^3} \sum_{n=1,3,5,\ldots}^{\infty} \frac{1}{n^3} exp\left(1 - \frac{kn^2\pi^2}{400}\right) t \sin\left(\frac{n\pi x}{20}\right)$$

Ex. 8.
$$\frac{1}{1250\pi^2} \sum_{n=1,3,5,\ldots}^{\infty} \frac{1}{n^2} sin\left(\frac{n\pi x}{2}\right) \sin 5000 n\pi t$$

Ex. 9.
$$\frac{80L}{\pi c} \sum_{n=1,3,5,\ldots}^{\infty} \frac{1}{n} \sin\left(\frac{n\pi x}{2L}\right) \sin\left(\frac{n\pi ct}{2L}\right)$$

Ex. 12.
$$\frac{4}{a\pi^2} \sum_{n=1,3,5,\ldots}^{\infty} \frac{(-1)^{(n-1)/2}}{n^2} \sin(0.01) n\pi \sin n\pi x \sin an\pi t$$

Sec 1.2

Ex. 1. $\frac{1}{3} \sum_{n=1}^{\infty} \frac{(-1)^{(n+1)}}{n} \left(\frac{\rho}{c}\right)^{6n} \sin 6n\theta$

Ex. 3. $\frac{40}{\pi} \sum_{n=1,3,5,\ldots}^{\infty} \left(\frac{r}{20}\right)^n \frac{\sin n\theta}{n}$

Ex. 12. $\frac{200}{a^2\pi^4} \sum_{n=1}^{\infty} \frac{1+n^2\pi^2}{n^4} \left[\int_0^{100} \sigma h(\sigma) \sin\frac{n\pi\sigma}{100} d\sigma\right] \frac{\sin(n\pi\rho/100)}{\rho} \sin\frac{an\pi t}{100}$

Ex. 13. $\frac{1}{\rho^2}(\rho^2 p_\rho)_\rho = \frac{1}{a^2} p_{tt}, \quad p(a,t) = 1, \quad \frac{\partial p}{\partial \rho}(b,t) = 0, \quad p(\rho,0) = 0, \quad \frac{\partial p}{\partial t}(\rho,0) = \rho$

Sec 1.3

Ex. 1. $\sum_{n=1,3,5,\ldots}^{\infty} \sin\frac{n\pi x}{L}\left[\frac{40}{n\pi}\cos\frac{an\pi t}{L} - \frac{20L}{an^2\pi^2}\sin\frac{an\pi t}{L}\right]$

Ex. 3.
$$\frac{L^2}{6} + 2t + \frac{2L^2}{\pi^2}\sum_{n=1}^{\infty}[(-1)^n(1_L) - 1]\cos\frac{n\pi x}{L}\cos\frac{an\pi t}{L}$$

Ex. 6.

$$2 - 0.0019x + \frac{1}{5\pi} \sum_{n=1}^{\infty} \frac{(-1)^n - 20}{n} \sin\frac{n\pi x}{100} exp\left(\frac{-n^2\pi^2 t}{10^6 RC}\right)$$

Ex. 8.

$$\Phi(x) = \frac{x}{2a^2}\left[L^2 - \frac{x^2}{3}\right], \quad z_{tt} = a^2 z_{xx}, \quad z(0,t) = 0, \tag{13.4}$$

$$\frac{\partial z}{\partial x}(L,t) = 0, \quad z(x,0) = f(x) - \Phi(x), \quad \frac{\partial x}{\partial t}(x,0) = 0$$

Ex. 10.

$$\Phi(x) = \frac{\sin x}{a^2} - x\left(1 + \frac{\cos L}{a^2}\right), \quad w_{tt} = a^2 w_{xx}, \tag{13.5}$$

$$w(0,t) = 0, \quad w_x(L,t) = 0, w(x,0) = -\Phi(x), \quad w_t(x,0) = f(x)$$

Sec. 4.2

Ex. 1. Use the analog of Equation (8.4.16).

Ex. 2. To remove the singularity at $r = 0$ multiply the equation by r^2. Avoid $r = 0$ by using a circle of radius $\epsilon \ll 1$ around this point.

Ex. 3. A particular solution of the equation is

$$r^3\left(\frac{1}{18} + \frac{\cos 2\theta}{10}\right)$$

Sec. 4.3

Ex. 3. Use the same finite difference formulas used in this section.

Ex. 5. To obtain a finite difference formula for the mixed derivative at the grid point (i,j) use the following strategy

$$\left(\frac{\partial^2 u}{\partial t \partial x}\right)_{i,j} = \frac{\partial}{\partial t}\left(\frac{\partial u}{\partial x}\right)_{i,j} = \frac{\partial}{\partial t}\left[\frac{(u_{i+1,j} - u_{i-1,j})}{2h}\right] = \tag{13.6}$$

$$\frac{1}{2h}\left[\frac{\partial}{\partial t}u_{i+1,j} - \frac{\partial}{\partial t}u_{i-1,j}\right] = \frac{1}{4h^2}(u_{i+1,j+1} - u_{i+1,j-1} + u_{i-1,j+1} - u_{i-1,j-1}).$$

ANSWERS TO CHAPTER 9 EXERCISES

Sec. 2

Ex. 2. First observe that we can choose to maximize the square of the distance from the origin rather that the distance itself since when the square of the distance is maximum the distance is also at a maximum.

The problem is then to maximize $z = x^2 + y^2$ subject to the constraint $w = x^2 - 2xy + 2y^2 = 1$. To this end we consider the function

$$f(x, y, \lambda) = x^2 + y^2 - \lambda(x^2 - 2xy + 2y^2).$$

Hence for extremum we must have

$$2x - 2\lambda x + +2\lambda y = 0, \quad 2y + 2\lambda x + 4\lambda y = 0, \quad x^2 - 2xy + 2y^2 = 1$$

which can solved for x, y (and λ).

Ex. 3. The surface area and volume of the cylinder are respectively

$$A = 2\pi r^2 + 2\pi rh, \quad V = \pi r^2 h$$

where r is the radius of the bottom and h the height. since A is constant we can solve for h and substitute in the expression for V. The problem is then to find the max of a one variable function. Otherwise one can use Lagrange multipliers by forming the function

$$f(r, h, \lambda) = \pi r^2 h - \lambda(2\pi r^2 + 2\pi rh).$$

Ex. 4. Form the function

$$f(x, y, z, \lambda, \mu) = xz + yz - \lambda(x^2 + y^2) - \mu yz.$$

Ex. 5(b). The perimeter and area of the rectangle are $P = 2(a + b)$, $A = ab$. Hence $b = P/2 - a$ and therefore

$$A = a(P/2 - a).$$

The max of this function is obtained when $a = P/4$ and therefore $a = b$.

Sec. 3

Ex. 1. According to Fermat principle we want to minimize the time it takes light between $\mathbf{x_1}$ and $\mathbf{x_2}$. This time is given by the integral

$$T = \int_{t_1}^{t_2} dt.$$

However $dt = \frac{ds}{v(\mathbf{x})}$ where ds is the distance traveled during the time interval dt and $v(\mathbf{x})$ is the light speed at the point. Hence we want to minimize

$$T = \int_{t_1}^{t_2} dt = \int_{\mathbf{x_1}}^{\mathbf{x_2}} \frac{ds}{v(\mathbf{x})}.$$

If we define the index of refraction in the medium as $n(\mathbf{x}) = \frac{c}{v(\mathbf{x})}$ (where c is the velocity of light in the vacuum) then

$$T = \frac{1}{c} \int_{\mathbf{x_1}}^{\mathbf{x_2}} n(\mathbf{x}) ds.$$

Sec. 4

Ex. 2. Using Equation (9.4.5) it follows that (since f is independent of $\mathbf{x} = (x, y, z)$ that

$$\dot{\mathbf{x}} = \mathbf{C} = (C_1, C_2, C_3).$$

Hence

$$\mathbf{x} = (C_1 x + D_1, \; C_2 y + D_2, \; C_3 z + D_3),$$

where $\mathbf{C}, \mathbf{D}$ are constants of integration.

Ex. 3. The differential equation for the geodesics is

$$\frac{d^2 x^\lambda}{dt^2} + \Gamma^\lambda_{\mu\nu} \frac{dx^\mu}{dt} \frac{dx^\nu}{dt} = 0,$$

where $\Gamma^\lambda_{\mu\nu}$ are the Christoffel symbols of the metric tensor g_{ij}. (For the definition of the Christoffel symbols consult a book on differential geometry.)

Sec. 5

Ex 1. The potential $V(w)$ has to be modified to include (in the integral) an additional term $-\rho g w$.

Exs. 2 and 3. Solutions are given in the text.

Sec. 6

Ex. 2. Following the example in the text (Equations. $(9.6.4) - (9.6.6)$ we consider small variations near the optimal solution ψ

$$\bar{\psi}(x) = \psi(x) + \epsilon \eta(x).$$

Substituting this in the expressions for I and J yields,

$$I(\epsilon) = \int_{x_1}^{x_2} [(\psi + \epsilon\eta)^2 (\psi_x + \epsilon\eta_x) + (\psi_x + \epsilon\eta_x)^2] dx,$$

$$J(\epsilon) = \int_{x_1}^{x_2} (\psi + \epsilon\eta)^2 dx = 1.$$

Using Lagrange multipliers we generate the function

$$K(\epsilon) = I(\epsilon) + \lambda J(\epsilon).$$

The conditions for extremum become

$$\frac{\partial K(\epsilon, \lambda)}{\partial \epsilon}\bigg|_{\epsilon=0} = 0, \quad J(\epsilon = 0) = 1.$$

This leads to

$$\int_{x_1}^{x_2} ((2\psi\psi_x + 2\lambda\psi)\eta(x) + (\psi^2 + 2\psi_x)\eta_x))dx = 0, \quad \int_{x_1}^{x_2} \psi^2 dx = 1.$$

Now use integration by parts to convert the integral containing η_x to one over η to obtain a differential equation for ψ. The constraint J is a normalization for ψ.

Sec. 7

Ex. 2 With drag Equation (9.7.6) is replaced by

$$\dot{x}(\tau) = -g\tau - \alpha x(\tau) + F(\tau)$$

where

$$\int_0^\tau u(s)ds.$$

The solution of the equation for x is

$$x(t) = Ce^{-\alpha t} - \frac{gt^2}{2} + e^{-\alpha t} \int_0^t F(\tau)e^{-\alpha\tau}d\tau.$$

Then follow the steps in the text.

Ex. 3 This requirement replaces the constraint given by Equation (9.7.2) by requiring that

$$J(T) = \int_0^T f(t)dt$$

is minimum, where $f(t)$ is the fuel consumed at time t.

ANSWERS TO CHAPTER 10 EXERCISES

Sec. 3.

Ex. 1. Let V be the volume occupied a body of arbitrary shape. The total force acting on the body due to the pressure at at each point on its surface S is

$$\mathbf{F} = -\int_S P\mathbf{n}dS$$

where $\mathbf{n}$ is the normal to the surface (The minus sign is due to the fact that $\mathbf{n}$ is in the outward direction from the surface). However using Gauss Theorem Using Equation (A.2)in the appendix of this chapter we have

$$\mathbf{F} = \int_V gradPdV = \rho gV\mathbf{k},$$

since $gradP = (0, 0, -\rho g)$ (remember that $gradP$ is in the direction in which the pressure is increasing).

Ex. 2. Use the vector identity

$$\nabla \times (\phi\nabla\psi) = \phi\nabla \times (\nabla\psi) + \nabla\phi \times \nabla\psi.$$

Ex. 3. A well known theorem from vector calculus states that if the curl of a vector field $\mathbf{u}$ is zero then (subject to some restrictions on the geometry of the domain) the vector field is conservative and there exists a function ϕ so that $\mathbf{u} = grad\phi$

Ex. 4. Show that ψ satisfies the continuity equation.

ANSWERS TO CHAPTER 12 EXERCISES

Sec. 2.

Ex. 1.

Modify Equation (12.1.5) to read

$$P(N \to N - 1) = Nk\Delta t + p(\Delta t)$$

Eq (12.1.7) becomes

$$P_N(t + \Delta t) = P_N(1 - kN\Delta t) + P_{N+1}k(N + 1)\Delta t.$$

Observe however that $N \geq 0$.

Ex. 2.

Assume that the processes of birth and death are independent. On the time interval Δt the probability of birth is

$$P(N \to N+1) = Nk_1\Delta t + p_1(\Delta t),$$

and the probability of death is

$$P(N \to N-1) = Nk_2\Delta t + p_2(\Delta t).$$

Eq (12.1.7) becomes then

$$P_N(t+\Delta t) = P_N(1-(k_1+k_2)N\Delta t) + P_{N-1}k_1(N-1)\Delta t + P_{N+1}k_2(N+1)\Delta t$$

Ex. 4.

Observe that σ^2 can be rewritten as

$$\sigma^2 = \sum_{N>N_0} N^2 P_N(t) - \mu(t)^2.$$

Now differentiate this formula with respect to t and use Equations (12.1.8), (12.1.9) and (12.1.18).

Sec. 3.

Ex. 1. Differentiate Equation (12.3.7) and use Equations. (12.3.4), (12.3.5).

Ex. 3.

Let the number of males and females in the colony at time t be denoted by $M(t)$ and $F(t)$ respectively. Assume that on the time interval $[t, t+\Delta t]$ the probability for one mating in the colony is $kM(t)N(t)\Delta t$ and the probability of more than one mating is $O((Deltat)^2)$. The probability that at time $t+\Delta t$ the number of males in the population is $M(t)+1$ is

$$P_{M+1}(t+\Delta t) = P_M(t) + \frac{1}{2}kM(t)F(t)\Delta t + O((Deltat)^2)$$

Similarly for the female population we have

$$P_{F+1}(t+\Delta t) = P_F(t) + \frac{1}{2}kM(t)F(t)\Delta t + O((Deltat)^2)$$

This leads to a system of two coupled equations for the probability densities which can be solved subject to initial conditions on the male and female populations at the initial time.

Sec. 4

Ex. 1.

Equation (12.4.2) is modified as follows:

$$P_\ell(t + \Delta t) = P_\ell(t)(1 - k\Delta t)(1 - Ns\Delta t) + \tag{13.7}$$
$$P_{\ell-1}(t)(k\Delta t)(1 - Ns\Delta t) + P_{\ell+1}(t)(Ns\Delta t)(1 - k\Delta t) + O((\Delta t)^2)., \quad \ell \geq N.$$

Similarly for $\ell < N$ we have

$$P_\ell(t + \Delta t) = P_\ell(t)(1 - k\Delta t) + P_{\ell+1}(t)(Ns\Delta t)(1 - k\Delta t) + O((\Delta t)^2).$$

Sec. 5

Ex. 1.
$$P^{2k+1} = P, \quad P^{2(k+1)} = P^2, \quad k = 0, 1, \ldots$$

Ex. 2. Make one single bet with all the money the gambler has.